T0418797

Handbook of Computational Intelligence in Biomedical Engineering and Healthcare

HANDBOOK OF COMPUTATIONAL INTELLIGENCE IN BIOMEDICAL ENGINEERING AND HEALTHCARE

Edited by

Janmenjoy Nayak
Bighnaraj Naik
Danilo Pelusi
Asit Kumar Das

Academic Press is an imprint of Elsevier
125 London Wall, London EC2Y 5AS, United Kingdom
525 B Street, Suite 1650, San Diego, CA 92101, United States
50 Hampshire Street, 5th Floor, Cambridge, MA 02139, United States
The Boulevard, Langford Lane, Kidlington, Oxford OX5 1GB, United Kingdom

Notices

Knowledge and best practice in this field are constantly changing. As new research and experience broaden our understanding, changes in research methods, professional practices, or medical treatment may become necessary.

Practitioners and researchers must always rely on their own experience and knowledge in evaluating and using any information, methods, compounds, or experiments described herein. In using such information or methods they should be mindful of their own safety and the safety of others, including parties for whom they have a professional responsibility.

To the fullest extent of the law, neither the Publisher nor the authors, contributors, or editors, assume any liability for any injury and/or damage to persons or property as a matter of products liability, negligence or otherwise, or from any use or operation of any methods, products, instructions, or ideas contained in the material herein.

Library of Congress Cataloging-in-Publication Data
A catalog record for this book is available from the Library of Congress

British Library Cataloguing-in-Publication Data
A catalogue record for this book is available from the British Library

ISBN: 978-0-12-822260-7

For information on all Academic Press publications visit our website at https://www.elsevier.com/books-and-journals

Publisher: Mara Conner
Acquisitions Editor: Chris Katsaropoulos
Editorial Project Manager: Emily Thomson
Production Project Manager: Selvaraj Raviraj
Cover Designer: Miles Hitchen

Typeset by TNQ Technologies

Contents

6. Application of soft computing techniques to calculation of medicine dose during the treatment of patient: a fuzzy logic approach

Ramjeet Singh Yadav

7. Multiobjective optimization technique for gene selection and sample categorization

Sunanda Das and Asit Kumar Das

8. Medical decision support system using data mining semicircular-based angle-oriented facial recognition using neutrosophic logic

R.N.V. Jagan Mohan

9. Preservation module prediction by weighted differentially coexpressed gene network analysis (WDCGNA) of HIV-1 disease: a case study for cancer

Ria Kanjilal, Bandana Barman, and Mainak Kumar Kundu

10. Computational intelligence for genomic data: a network biology approach

Parameswar Sahu, Fahmida Khan, and Subrat Kumar Pattanayak

11. A Kinect-based motor rehabilitation system for stroke recovery

Sriparna Saha and Neha Das

12. Empirical study on Uddanam chronic kidney diseases (UCKD) with statistical and machine learning analysis including probabilistic neural networks

T. PanduRanga Vital

13. Enhanced brain tumor detection using fractional wavelet transform and artificial neural network

Bhakti Kaushal, Mukesh D. Patil, and Gajanan K. Birajdar

14. A study on smartphone sensor-based Human Activity Recognition using deep learning approaches

Riktim Mondal, Dibyendu Mukhopadhyay, Sayanwita Barua, Pawan Kumar Singh, Ram Sarkar, and Debotosh Bhattacharjee

Contributors

David Al-Dabass School of Science and Technology, Nottingham Trent University, Nottingham, United Kingdom

Nagaraj Balakrishnan Department of Electronics and Communication Engineering, Rathinam Technical Campus, Coimbatore, Tamil Nadu, India

Valentina E. Balas Automation and Applied Informatics, Aurel Vlaicu University of Arad, Romania, Arad, Romania

Bandana Barman Department of Electronics & Communication Engineering, Kalyani Government Engineering College, Kalyani, West Bengal, India

Sayanwita Barua Department of Computer Science and Engineering, Jadavpur University, Kolkata, West Bengal, India

Debotosh Bhattacharjee Department of Computer Science and Engineering, Jadavpur University, Kolkata, West Bengal, India

Gajanan K. Birajdar Department of Electronics Engineering, Ramrao Adik Institute of Technology, Navi Mumbai, Maharashtra, India

Asit Kumar Das Department of Computer Science and Technology, IIEST, Howrah, West Bengal, India

Neha Das Department of Computer Science and Engineering, Maulana Abul Kalam Azad University of Technology, Kolkata, West Bengal, India

Sunanda Das Department of Computer Science and Engineering, SVCET, Chittoor, Andhra Pradesh, India

L. Jegatha Deborah Department of Computer Science and Engineering, University College of Engineering, Tindivanam, Tamil Nadu, India

Susovan Jana School of Education Technology, Jadavpur University, Kolkata, West Bengal, India

Ria Kanjilal Department of Electronics & Communication Engineering, Kalyani Government Engineering College, Kalyani, West Bengal, India

Bhakti Kaushal Department of Electronics and Telecommunication Engineering, Ramrao Adik Institute of Technology, Navi Mumbai, Maharashtra, India

Fahmida Khan Department of Chemistry, National Institute of Technology, Raipur, Chhattisgarh, India

Mainak Kumar Kundu Department of Electronics & Communication Engineering, Brainware Engineering College, Barasat, Kolkata, West Bengal, India

Raffaele Mascella Faculty of Communication Sciences, University of Teramo, Teramo, Italy

Lela Mirtskhulava Iv. Javakhishvili Tbilisi State University/San Diego State University, Tbilisi, Georgia

Manohar Mishra Department of Electrical and Electronics Engineering, Institute of Technical Education and Research, Siksha 'O' Anusandhan (Deemed to be) University, Bhubaneswar, Odisha, India

R.N.V. Jagan Mohan Department of Computer Science and Engineering, Sagi Rama Krishnam Raju Engineering College, Bhimavaram, Andhra Pradesh, India

Subhashree Mohapatra Department of Computer Science and Engineering, Institute of Technical Education and Research, Siksha 'O' Anusandhan (Deemed to be) University, Bhubaneswar, Odisha, India

Riktim Mondal Department of Computer Science and Engineering, Jadavpur University, Kolkata, West Bengal, India

Dibyendu Mukhopadhyay Department of Computer Science and Engineering, Jadavpur University, Kolkata, West Bengal, India

Ranjan Parekh School of Education Technology, Jadavpur University, Kolkata, West Bengal, India

Mukesh D. Patil Department of Electronics and Telecommunication Engineering, Ramrao Adik Institute of Technology, Navi Mumbai, Maharashtra, India

Subrat Kumar Pattanayak Department of Chemistry, National Institute of Technology, Raipur, Chhattisgarh, India

Arunkumar Rajendran Department of Electronics and Communication Engineering, Rathinam Technical Campus, Coimbatore, Tamil Nadu, India

S.C. Rajkumar Department of Computer Science and Engineering, University College of Engineering, Panruti, Tamil Nadu, India

Sriparna Saha Department of Computer Science and Engineering, Maulana Abul Kalam Azad University of Technology, Kolkata, West Bengal, India

Parameswar Sahu Indian Council of Agricultural Research (ICAR-CIFRI), Barrackpore, West Bengal, India

Bijan Sarkar Department of Production Engineering, Jadavpur University, Kolkata, West Bengal, India

Ram Sarkar Department of Computer Science and Engineering, Jadavpur University, Kolkata, West Bengal, India

Pawan Kumar Singh Department of Information Technology, Jadavpur University, Kolkata, West Bengal, India

Tripti Swarnkar Department of Computer Application, Institute of Technical Education and Research, Siksha 'O' Anusandhan (Deemed to be) University, Bhubaneswar, Odisha, India

P. Vijayakumar Department of Computer Science and Engineering, University College of Engineering, Tindivanam, Tamil Nadu, India

T. PanduRanga Vital Department of Computer Science and Engineering, Aditya Institute of Technology and Management, Srikakulam, Andhra Pradesh, India

Ramjeet Singh Yadav Department of Computer Science and Engineering, Ashoka Institute of Technology and Management, Varanasi, Uttar Pradesh, India

Biographies

Janmenjoy Nayak is working as an associate professor at the Department of Computer Science and Engineering, Aditya Institute of Technology and Management (AITAM), Tekkali, K Kotturu, AP- 532201, India. He has published more than 110 research papers in various reputed peer reviewed referred journals, international conferences, and book chapters. He is the recipient of the best researcher award from Jawaharlal Nehru University of Technology, Kakinada, Andhra Pradesh for 2018–2019, and many other awards. His area of interest includes data mining, nature inspired algorithms and soft computing. He has edited nine books from various publishers such as Elsevier and Springer.

Bighnaraj Naik is an assistant professor in the Department of Computer Application, Veer SurendraSai University of Technology (Formerly UCE Burla), Odisha, India. He has published more than 90 research papers in various reputed peer reviewed international journals, conferences, and book chapters. He has edited eleven books from various publishers such as Elsevier, Springer, and IGI Global. At present, he has more than 10 years of teaching experience in the field of Computer Science and IT. He is a member of IEEE. His area of interest includes data mining, computational intelligence, soft computing, and its applications.

Danilo Pelusi is working as an associate professor at the Faculty of Communication Sciences, University of Teramo. As Associate Editor of IEEE Transactions on Emerging Topics in Computational Intelligence, IEEE Access, International Journal of Machine Learning and Cybernetics (Springer), and Array (Elsevier), he served as guest editor for Elsevier, Springer, and Inderscience Journals and as program member of many conferences, as well as editorial board member of many journals. Reviewer of reputed journals such as *IEEE Transactions on Fuzzy Systems* and *IEEE Transactions on Neural Networks and Machine Learning*, his research interests include fuzzy logic, neural networks, information theory, and evolutionary algorithms.

Asit Kumar Das is a professor of the Department of Computer Science and Technology, Indian Institute of Engineering Science and Technology Shibpur, Howrah, and currently acting as the head of the Center of Healthcare Science and Technology of his institute. He has published one research monograph, three edited books, many book chapters, and over 100 research articles in peer-reviewed journals and international conferences. His current research interests include data mining and pattern recognition, social network analysis, evolutionary computing, text, audio, and video processing. Prof. Das has already guided five PhD scholars, and seven more scholars are currently working under him.

Preface

In some coming decade, the expansion of medical data (both structured and unstructured) will present issues as well as prospects for hospitals and academics. The current situation of storage of amount of data is quite huge in contemporary databases due to the availability and popularity of the internet as well as cloud sharing. Healthcare has always been a challenging field and needs more focus. The advantages of intelligent computing have been extensively discussed in the medical literature. Computational intelligence includes the ability to use refined algorithms to "learn" correct features from a bulky healthcare data, which may be used to obtain insights to assist clinical practice. Automated systems can aid physicians by providing current medical information from news articles, journals, textbooks, and clinical practices to inform proper patient care. For the past decade, there is an aggressive increment in the development of computational intelligence methods. Data produced by the medical organizations is extremely huge and multifaceted due to which it is hard to examine the information in order to make significant conclusion regarding the health of a patient. This information holds particulars concerning hospitals, patients, medical claims, cost of treatments, etc. So, there is a demand to generate a commanding tool for examining and removing vital information from this difficult data. Thus the information needs to be reviewed and structured in order to preserve efficient decision-making. In the present scenario of computing, computational intelligence tools present adaptive mechanisms that permit the understanding of difficult data and altering environments. The results of computational intelligence technologies are to present profits to medical fields for assembling the patients having same type of diseases or fitness problems, so that medical organization gives them effectual treatments.

In spite of much advancement, we believe that machines (intelligent computing methods) cannot replace human physicians in the upcoming future, but it can definitely assist physicians to make better clinical decisions or even replace human judgment in certain functional areas of healthcare. Healthcare is a multifaceted domain, which includes advanced decision-making, remote monitoring, healthcare logistics, operational excellence, and modern information systems. This proposed book is more focused and intricate toward complex problem solving with integration of computational intelligence, biomedical, and healthcare applications. The book is a voluminous collection for the state of the art as well as advances of different computational intelligence methods and will provide effective solutions for the confront faced by the healthcare institutions and hospitals for effective analysis, storage and analysis of data. Various methods such as Dynamical Systems based on Deep Learning, Biosignal Processing, Decision Support System using Data-Mining, Fuzzy Logic, Multiobjective Optimization Technique, Neutrosophic Logic, Probabilistic Neural Networks, Fractional Wavelet

Transform, etc. are discussed with several applications to meet the requirements of the latest biomedical and healthcare challenges.

Chapter 1 discusses the application of dynamical systems based on deep learning algorithms to model emergent behavior for healthcare diagnostics. The algorithms and methodologies based on dynamical-system deep learning are used to formulate models for healthcare variables. Further, hybrid recurrent nets are proposed to construct deep learning models of observed trajectory patterns of these variables. Each observed trajectory is subjected to a deep learning mining process to determine its dynamical behavior parameters. Results obtained from this study using the simulation demonstrated the quality of the algorithms in dealing with the range of difficulties inherent in the problem.

Chapter 2 focuses on the enhancement of the behavior and nature of the deep learning method with clustering algorithm. The major objective is to analyze the impacts of unsupervised algorithms in context to deep learning. Methods such as autoencoder with clustering, Deep Embedded Clustering, Discriminately Boosted Clustering, image clustering, Deep Embedded Regularized Clustering, Variation Deep Embedding, Information Maximizing Generative Adversarial Network, Joint unsupervised Learning, Deep Adaptive Image Clustering, and the Deep Clustering framework based on Orthogonal AutoEncoder are thoroughly discussed with their challenges and issues in processing medical image data.

Chapter 3 proposes an automatic optic disc detection and segmentation technique to address the task of optic disc in retina. With the converted grayscale image and noise removal steps, the authors have used edge detection operation to find the edges of the optic disc. The possible optic disc center from the edges is calculated using Circular Hough Transform technique. With a supervised machine learning algorithm, the actual and perfect optic disc region among the candidate regions for the optic disc has been computed.

Chapter 4 is about the development of an intelligent healthcare system to monitor the driver's heart rate that utilizes an artificial recurrent neural network (RNN) model, which continually monitors driver's health condition and persistently update to the cloud server. To accomplish a reliable emergency service, the proposed proximity based communication model transmits the critical condition record to the service providers even without internet capability. The Long-Short Term Memory (LSTM) is capable of learning the driver's behavior continually in long-term decencies of his/her activity which converts sensor reading into optimized EHR. The proposed work would be noticed as one step toward guaranteed guarded journey.

Chapter 5 discussed the application of AI as a solution to Gastroenterology-based medical diagnosis. In Gastroenterology-based research area, general practitioner handles huge amounts of medical data and numerous varieties of imaging instruments. The artificial intelligence has been successfully applied in the field of GE for effective diagnosis and analyzing gastrointestinal images. The chapter summarize the AI applications in Gastroenterology with several key issues/challenges. It briefly enlighten about the Endoscopy, X-ray, Ultrasound, CT-scan, MRI, PET, etc. using machine learning and deep learning based approaches.

Chapter 6 proposed a fuzzy logic based intelligent system calculation of medicine doses for chronic intestine illness symptoms like sedimentation and prostate antigen. As the medicine doses play a vital role during the procedure and observation of the patient, appropriate medicine dose require for the long-duration treatment of any patient. The chapter briefly discusses about the components of fuzzy membership functions, rules and inference system used for the calculation of medicine dose. With several experiments the authors claimed that, their proposed method is a suitable intelligent based framework for determining the dose of medicine given to patients with chronic intestinal infections.

Chapter 7 developed a multiobjective optimization technique based on an improved strength Pareto evolutionary algorithm to select only the few important genes responsible for disease identification. The method explores the whole search space for approximating the paretooptimal front that provides the optimal solution. Each chromosome in the population is evaluated using three different objective functions considering the external cluster validation index, number of genes in a sample and mutual correlation between the objects separately. The experimental results of the proposed method prove the effectiveness and acceptability of the improved strength Pareto evolutionary algorithm for the purpose of important genes selection and sample categorization.

Chapter 8 is about the medical decision support system using data-mining semicircular based angle oriented facial recognition using neutrosophic logic. By adopting two-level optimizations such as micro and macro, undesirable attitude-oriented images are deleted from the input dataset and identified the similar face from specific angle-oriented images. The experimental effects records are compared to perspective-based images of various large databases such as Yale, MIT, FERET, and College Academic for identification and found to be the efficient approach than others.

Chapter 9 discussed Weighted Differentially Coexpressed Gene Network Analysis (WDCGNA) of HIV-1 Disease for Preservation Module Prediction. The preservation patterns of differentially coexpression modules are also determined for better understanding, and eigengene network of differentially coexpression modules are built to represent the characteristic expressions of modules. The analyses have discovered the strongest preservation in Nonprogressor-Acute network and identified the significant genes in HIV-1 disease progression as well as in cancer progression.

Chapter 10 has established a building block between big data approaches with computational biology to identify/target the proteins from the available genomic datasets. This chapter is an overview of discussion of brief history of advancement of DNA sequencing and associated technologies as well as characteristics features of different sequencing platforms. Next Generation Sequencing datamining and analysis provided the genetic information related with different functions and involvement in various pathways. Network biology approach contributed for the better understanding of functional values and pathway enrichment architecture of cancer encouraging proteins from the genetic datasets.

Chapter 11 developed a Multilayer Perceptron as the learning neural network to learn and map the exercises done by stroke patients based on the feature space of the training exercise set. The proposed method is useful for day-to-day monitoring of improvement of the patient condition measurement while performing the exercises without the need to regular visit to the doctor. The Multilayer Perceptron along with back propagation algorithm is used to compute the extent of correctness of the performed exercise to measure the improvement of the patient on a daily basis.

Chapter 12 proposed a Probabilistic Neural Network model for an empirical

Study on Uddanam Chronicle Kidney Diseases (UCKD) with statistical analysis. Chronic kidney disease is a threatening state of living, can be induced due to malfunctioning of kidney or its pathology. It is critical to recognize factors that precipitate risk for CKD, even in people with typical Glomerular Filtration Rate. The proposed method is able to predict the CKD in Uddanam area with an accuracy of 100%. This study will be very useful to analysts and government to make decisions for further steps about Uddanam Chronic Kidney Diseases.

Chapter 13 developed a method for detecting enhanced brain Tumor using Fractional Wavelet Transform and Artificial Neural Network. The intention of the proposed method is for the early detection of brain tumor with high accuracy, sensitivity, and specificity by using large dataset of transverse relaxation time weighted MRI scans. The authors have extended the approach of 2D-discrete wavelet transform (DWT) to 2D-fractional wavelet transform (FrDWT) and the features are reduced using PCA. Further, they classified the brain MR images with the help of artificial neural network and achieved better performance.

Chapter 14 discussed on the performance analysis of various deep learning approaches based classification models for Smartphone Sensor based Human Activity Recognition. Human Activity Recognition (HAR) is a rapidly growing research field in the domain of computer vision where sequence of data for a specified time span is collected from the sensors like accelerometer and gyroscope present in these smart devices. HAR plays a crucial role in detecting a user interaction with environment which helps in surveillance, health care, building smart environment based on human-computer interaction. The authors have used five deep learning methods such as Convolutional Neural Network (1D-CNN), Recurrent Neural Network with Long-Short Term Memory (RNN-LSTM), CNN-LSTM, ConvLSTM, and Stacked-CNN for automatic extraction of meaningful information from raw sensor data. With the effective simulation results, the authors are confident about the methods in HAR applications.

We would like to thank all the contributors and the reviewers for their contributions and dedicated efforts for the successful completion of this book. We want to specially thank to the editorial team of Elsevier for their valuable technical support and superior efforts. We hope that the work reported in this volume will motivate for the further research and development efforts in the performance evaluation of biomedical and medical domain.

Editors

Dr. Janmenjoy Nayak
Dr. Bighnaraj Naik
Dr. Danilo Pelusi
Dr. Asit Kumar Das

CHAPTER

1

Application of dynamical systems based deep learning algorithms to model emergent characteristics for healthcare diagnostics

David Al-Dabass[1], *Lela Mirtskhulava*[2]

[1]School of Science and Technology, Nottingham Trent University, Nottingham, United Kingdom; [2]Iv. Javakhishvili Tbilisi State University/San Diego State University, Tbilisi, Georgia

1. Introduction

Healthcare diagnostics is a vital area that can benefit enormously from progress in computer science, including advances made within computational intelligence in deep learning, illustrated and reported by available statistical data [1], Karam et al. [2], Mirtskhulva et al. [3], and research on brain-computer interface [4]. Brain monitoring to diagnose deficiencies in biological intelligence functions is one such area that is undergoing wide research and progress such as: EEG research [5], neurotechnology to enhance brain-computer interfaces [6], and projects reported by Mirtskhulava et al. [7,8], Yalung et al. [9], Padierna Sosa et al. [10], Michalopolous et al. [11], and Bhattacharya et al. [12]. Within advances being made in computational intelligence, feedback inference networks are emerging as an active area to represent healthcare variables to model dynamic intelligent systems. The input parameters of the healthcare variables are resolved from measurements by using a differential estimation mechanism for representing the knowledge embedded within the patient. The usage of dynamic knowledge mining processes guarantees that growth of knowledge is repeatedly traced as reported in projects by Al-Dabass et al. [13–16], Baily et al. [17], and Berndt [18] to find patterns in Time Series using Dynamic Programming. Deep Knowledge, represented by means of the input variables of the patterns of evolution of knowledge first layer, is then resolved using implementation of second layer dynamic procedures such as complexity theory by Bovet et al. [19], estimation of transfer function utilizing FFT based analyzers by

Handbook of Computational Intelligence in Biomedical Engineering and Healthcare
https://doi.org/10.1016/B978-0-12-822260-7.00009-1

Crawley [20], modeling and analysis of dynamic systems by Close et al. [21], and Dewolf et al. technique for continuous time parameter estimation [22]. In the applications of deep learning, for instance, there is a necessity to find the sources of specific style sequences. Alternative functions of deep learning comprise of cyclic trends in the values of stock market and trade value of retailing and commerce, alterations in the restoration aspects of patient, and prediction of motion uncertainties in compound engineering systems as reported by Gersch [23], Kailath [24], Kalman [25], Manila et al. [26], Man [27], Schank et al. [28] and Seveance [29]. In this chapter we develop complete mathematical derivation along with simulations and instances to demonstrate the techniques that have been included as follows:

A. Hybrid Inference Networks: A multilayer analytical pyramid is proposed to mathematically express knowledge contained in biological systems. This knowledge is fundamentally time varying and require differential models to express it and obtain its movement parameters from data in the output patterns. Where regular artificial neural nets (ANNs) are employed to carry out inference, the mathematical function between input (cause) and output (effect) is time invariant and is determinable by deduction listing all input variables (causes) then using a gradual stepwise process progressing through the network layers to compute the output values. Computing from output back to input to estimate the causes that generated the output pattern, e.g., diagnostics scenarios, the procedure needs to analyze the output pattern and progress backward stepwise layer-by-layer to calculate input values (causes).

B. Deep Learning and Knowledge Mining for Healthcare Models: The concepts described above are applied to mathematical models of bioengineering systems to cover the cases where knowledge contained with the data is not fixed but varies in time. As expected the output pattern is no longer fixed with time, and is used to drive a dynamical system to discover the embedded knowledge represented by the parameters which are now dynamic. There is deeper knowledge contained within these parameters which are discovered by computation of the input driving a dynamic process as layer-2. As before, these procedures embody a dynamic section to evaluate higher derivative in time driving static expressions for parameter evaluation.

2. Deep learning applications for brainwaves monitoring

A typical example of the application of deep learning technology occurs in medicine, where monitoring involves the observation of one or several medical parameters over time focused on a specific disease. Medical monitoring is classified by the target of interest. Here we focus on a kind of neurological monitoring dealing with the brainwave-monitoring with the aid of Brainwave Computer Interface (BCI) through Electroencephalography (EEG). Brainwave-monitoring using EEG is most popular method in neurological monitoring for monitoring and diagnosing a series of neurological diseases like Autism, Alzheimer's, epilepsy, stroke and etc. EEGLAB is the powerful tool of MATLAB allowing us to process high-density EEG dataset through an interactive graphic user interface (GUI) through independent component analysis (ICA), time/frequency analysis (TCA), and the standard method of averaging.

2.1 Brain disorder

A global rise of brain disorders and the increasing government funding for improving healthcare systems can supplement the growth of the global market of BCI (Brain Computer Interface). The noninvasive BCI technology has accounted about 85% of revenue of the total global BCI market in 2013, and demonstrated a steady growth that eliminates the need to perform surgery and implantation of BCI devices (Fig. 1.1). The worldwide statistics showed that, an income of market of BCI was about 125.21 million US-dollars in 2018. According to a forecast the market will grow at a Component Annual Growth Rate (CAGR) of 12.43%, reaching 283 million US dollars in 2025 [1]. The neural activity of the cerebral cortex is usually retrieved and used for controlling artificial (prosthetic) limbs. A noninvasive technique is becoming popular, allowing doctors to replace the invasive method by wearable devices capable to easily measure the neural activity. After gaining the data from EEG they will be preprocessed using EEGLAB. An EEG signal measurement gives us a value of currents in dendrites as a result of excitations of synapsis through the neurons inside the cerebral cortex. The signal is not capable to measure each of the neurons but capable to measure many of them. Signals are recorded using the electrodes. Dr. Jalal Karam and Dr. Lela Mirtskhulava showed that the brain parts can be specified by the electrodes positions [2,3]. BCI cannot read the mind but can detect the frequency patterns in the brain. Our thoughts, behaviors

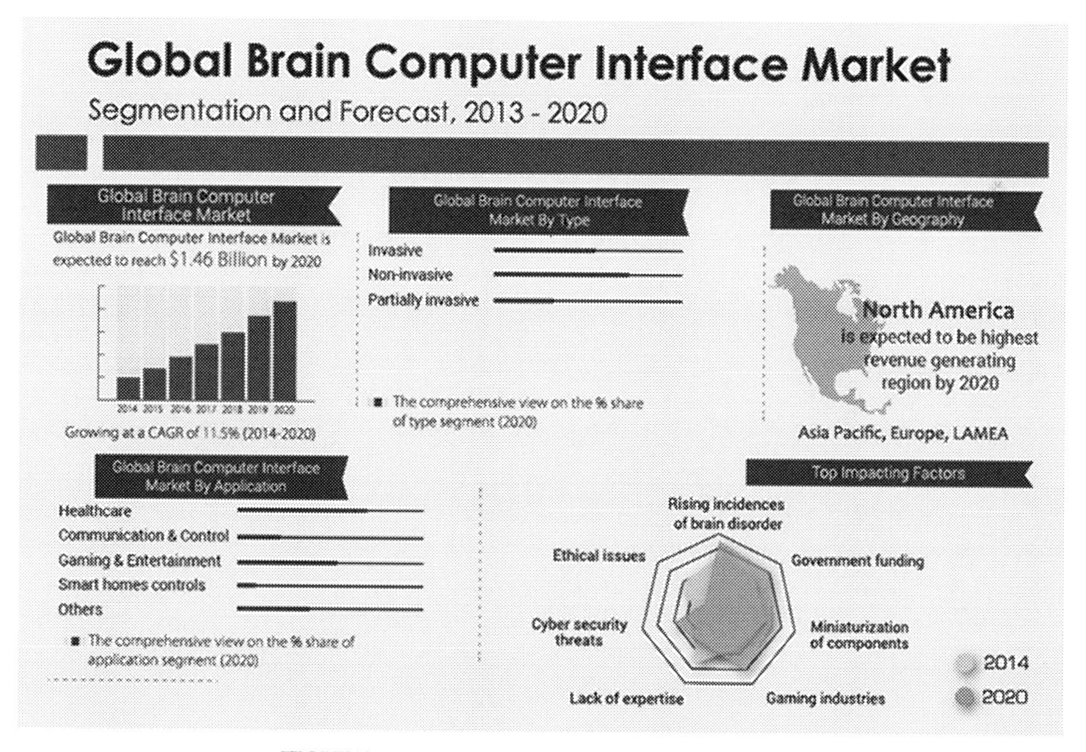

FIGURE 1.1 Global brain computer interface market.

or emotions are result of the communication between neurons in our brain. Brainwaves are generated by synchronized electrical pulses from plenty of neurons communicating with each other.

2.2 Brain-computer interface (BCI)

Brain is the central control unit of the human body controlling our thoughts, speech, movements and memory. Brain is responsible to regulate the function of the human organs. Brain works efficiently and automatically when it is healthy but brain disorders can cause devastating results. We differentiate three main groups of brain computer interfaces like invasive, semiinvasive and noninvasive. Invasive technique implies to use devices for capturing the signals generated by brain where surgery is required to insert them into the brain. Semiinvasive techniques implies to insert devices into the skull not into the brain [4–6].

Brain computer interface (BCI) (Fig. 1.2) is a device enabling the users to have an interaction with computers by mean of brain activity generally measured by electroencephalography (EEG), which is a physiological method for recording the brain activities through electrodes placed on the scalp surface, and is more widely used due to its noninvasive nature. In noninvasive techniques electrodes are placed on the scalp (Fig. 1.3). EEG-based interfaces are very easy to wear and avoid surgery but have poor spatial resolution. They cannot use high frequency signals because the skull can dampens signals. After capturing EEG signals can be processed to get control signals to be readable by the computers. The processing of EEG signals is a difficult task for building a high quality BCI. The EEG recording shows the mental or physical state of the subject. If EEG shows alpha brainwaves with high amplitudes in the occipital area, it means that the subject is relaxed and the eyes are closed. If the eyes are opened, alpha-waves disappear.

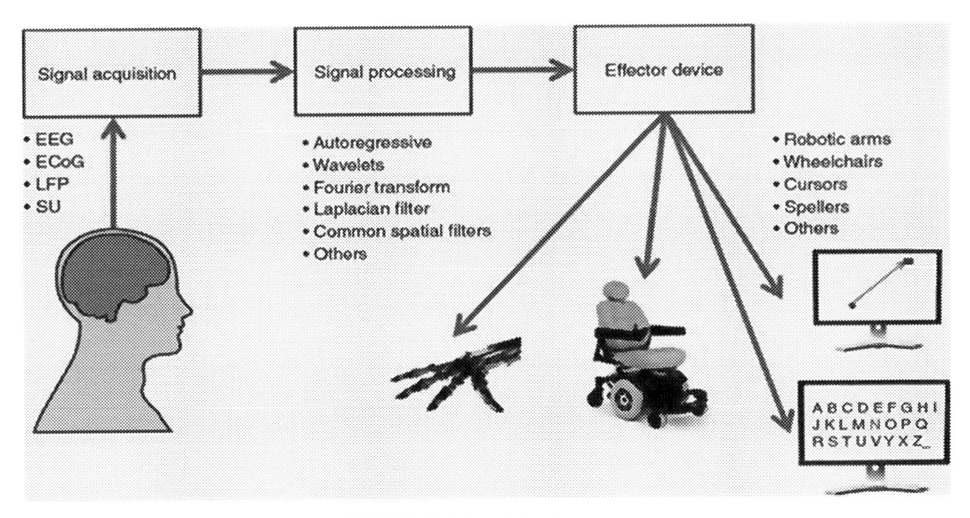

FIGURE 1.2 BCI diagram.

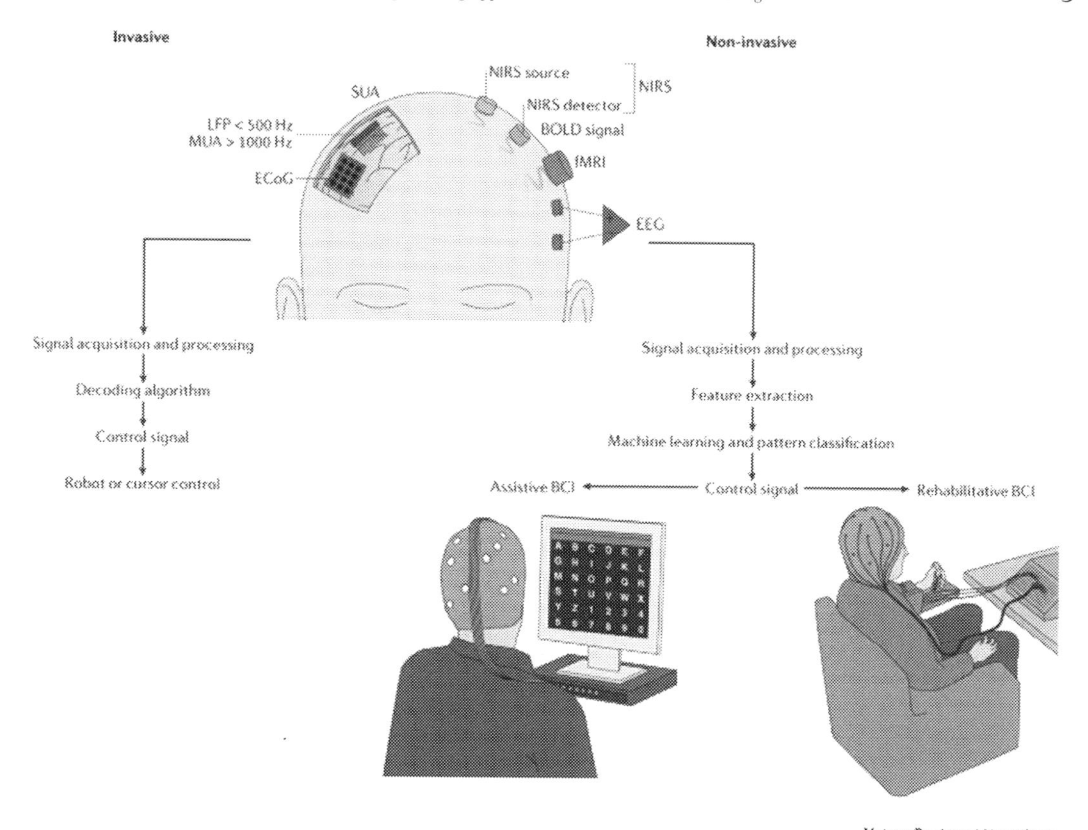

FIGURE 1.3 Invasive and noninvasive BCI [4].

2.3 Brainwaves studies

Our brain has interconnections of billions of neurons communicating with each other through electrical currents traveling throughout our neural network acting as electrical impulses. During an activation of neurons, electrical pulses are generated forming brainwaves as a result. Here are different brainwaves according to different states of activities and thoughts. Analysis and monitoring of brainwaves play crucial roles to treat neurological malfunctions and may prevent fatality. Strokes can occur during any time, even during sleep in some cases. The time of onset of a stroke at night time is less prone to be discovered. It is vital to detect the onset of a stroke because many of them emerge due to a blood lump in a vessel inside the brain, and are suitable for treatment. Mirtskhulava et al. [7,8,30] showed that the time of onset of a stroke is crucial because limbs can be affected and therefore cause loss of motor function.

Thus, monitoring brainwaves through EEG by recording any frequency changes would enable the time of beginning of the stroke to be determined, even occurring during sleep. A dataset in BDF format was used in sleep research where brainwaves were characterized by their frequencies and amplitudes measured in Hertz. According to these frequency values, it is possible to determine the symptoms of diseases. Major research projects carried by Holewa et al. [31], Wolpaw et al. [32], Yalung et al. [9], Padierna Sosa et al. [10], Michalopolous et al. [11], Bhattacharya et al. [12], Yuan [33], Caminiti, et al. [34], Nolan et al. [35], Gurumurthy et al. [36], together with stroke dataset [37], amply illustrated that fast and slow activities of the brain are intrinsically related to high frequency and small amplitude, and low frequency and high amplitude respectively. There are five main types of brainwaves (Fig. 1.4).

2.4 EEGLAB preprocessing

EEGLAB is an interactive toolbox built in Matlab to proceed data processing obtained from EEG [2]. EEGLAB is equipped with an interactive GUI giving the possibility to process high-density brain data obtained from EEG in two ways through ICA and TFA. EEG data out of the recording device is a continuous process signal. It is like measuring a difference of potential on an oscilloscope (Fig. 1.5). To make sense of data, we need to:

(1) Extract meaningful measures from brain oscillations.
(2) Compare brain data in different conditions.
(3) Assess reliable changes due to external stimuli.

Which involves the following Preprocessing steps:

(1) Collect EEG data.
(2) Import into EEGLAB.
(3) Import event markers and channel locations.
(4) Rereference/download-sample.
(5) High pass filter (~0.5–1 Hz).
(6) Examine raw data.
(7) Identify/reject bad channels.
(8) Reject large artifact time points.
(9) Run ICA and reject components.

EEGLAB supports different data formats. We used the dataset generated in the BDF file (Fig. 1.6) for a number of 16 channels at each time-point. There is only one Epoch before starting to process data. The Epoch gives a duration in seconds in the entire recording with the specific sampling rate measured in Hertz (Hz) and defined as a series of time points obtained in a second. EEG data consists of time-points representing the voltage samples measured in microvolts (μV). Sampling rate values varied from 250 to 2000 Hz. When the sampling rates are high the signal's dynamics' resolution is better but the data processing time is long (Fig. 1.7).

We filtered continuous EEG data using linear finite impulse response (FIR) filtering method in EEGLAB which involved the inverse Fourier transform. We updated dataset after applying filtering. Digital filters improved the signal-to-noise ratio or line noise at 50/60 Hz. Human EEG waves in sleep state are typically in a range of 100–500 μV in amplitude, we used a signal with a 200 μV (Fig. 1.8). The filters commonly used by EEG can be low pass and high pass for attenuating high and low frequencies respectively. For example, a high

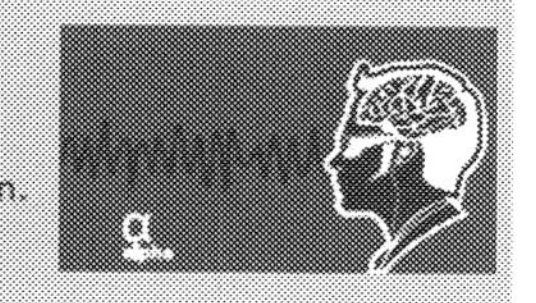

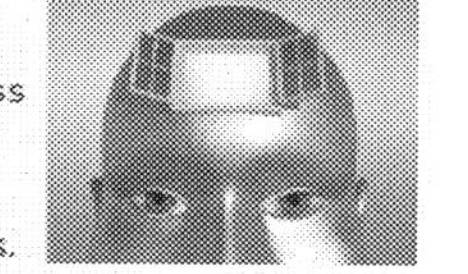

FIGURE 1.4 The frequency patterns in the brain.

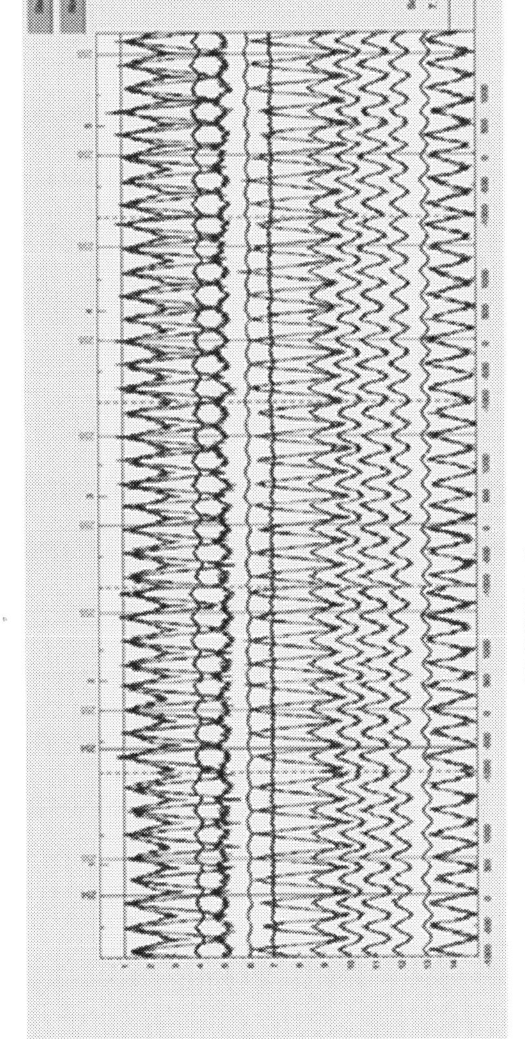

FIGURE 1.5 Component activities.

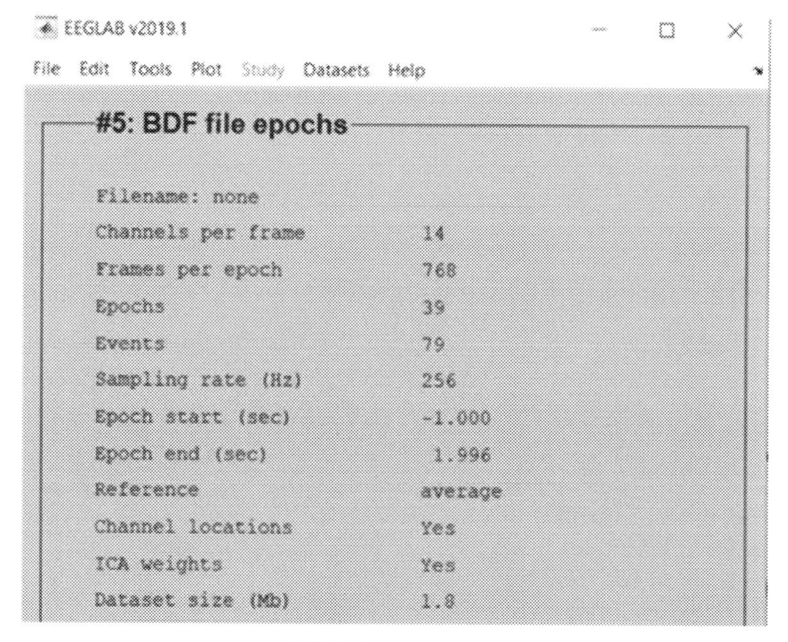

FIGURE 1.6 BDF file.

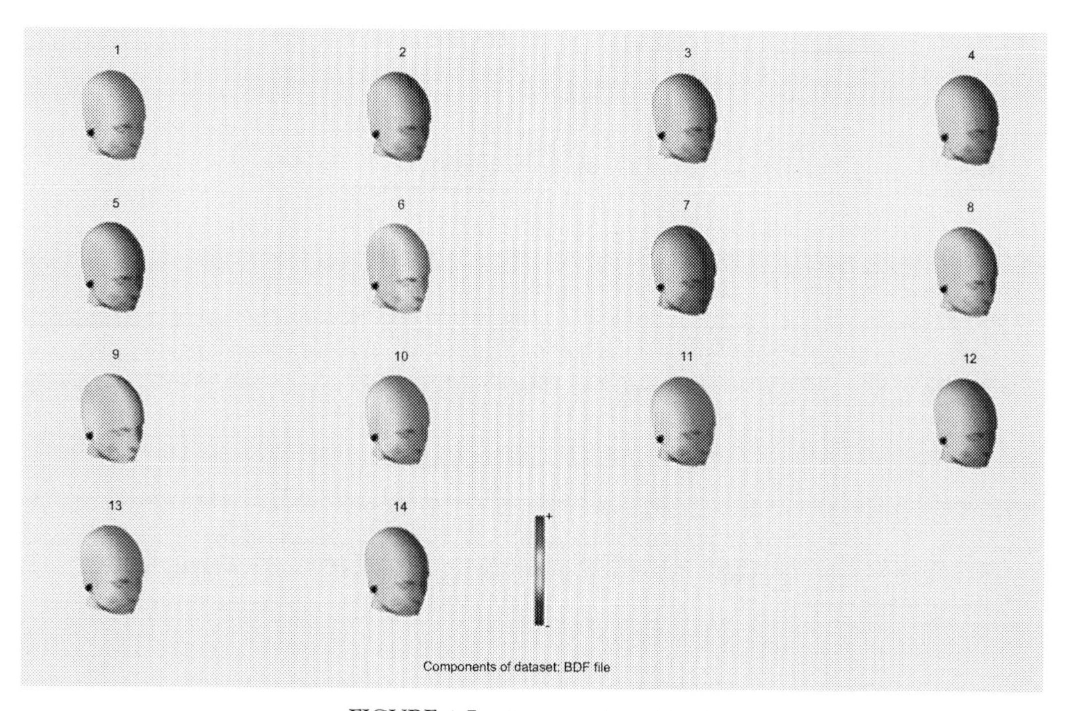

FIGURE 1.7 Components of dataset.

pass filter with a cut-off, 0.5 Hz frequency can be used for attenuating the components of lower frequency like 0.5 Hz and pass the components of higher frequency higher than 0.5 Hz. The aim of an ICA is to generate the available independent signals in the channel data. When the data were full rank the channels numbers and independent components (ICs) numbers are quite similar (Figs. 1.9 and 1.10). We used EEGLAB to visualize and model brain dynamics based on events and an individual dataset.

3. Healthcare Modeling and simulation using feedback hybrid artificial neural networks

As shown in Section 2, many systems in healthcare display complicated patterns that cannot be modeled mathematically with basic ANNs. Here we remodel this problem in the form of the feedback ANNs, which are hybrid to contain a mix of stationary nodes (arithmetic or logical or both), and feedback nodes (recurrent). The output of a sample feedback element is expressed as a second-order differential process. The nonvariable elements (parameters) that generate the output of this feedback node may in turn show dynamic time-varying pattern that can be expressed mathematically with more feedback nodes. Further levels of feedback elements can then be incorporated to generate a full display of the output pattern. We derive mathematical procedures to accurately evaluate the parameter values of the mathematical equations from output patterns of the systems such as biological brains. We propose a new process based on a multilayer system of six level simplex system to emulate complex output pattern seen in numerous practical applications including biological systems as shown in Section 2 as well as other AI applications such as those in economy and technology.

3.1 Static-dynamic hybrid mix of ANN models

Output patterns of measurements in bioengineering show time variation even when the nonvariable elements (parameters) of the model has fixed nontemporal values; see Fig. 1.11. Such oscillatory pattern can be emulated by order-2 mixed model; see Fig. 1.12 and see Al-Dabass et al. [15], which is formed on the well-known differential process of order-2 as shown in Eq. (1.1) below:

$$\alpha^{-2}x'' + 2 \cdot \lambda \cdot \alpha^{-1} \cdot x'' + x = u \tag{1.1}$$

Here x is the node output, α (Alpha) represents the natural frequency, λ (Lambda) is used to the represent the damping ratio, and u represents input, respectively, which expresses the three input constants to represent the input vector. To build this dynamic paradigm as a feedback net, a double integral steps are utilized to configure a mixed integral-feedback network, illustrated in Fig. 1.12 below.

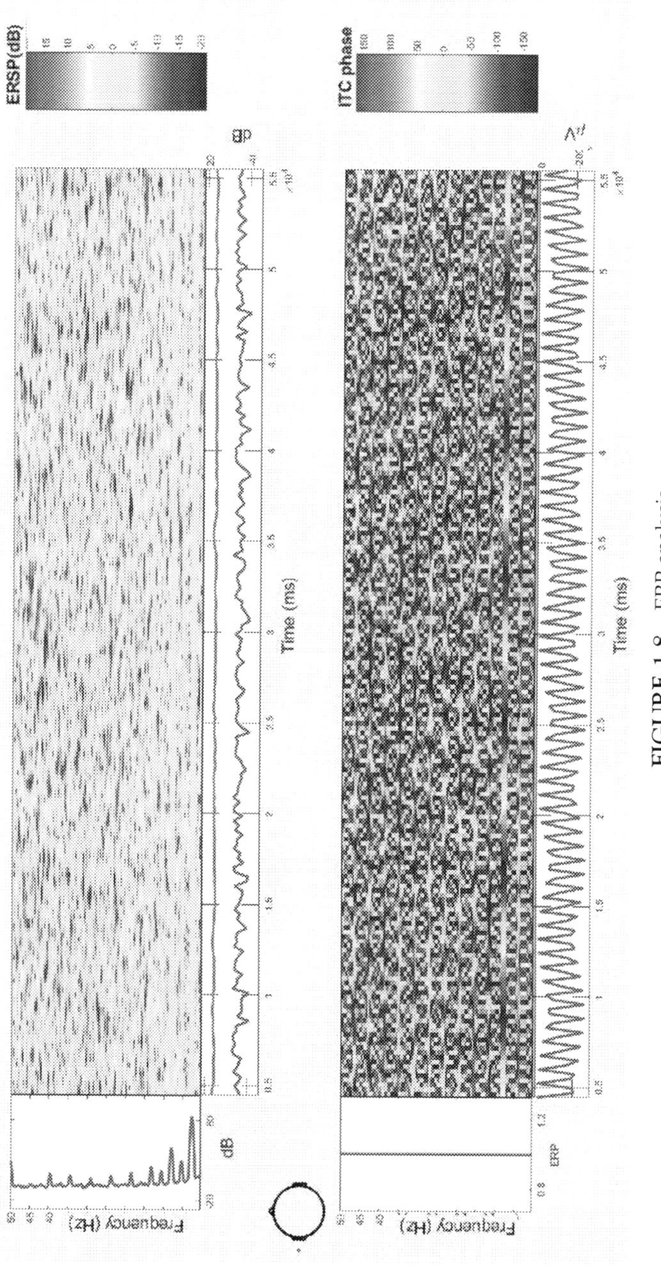

FIGURE 1.8 ERP analysis.

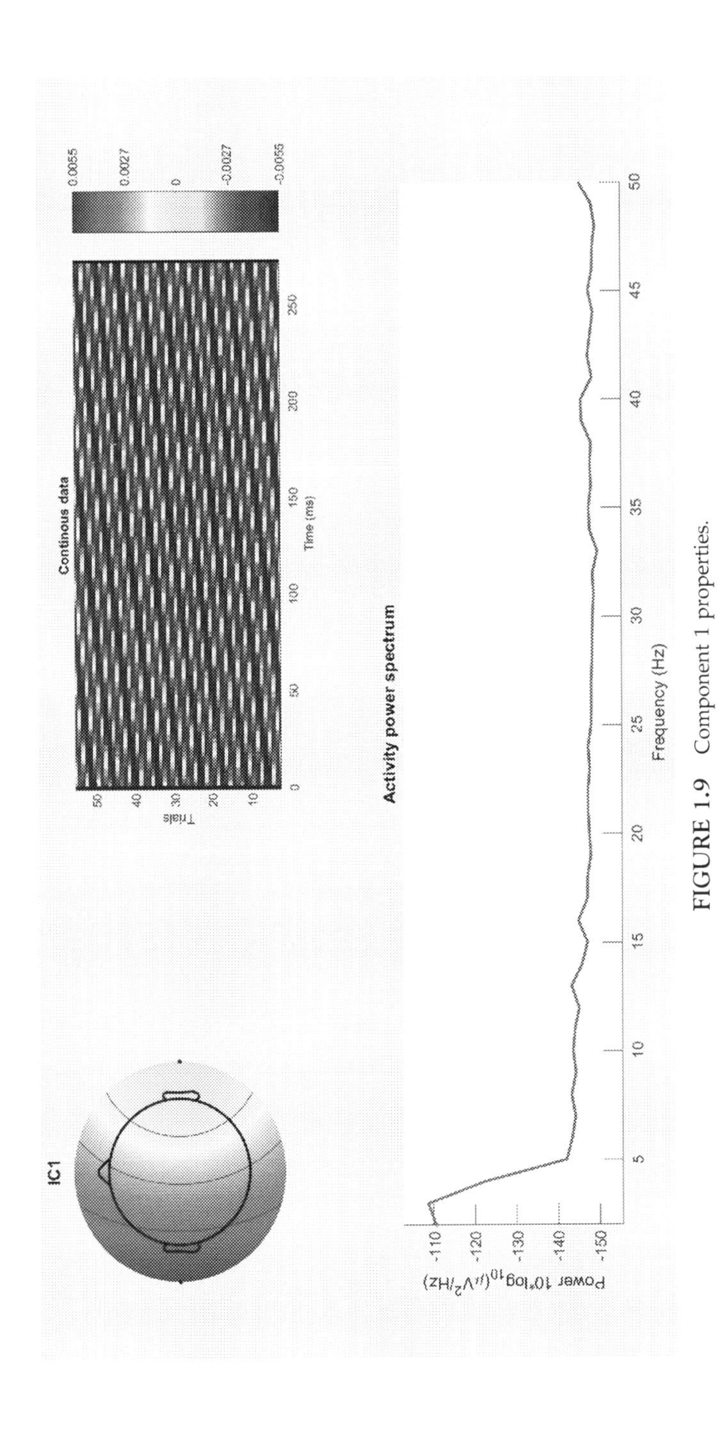

FIGURE 1.9 Component 1 properties.

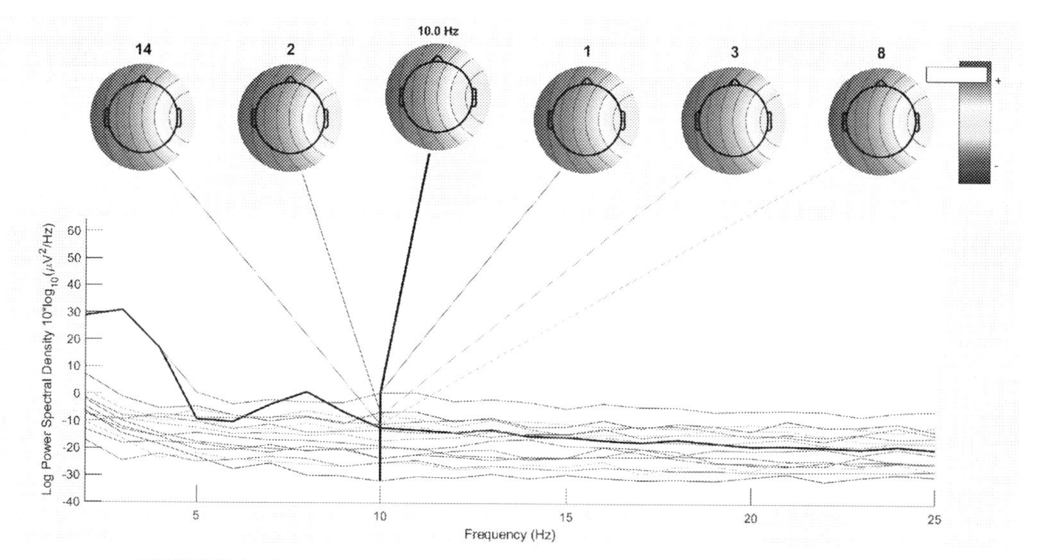

FIGURE 1.10 The mean log spectrum of a set of data epochs at all channels.

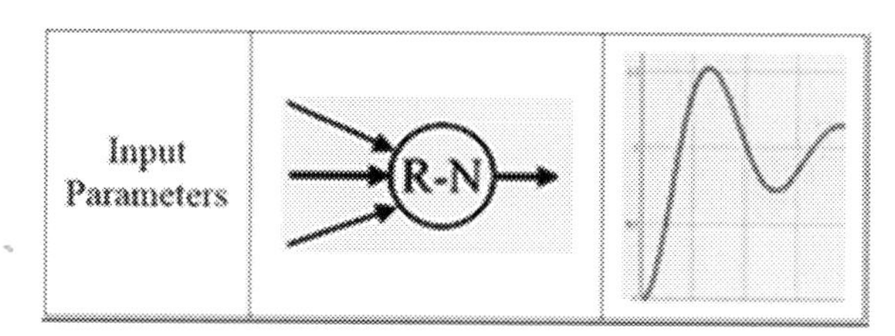

FIGURE 1.11 A Feedback or Recurrent-Node (R-N) shows a dynamic pattern at the output node even containing fixed time input constants (parameters).

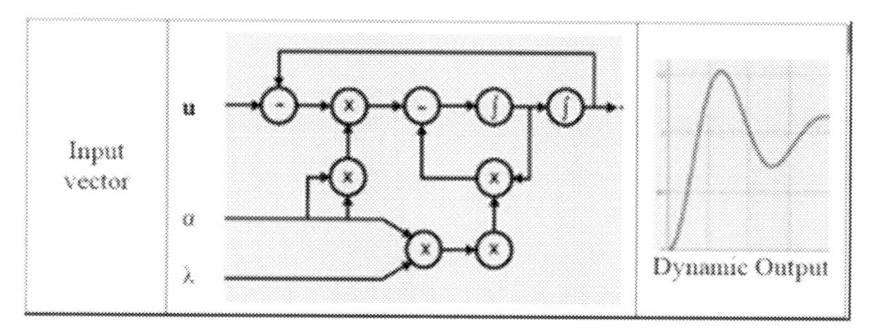

FIGURE 1.12 Hybrid integral-feedback network to design the dynamic output of the element in Fig. 1.11.

3.2 Architecture of the hybrid integral-feedback network

The process illustrated in Fig. 1.12 represents directly the structure of Eq. (1.1), the derivation is simply carried in the following steps:

(1) To obtain Eq. (1.1A), multiply both sides of Eq. (1.1) using Alpha-squared (α^2) as shown:

$$x'' + 2\cdot\lambda\cdot x'' = \alpha^2\cdot(u-x) \tag{1.1A}$$

The Eq. (1.1B) can be obtained by rearranging of x''

$$x'' = \alpha^2\cdot(u-x) - 2\cdot\lambda\cdot\alpha\cdot x'' \tag{1.1B}$$

For the remaining steps, we quote below the author's previous work Al-Dabass et al. [15]:

(2) The first subtraction node that appears on the left side is provided with the output of the net x. Therefore, the output is $(u - x)$ since u is accepted as the input from the left side of this node.

(3) α is considered as value of intermediate input from the left in the entire network. The multiplication node x is provided with two separate value of the middle input value α to form α^2 which is considered as the output. This is represented using an up arrow, which is provided as input to the lower multiplier node above it as shown in Fig. 1.12. It is represented as node two from the left in the top chain of nodes.

(4) Therefore, the multiplier node output is expressed as α^2 $(u - x)$ which is considered as the RHS of Eq. (1.1A).

(5) The λ is considered as the bottom input from the left that is provided as the lower input value to the first two multiplier nodes in the bottom chain nodes. The output is considered as $\lambda.\alpha$ because α is the top input value of this node. Further, this node is multiplied with 2 to generate $2\cdot\lambda\cdot\alpha$ in the second node.

(6) As specified in step (2), the final node from the right side of the top node chains is an integrator node used to produce x. The input must be provided as derivative of x i.e., x' as last node is an integrator node. Then it is multiplied with $2\cdot\lambda\cdot\alpha$ which is the right node of the bottom 2-node chain to generate the output $2\cdot\lambda\cdot\alpha\cdot x'$. The x' is the LHS term in Eq. (1.1A) while in Eq. (1.1B) it is the RHS term.

(7) In order to obtain full RHS term of Eq. (1.1B) which is represented as $\alpha^2.(u - x) - 2\cdot\lambda\cdot\alpha\cdot x'$, subtract the output of the integrator node form the output of the middle node of the top row node chains.

(8) As the input of the final integrator (on the right in the top chain) is the first derivative of x (x'), and as it is also the output of the first integrator, therefore the input to the first integrator node must be x'', i.e., the LHS of Eq. (1.1B).

(9) Finally, the complete equation can be obtained by merely feeding the output of the middle node to the input of the second integrator node form the right in the row of the top chain nodes.

3.3 Prototypes of hierarchical feedback nodes

The pattern of the output of a typical healthcare variable may be more complex than can be expressed by an elementary dynamic model of order-2. Here each input constant (parameter) can itself be assumed to possess differential pattern, which can possess oscillation. We illustrate in Fig. 1.13 an example with two of the three constant values in the input (parameters) possess differential behavior of order-2.

The second-order paradigm of an element in a particular level in the pyramid is expressed using Eq. (1.1) specified above.

We begin with the last, output, element, and allow both the input parameters u and Alpha α to possess their individual second-order dynamics.

The input represented by u is the result of second-order process given in Eq. (1.2) below:

$$\alpha_u^{-2}\mathbf{u}'' + 2\cdot\lambda_u\cdot\alpha_u^{-1}\cdot\mathbf{u}' + \mathbf{u} = u_u \tag{1.2}$$

The input parameter Alpha α, which is used for representing natural frequency is the result of the second-order process expressed in Eq. (1.3) below:

$$\alpha_\alpha^{-2}\alpha'' + 2\cdot\lambda_\alpha\cdot\alpha_\alpha^{-1}\cdot\alpha' + \alpha = u_\alpha \tag{1.3}$$

Hence, the output pattern is accomplished by a six-order differential equation vector (utilizing the easily available Runge-Kutta in Mathcad for the current example).

Vector Form of First-order Differential Equations: The second-order equations are transformed into differential equations of first-order vector that can be quickly computed to produce a simulation output trajectory of the node.

$x_1 = x$, and $x_2 = x'$.

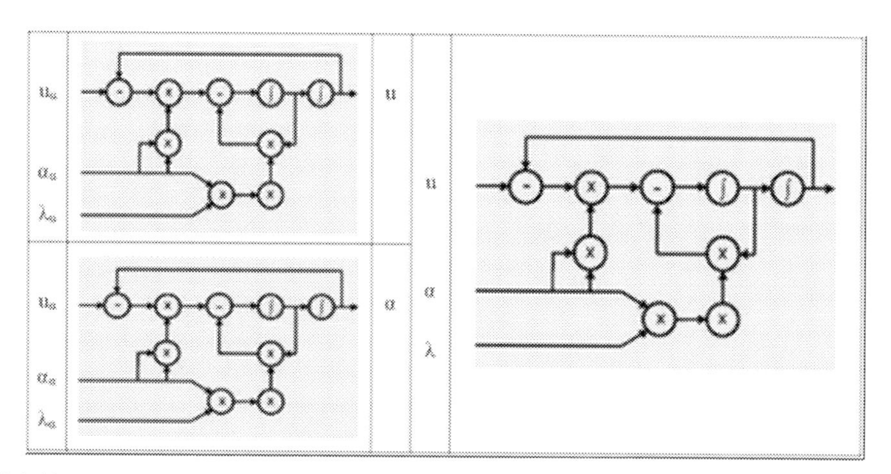

FIGURE 1.13 The two input parameters, u and α, of the end element possess time-varying output modeled as mixed second-order integral recurrent nets.

$$D(t,x) = \begin{bmatrix} x_2 \\ x_5 \cdot x_5 \cdot x_3 - 2 \cdot z \cdot x_5 \cdot x_2 - x_5 \cdot x_5 \cdot x_1 \\ x_4 \\ (wu \cdot wu \cdot uu) - 2 \cdot zu \cdot wu \cdot x_4 - wu \cdot wu \cdot x_3 \\ x_6 \\ (ww \cdot ww \cdot uw) - 2 \cdot zw \cdot ww \cdot x_6 - ww \cdot ww \cdot x_5 \end{bmatrix}$$

FIGURE 1.14 The derivative vector for producing the healthcare variable computed-result utilizing subsystems for the parameters u and α (Alpha).

In the above Fig. 1.14, x and x′ are represented by variables x_1 and x_2, u and u′ are depicted by x_3 and x_4 and α and α' are expressed by x_5 and x_6 respectively. The following parameter values are used to produce the output pattern shown in Fig. 1.5. In subsystem u, the value of u starts from 0 and aims at $u_u = 1$ with $\alpha_u = 5$ rad/s and $\lambda_u = 0.3$. while in the subsystem α, the values of α starts from 4 rad/s and aims at $u_\alpha = 32$ rad/s with $\alpha_\alpha = 4$ rad/s and $\lambda\alpha = 0.1$. The estimated amalgamated behavior pattern of x (red [dark gray in printed version] oscillatory trace), u (blue [black in printed version] upper trace) and α (green [light gray in printed version] lower trace) has been depicted as shown in Fig. 1.15.

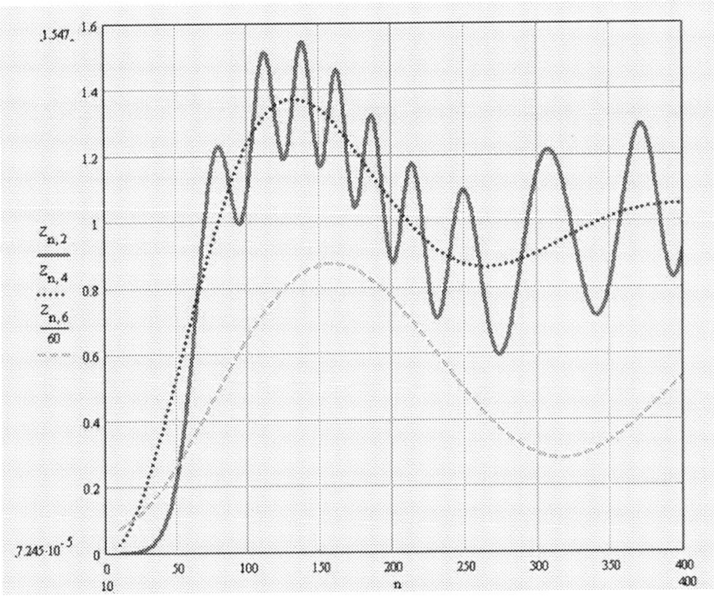

FIGURE 1.15 Simulated output pattern of a hierarchical feedback element (oscillatory trace), with two variable inputs: u (*dotted-gray* upper trace) and α (*dashed-gray* lower trace).

4. Derivative estimation using feedback networks

To determine the variables of second-order derivative models that resembles the nature of the composite high-order structures, a number of precise algorithms that are static and dependent on the behavior of biological and healthcare schemes have been developed. These algorithms depend on the availability of the trajectory time derivatives. In this section, a framework of cascaded feedback network is suggested to estimate these derivatives in the succeeding phase. The technique is tested successfully on parameter tracking algorithms that varying from the constant parameter algorithm that only makes use up to fourth-order derivatives to an algorithm that pursues two variable parameters and makes use up to the eighth-order derivatives.

4.1 Algorithm for constant parameters using single point data

Consider using the first-order to fourth-order derivatives at a single point. Given the second-order system in Eq. (1.4):

$$\alpha^{-2}x'' + 2\cdot\lambda\cdot\alpha^{-1}\cdot x' + x = u \tag{1.4}$$

Eq. (1.5) is obtained by differentiating Eq. (1.4) with respect to t.

$$\alpha^{-2}x''' + 2\cdot\lambda\cdot\alpha^{-1}\cdot x'' + x' = 0 \tag{1.5}$$

Eq. (1.6) is derived by dividing Eq. (1.5) with x''.

$$\alpha^{-2}x'''/x'' + 2\cdot\lambda\cdot\alpha^{-1} + x'/x'' = 0 \tag{1.6}$$

Then Eq. (1.6) is differentiated using t to give Eq. (1.7).

$$\alpha^{-2}\cdot\left[\left(x''\cdot x'''' - x'''^{2}\right)/x''^{2}\right] + 0 + \left[\left(x''^{2} - x'\cdot x'''\right)/x''^{2}\right] = 0 \tag{1.7}$$

The Eq. (1.8—1.10) is obtained by using the expressions of estimated α and λ, using Eq. (1.5) and estimated u

$$E\alpha^{2} = \left[x''\cdot x'''' - x'''^{2}\right]/\left[x''\cdot x''' - x''^{2}\right] \tag{1.8}$$

$$E\lambda = -\left[E\alpha^{-2}x''' + x'\right]/\left[2\cdot E\alpha^{-1}\cdot x''\right] \tag{1.9}$$

$$Eu = E\alpha^{-2}\cdot x'' + 2\cdot E\lambda\cdot E\alpha^{-1}\cdot x' + x \tag{1.10}$$

4.2 High-order algorithms

The first and higher-order derivative of u is considered as nonzero value. For modesty, the two parameters a and b (the coefficients of x'' and x' are used for making symbol manipulation simple) are assumed as constant. Due to this, both the parameters a and b disappear on performing the first differentiation. The additional information required for u', u'', u''' and u'''' be as nonzero are obtained from the fifth to eighth-order derivative of the pattern. In this paper, only the u' case is explained as displayed in Eqs. (1.11–1.16). The remaining cases are the simple extension of the u' case and are left to the reader for practice.

$$a \cdot x'' + b \cdot x' + x = u \tag{1.11}$$

Differentiating Eq. (1.11) with t and assuming u' as nonzero gives Eq. (1.12).

$$a \cdot x''' + b \cdot x'' + x' = u' \tag{1.12}$$

Further, differentiating Eq. (1.12) and assigning $u'' = 0$ gives Eq. (1.13).

$$a \cdot x'''' + b \cdot x''' + x'' = 0 \tag{1.13}$$

For isolating b, Eq. (1.11) is divided with x'''

$$a \cdot x''''/x''' + b + x''/x''' = 0 \tag{1.14}$$

Further, to exclude b differentiate Eq. (1.14).

$$a \cdot \left(x''''' \cdot x'' - x''''^2\right)/x'''^2 + \left(x'''^2 - x'' \cdot x''''\right)/x'''^2 = 0 \tag{1.15}$$

Then re-arrange the Eq. (1.15) to obtain Estimated a, E(a) as shown.

$$E(a) = \left(x'' \cdot x'''' - x'''^2\right)/\left(x''''' \cdot x''' - x''''^2\right) \tag{1.16}$$

Inserting the value of a from Eq. (1.16) into Eq. (1.14) gives b, which is used to obtain Estimated b, E(b)

$$E(b) = -x''/x''' - a \cdot x''''/x'''$$

And the substitution of a and manipulation gives Eq. (1.17).

$$E(b) = (x'' \cdot x'''' - x''' \cdot x''''')/\left(x''''^2 - x''' \cdot x'''''\right) \tag{1.17}$$

The u is solved by substituting these values of a and b in Eq. (1.17)

$$u = a \cdot x'' + b \cdot x' + x$$

4.3 A feedback architecture to estimate time derivatives

Fig. 1.16 depicts the framework of each cell of the feedback subnet. The next higher-order time derivative is obtained by providing the output of every cell as input value to the next layer. By utilizing the fourth-order Runge-Kutta method in Mathcad, the simulations of the cascade of first-order feedback network filters and output of the system are generated. Figs. 1.17 and 1.18 represents a cascade of five feedback cells and the trajectory model of the second-order derivative. An essential set of time derivates which are resulted from the trajectory of an oscillatory second-order dynamical system is depicted in Fig. 1.19.

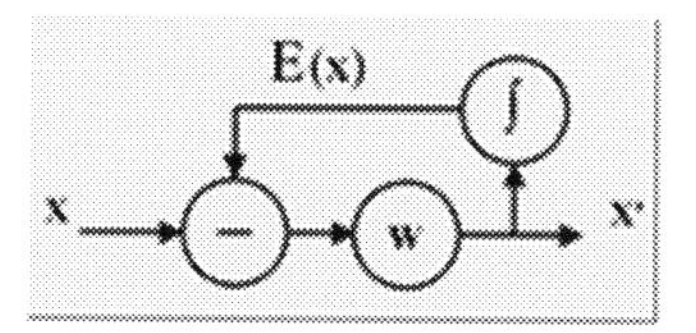

FIGURE 1.16 A feedback system having single stage subnetwork used for determining the derivative $x' = w(x - E(x))$ in the path of feedback by utilizing an integrator; the subnetwork uses a low pass filter with w cut off frequency.

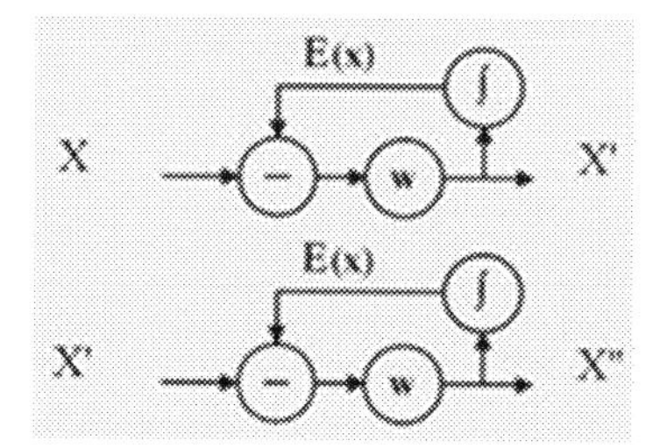

FIGURE 1.17 A feedback system having second-order derivative network for determining first and second time derivatives.

$$D(t,x) := \begin{bmatrix} x_2 \\ -\omega^2 \cdot x_1 - 2\zeta \cdot \omega \cdot x_2 + \omega^2 \cdot u \\ G\cdot(x_1 - x_3) \\ G\left[G\cdot(x_1 - x_3) - x_4\right] \\ G\left[G\left[G\cdot(x_1 - x_3) - x_4\right] - x_5\right] \\ G\left[G\left[G\left[G\cdot(x_1 - x_3) - x_4\right] - x_5\right] - x_6\right] \\ G\left[G\left[G\left[G\left[G\cdot(x_1 - x_3) - x_4\right] - x_5\right] - x_6\right] - x_7\right] \end{bmatrix}$$

$$Z := \mathrm{Rkadapt}(x, t0, t1, N, D)$$

FIGURE 1.18 A cascade of five feedback cells and the second-order trajectory model.

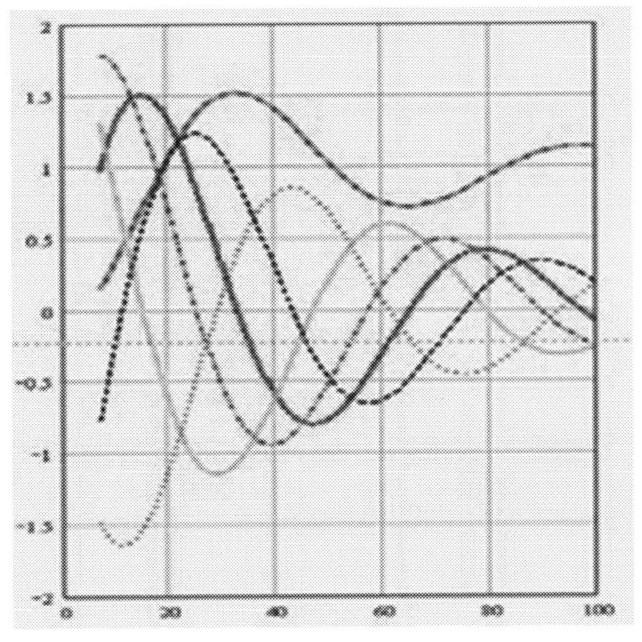

FIGURE 1.19 An essential set of time derivatives resulted from the trajectory of an oscillatory second-order dynamical system.

4.4 Results and discussion

The results and discussions are described as follows.

D1. Utilization of constant parameters in First Algorithm: This algorithm uses single point data and four time derivatives of higher-order. The continuous finding of the first to fourth time derivative represented by x', x'', x''' and x'''' are provided by filter cascade. These estimates are further used to produce a continuous determination of all the parameters at every point on the trajectory. Fig. 1.20 depicts the approximated results of Alpha, zeta, and u, which represents fast and precise convergence.

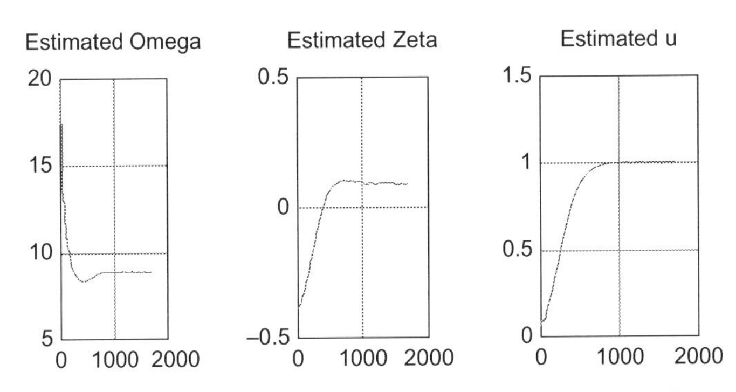

FIGURE 1.20 Results of estimated constant Alpha, zeta and u.

***D2. Discussion*:** The approximated values of constant parameters are almost nearer to the values of the desired set. An adequate series of values such as α from 1 to 10, λ between +/− (0.01−1), and u between +/− (0.5−40) can be estimated using derived algorithms. This range can be used to produce accurate estimates. Estimation errors decrease with increase in the value of α, specifically for small λ (less than 0.5), where vibrations generate ample deviation in the parameters to minimize the errors. The discrepancy between the time derivatives of simulated system such as x, x′ and x″ and their estimates from the filter cascade mainly depends on the value of G, which represents the cut-off frequency. The high value of G provides more precise estimations but makes the algorithm more liable to noise and vice versa. Form the simulation point of view, the other disadvantage of using high G value is that simulation time enhances significantly because of the integration routine adjusting to forever smaller steps. Moreover, the algorithm also produces speedy convergence rate.

***D3. Results generated by Higher-Order Algorithm*:** By utilizing its own parameters of natural frequency represented by α, damping ratio represented by λ and input represented by u, Mathcad routines are used for generating the input u as second-order system. The damping ratio of the input subsystem was assigned to 0.05 for producing an oscillatory output that was sufficient for the thorough evaluation of the algorithm used for parameter tracing. The input frequency is initialized to 16 rad/s, which is equivalent to one quarter of the data natural frequency. Fifth time derivative is generated by increasing the value of derivative generation cascade to one. The results are depicted as shown in Fig. 1.21.

The original input is considered as the smooth trace as it produces 16 rad/s = 2.546 Hz, which is nearly equivalent to one and one-quarter cycles for a period of half a second as

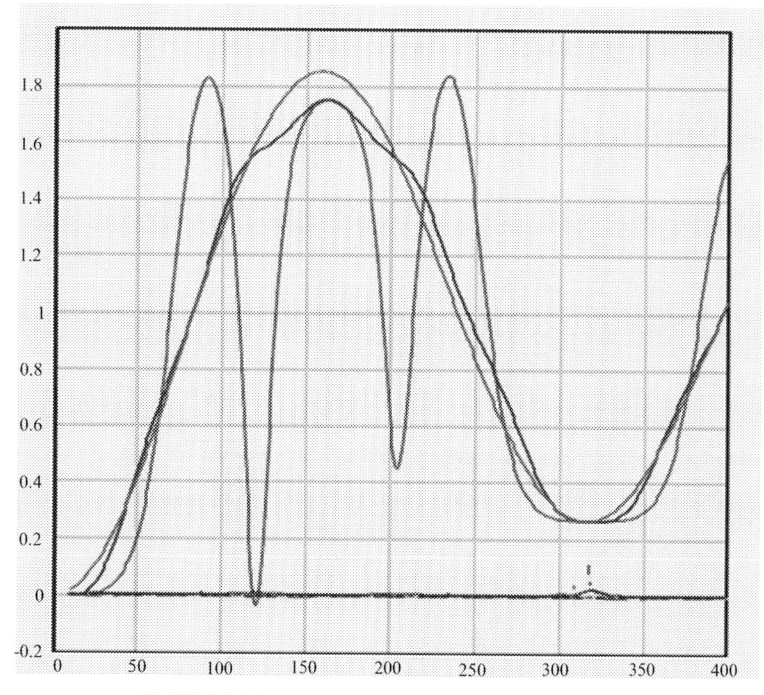

FIGURE 1.21 Results of the high-order algorithm.

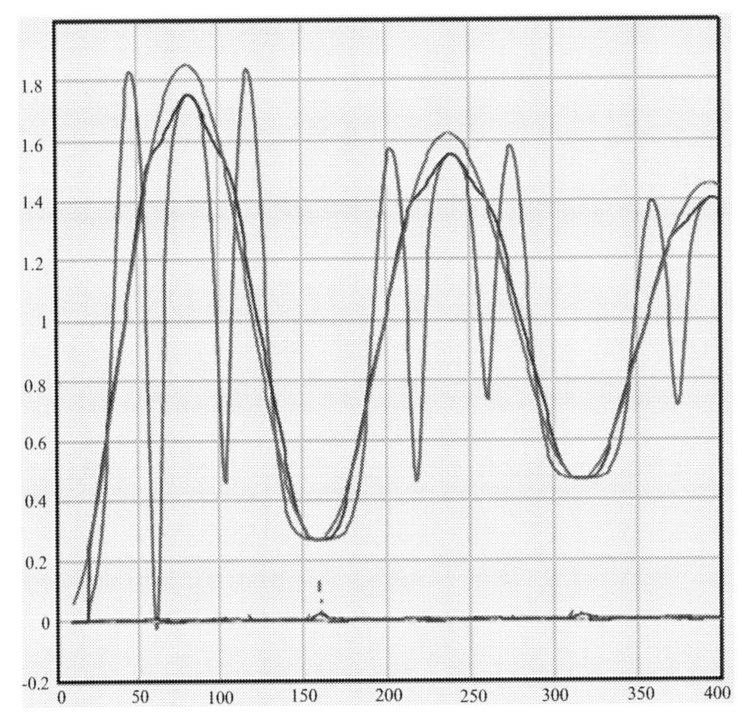

FIGURE 1.22 Results of the high-order algorithm for one second integration time.

expected. The large jagged indications represent the results of the former constant u derivation algorithm that completely fails in tracking the input parameters. The third trace represent the result obtained from the new algorithm that is mainly used in tracking the input parameters more closely. However, initially slight divergence occurs adjacent to the peak of the cycle but finally it gives back to trace the input trajectory properly through its right down and round the lower channel.

As time progresses, the quality of tracing can be ensured using a second set of results as depicted in Fig. 1.22. These results are acquired by extending one second of the integration time to generate two and half cycles. From the figure, it is obvious that the tracing of input parameters stays stable. It is also observed that the old algorithm that failed in tracking the upper half of the input patterns performs good in tracking the lower half of the input trajectory but comparatively less when compared with the performance of the new algorithm.

5. Usage of deep learning knowledge mining in Hybrid Inference Networks

The usage of deep learning knowledge mining in Hybrid Inference Network consists of the following:

***Inference Networks*:** To efficiently model the variables in biological healthcare systems, a multilevel framework is required. The knowledge entrenched in these systems is altering repeatedly because of it nature. Therefore, there is a need to develop dynamic models for the efficient parameter representation and acquisition from the noticed data. In a regular inference network model, the relationship between cause and effect is stationary that means the effects can be simply obtained by using a deduction process that takes into consideration all the causes in a stepwise manner which operates at all the levels of network to reach to the final effect. Alternatively, analysis in the opposite direction such as that utilized in analysis is defined as knowledge mining. That means in knowledge mining, it begins with identifying the effect and then work back using the network nodes to decide the causes for effect.

Dynamical Knowledge Mining Processes: These concepts are enforced on feedback or dynamic system networks where certain part or entire part of the data in the knowledge base varies according to time. Now, the effect used is based on time dependent behavior pattern. Then, the knowledge of the system by means of time differing input parameters can be obtained by using effect as an input value to the differential abduction process. Further, to generate second level parameters these input parameters will integrate knowledge by itself which is acquired through the process of second level mining. To compute the input parameters, the mining process incorporates differential part for finding the higher-order time derivative knowledge accompanied with a nonlinear algebraic part.

Hierarchical Input Parameters with Temporal Behavior: The pattern of the output system may be too composite than the one depicted by the model using second-order differentiation. In this instance, every input parameter is designed itself to have a dynamic nature that may or may not have oscillatory behavior. One example of such instance is shown in Fig. 1.13 where two of the three input parameters exhibit dynamic characteristics of the second-order system.

5.1 Knowledge mining algorithms

A few explicit algorithms have been derived by Al-Dabass et al. [38], depending on the information accessible from the systems time trajectory for characterizing the nature of second-order model using three input parameters. A set of three simultaneous algebraic equations are formed by leaving second-order model at its second derivative form and by using three points on the trajectory that provides data about position, velocity and acceleration. The estimates of u, α, and λ are further generated by using three algebraic equations. Further, an online dynamic algorithm was designed in real time to produce continuous parameter evaluation by merging the estimates of time derivative trajectory with these specific static nonlinear functions.

A1. Multipoint Algorithms: We derived a number of algorithms to compute values of input parameters using as many points from the trajectory that are necessary to form a set of simultaneous algebraic equations. The unknown variables are formed from the parameters to be estimated, while the constant parameters of the algebraic equations are formed from the values of trajectory and their time derivatives.

Algorithm 1. Three-Points in x, x′, and x″ are used for finding α, λ, and u by considering set of three algebraic expressions as shown in Eqs. (1.18–1.20)

$$\alpha^{-2}x_1'' + 2\cdot\lambda\cdot\alpha^{-1}\cdot x_1' + x_1 = u \qquad (1.18)$$

$$\alpha^{-2}x_2'' + 2\cdot\lambda\cdot\alpha^{-1}\cdot x_2' + x_2 = u \quad (1.19)$$

$$\alpha^{-2}x_3'' + 2\cdot\lambda\cdot\alpha^{-1}\cdot x_3' + x_3 = u \quad (1.20)$$

The Eq. (1.21) is obtained by subtracting Eq. (1.19) from Eqs. (1.18) and (1.22) is derived by subtracting Eq. (1.20) from Eq. (1.18)

$$\alpha^{-2}\cdot\left(x_1'' - x_2''\right) + 2\cdot\lambda\cdot\alpha - 1\cdot\left(x_1' - x_2'\right) + (x_1 - x_2) = 0 \quad (1.21)$$

$$\alpha^{-2}\cdot\left(x_1'' - x_3''\right) + 2\cdot\lambda\cdot\alpha - 1\cdot\left(x_1' - x_3'\right) + (x_1 - x_3) = 0 \quad (1.22)$$

Then dividing Eq. (1.21) with $\left(x_1' - x_2'\right)$ and Eq. (1.22) with $\left(x_1' - x_3'\right)$ gives Eqs. (1.23) and (1.24)

$$\alpha^{-2}\cdot\left(x_1'' - x_2''\right)/\left(x_1' - x_2'\right) + 2\cdot\lambda\cdot\alpha - 1 + (x_1 - x_2)/\left(x_1' - x_2'\right) = 0 \quad (1.23)$$

$$\alpha^{-2}\cdot\left(x_1'' - x_3''\right)/\left(x_1' - x_3'\right) + 2\cdot\lambda\cdot\alpha - 1 + (x_1 - x_3)/\left(x_1' - x_3'\right) = 0 \quad (1.24)$$

then Eq. (1.24A) is obtained by further subtraction

$$\alpha^{-2}\cdot\left[\left(x_1'' - x_2''\right)/\left(x_1' - x_2'\right) - \left(x_1'' - x_3''\right)/\left(x_1' - x_3'\right)\right] + \left[(x_1 - x_2)/\left(x_1' - x_2'\right) - (x_1 - x_3)/\left(x_1' - x_3'\right) = 0 \quad (1.24A)$$

The following are the notations used

$$\Delta_{12} = (x_1 - x_2), \Delta_{12}' = \left(x_1' - x_2'\right), \Delta_{12}'' = \left(x_1'' - x_2''\right)$$

$$\Delta_{13} = (x_1 - x_3), \Delta_{13}' = \left(x_1' - x_3'\right), \Delta_{13}'' = \left(x_1'' - x_3''\right)$$

to obtain expressions for estimated α and λ using Eq. (1.21) and estimated u

$$E\alpha^2 = \left[\Delta_{13}''\cdot\Delta_{12}' - \Delta_{12}''\cdot\Delta_{13}'\right]/\left[\Delta_{12}\cdot\Delta_{13}' - \Delta_{13}\cdot\Delta_{12}'\right]$$

$$E\lambda = \left[-E\alpha^{-2}\cdot\Delta_{12}'' - \Delta_{12}\right]/\left[2\cdot E\alpha - 1\cdot\Delta_{12}'\right]$$

$$Eu = E\alpha^{-2}\cdot x_1'' + 2\cdot E\lambda\cdot E\alpha^{-1}\cdot x_1' + x_1$$

Algorithm 2. This algorithm uses two points and one additional derivative. It uses two sets of x, x', x'' and x''' as shown in Eqs. (1.25) and (1.26).

$$\alpha^{-2}x_1'' + 2\cdot\lambda\cdot\alpha^{-1}\cdot x_1' + x_1 = u \quad (1.25)$$

$$\alpha^{-2}x_2'' + 2\cdot\lambda\cdot\alpha^{-1}\cdot x_2' + x_2 = u \tag{1.26}$$

The Eq. (1.27) is obtained by subtracting Eq. (1.26) from (1.25) and then by dividing with $(x_1' - x_2')$ as shown below.

$$\alpha^{-2}\cdot(x_1'' - x_2'')/(x_1' - x_2') + 2\cdot\lambda\cdot\alpha^{-1} + (x_1 - x_2)/(x_1' - x_2') = 0 \tag{1.27}$$

Then differentiating Eq. (1.27) with t gives Eq. (1.28):

$$\left[\alpha^{-2}\left[(x_1' - x_2')\cdot(x'''_1 - x'''_2)\right] - (x_1'' - x_2'')^2\right]/(x_1' - x_2')^2 + 0 + \left[(x_1' - x_2')^2 - (x_1 - x_2)\cdot(x_1'' - x_2'')\right]/ \times (x_1' - x_2') = 0 \tag{1.28}$$

The following are the notations used

$$\Delta_{12} = (x_1 - x_2),\ \Delta'_{12} = (x\prime_1 - x_2'),$$

$$\Delta''_{12} = (x_1'' - x_2'') \text{ and } \Delta'''_{12} = (x'''_1 - x'''_2)$$

to find expressions for estimated α and λ using Eq. (1.27) and estimated u

$$E\alpha^2 = \left[(\Delta'_{12})\cdot(\Delta'''_{12}) - (\Delta''_{12})^2\right]/\left[(\Delta'_{12})^2 - \Delta_{12}\cdot\Delta''_{12}\right]$$

$$E\lambda = \left[-E\alpha^{-2}\cdot\Delta''_{12}/\Delta'_{12} - \Delta_{12}/\Delta'_{12}\right]/\left[2\cdot E\alpha^{-1}\right]$$

$$Eu = E\alpha^{-2}\cdot x_1'' + 2\cdot E\lambda\cdot E\alpha^{-1}\cdot x_1' + x_1$$

A2. Single Point Algorithms: In single point algorithm, the constant parameter condition is made flexible by presuming a linear time variation which means it considers constant for first-order derivatives while zero for second and higher-order derivatives. In this new case, there is a requirement for more information, which can be retrieved from the system output pattern using higher-order time derivatives. It is possible to use explicit parameters as well as those of the first-order derivatives. To generate repetitive trajectories of the parameters in real time, a set of three expressions, one for each input parameter, is devised and numerically estimated along with the state estimation vector observer. This technique is a different technique when compared with state derivation vector. In this technique, instead of using some function of error to drive the derivatives among the system and model output, continuous tracking is provided by an explicit function that assist in speedy convergence of the actual parameter. Even when the parameters are changing very quickly with respect to systems natural frequency or time constant, this technique still performs better. Further, it is classified based on the order of the parameter deviation utilized in the derivation such as constant first-order polynomial (constant u' but $u'' = 0$), second-order polynomial (constant u'' but $u''' = 0$) etc.

Algorithm 3. It makes use of Time-Invariant Parameters. It considers the usage of first to fourth-order time derivatives at a single point. The second-order system considered is shown in Eq. (1.29):

$$\alpha^{-2}x'' + 2\cdot\lambda\cdot\alpha^{-1}\cdot x'' + x = u \tag{1.29}$$

Differentiating Eq. (1.29) using t and then dividing with x'' gives Eq. (1.30):

$$\alpha^{-2}x'''/x'' + 2\cdot\lambda\cdot\alpha^{-1} + x''/x'' = 0 \tag{1.30}$$

Further, differentiation using t again gives the following.
$\alpha^{-2}\cdot\left[\left(x''\cdot x'''' - x'''^{2}\right)/x''^{2}\right] + 0 + \left[\left(x''^{2} - x''\cdot x'''\right)/x''^{2}\right] = 0$, (see Eqs. 1.4–1.15).
The expressions for estimated α and λ, using Eqs. (1.3–1.14), and estimated u are

$$E\alpha^{2} = \left[x''\cdot x'''' - x'''^{2}\right]/\left[x''\cdot x''' - x''^{2}\right]$$

$$E\lambda = -\left[E\alpha^{-2}x''' + x''\right]/\left[2\cdot E\alpha^{-1}\cdot x''\right]$$

$$Eu = E\alpha^{-2}\cdot x'' + 2\cdot E\lambda\cdot E\alpha^{-1}\cdot x'' + x$$

Algorithm 4. Here we consider the more complicated case where the parameters change with time in the simplest way i.e., first order. Assume the first time-derivative of u to be nonzero. This algorithm was fully developed in sub-Section B in last Section 4, the reader may refer to it for the step-by-step derivation details.

5.2 Discussion of results

Algorithm 3 using Constant Parameters: This algorithm-3 utilizes a single point data and further two extra time derivatives when compared with Algorithm-1. The continuous evaluation of fourth-order time derivative which is represented by x'''' can be obtained by incrementing the filter cascade by one. The continuous evaluation of all parameters at every position of the trajectory can be generated through the disappearance of separation problem. Further, program three was executed and the results depicts that it achieves fast and precise convergence [14].

B1. Discussion: Refer to the discussion in paragraph D2 in Section 4 which applies here exactly. Furthermore, progressive faster convergence is contributed by both the algorithms with Algorithm-3 being the fastest to converge.

B2. Results: Further, results are displayed for the instance when there are changes in one or two parameters. The results in Fig. 1.23 are displayed when the parameter u is changing. Moreover, algorithm-3 and algorithm-4 are used to trace these results. From the results, it can

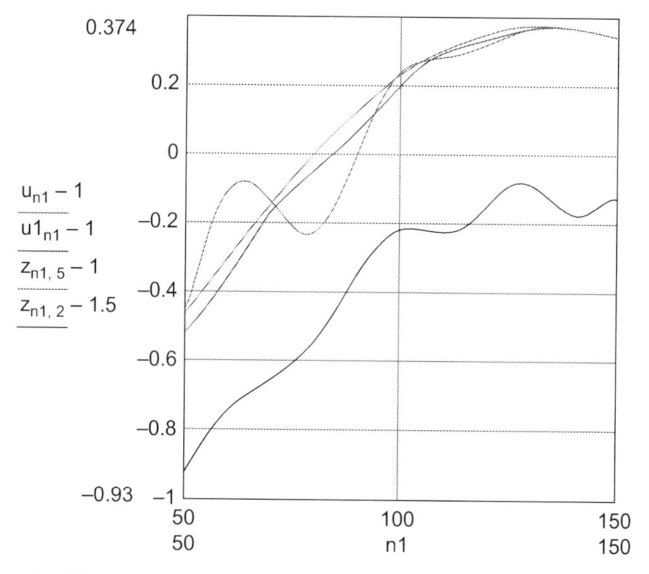

FIGURE 1.23 Using algorithm-3 (top sinusoid-like trace) and algorithm-4 (gentle wavy immediately below u trace) for tracking the u input parameter (smooth trace). The black trace represents the trajectory of the output node.

be noticed that algorithm-3 displays large oscillation while the algorithm-4 displays better smoother tracking. Fig. 1.24 displays the results of both algorithm-3 and algorithm-4 tracking the change in input parameter α. From these results also, it can be noticed that algorithm-4 produces better smoothing estimate when compared to algorithm-3 which generates large oscillations.

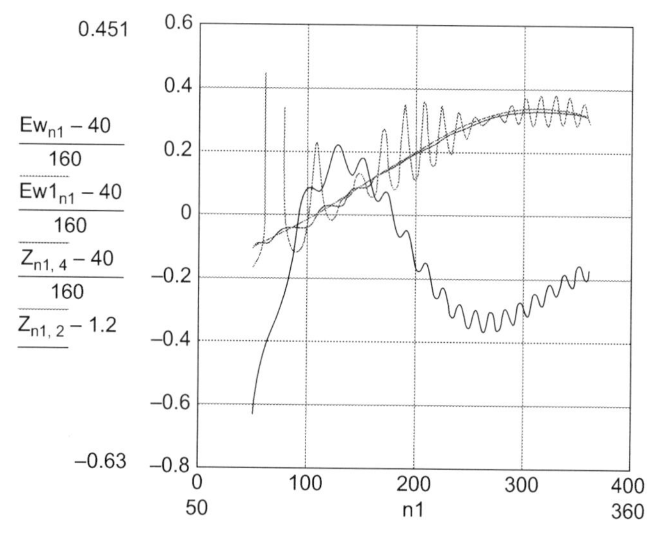

FIGURE 1.24 Using algorithm- 3 (top jagged trace) and the new algorithm (entwined with α trace) to track the input parameter α (smooth trace). The bottom trace represents trajectory of output node.

6. Conclusions

We examined brainwave monitoring as a typical application in a healthcare system with details of current techniques used in the process. We then put forward a system of hybrid deep-learning nets to express the complicated output of healthcare and biological processes. To compute the parameter values to emulate the system output, a number of mathematical procedures based on deep-learning knowledge mining were derived. Two of the procedural algorithms were based on several point values from the output pattern: three values for algorithm-1 and two values for algorithm-2. We then derived further two algorithms based on one point value: the first was based on the presumption that the parameters were time-invariant and utilizing higher time derivatives of the output pattern (first, second, third, and fourth time derivatives), while the second algorithm extracted additional knowledge from the fifth-order time derivative of the output pattern to account for one of the input parameters (i.e., u) to have a first-order time derivative of not zero. Algorithm-4 was evaluated to demonstrate its capability to emulate the input variables of a reduced-order model. The experiment included the simulation of a slightly damped dynamical second-order feedback/recurrent network. Further, the outcomes demonstrated the ability of the algorithm to keep close track during a long time period. It provided proof that this algorithm was far better than algorithm-3 which depended on stipulating in the derivation that the input must be time invariant.

References

[1] https://www.statista.com/statistics/1015013/worldwide-brain-computer-interface-market-value/#statisticContainer.

[2] J. Karam, S. Al Majeed, C.N. Yalung, L. Mirtskhulava, Neural network for recognition of brain wave signals, Int. J. Enhanc. Res. Sci. Technol. Eng. 5 (Issue 10) (October-2016) 2319–7463. ISSN.

[3] L. Mirtskhulva, Monitoring brain attacks using AI diagnosis, AIMed J. (2018). CA, USA. Featured Article, https://ai-med.io/ai-med-news/monitoring-brain-attacks-ai-diagnosis/?fbclid=IwAR0sfMiCnUZfYTPc0Rd-TzejVE8xi73iP9WW1ns9yuJMkLEQZr-PzHrztJc.

[4] https://towardsdatascience.com/a-beginners-guide-to-brain-computer-interface-and-convolutional-neural-networks-9f35bd4af948.

[5] https://sccn.ucsd.edu/eeglab/index.php.

[6] https://www.theguardian.com/technology/2018/jan/01/elon-musk-neurotechnology-human-enhancement-brain-computer-interfaces.

[7] L. Mirtskhulava, J. Wong, G. Pearce, S. Al-Majeed, Artificial neural network model in stroke diagnosis, Conference: 2015 17th UKSIM-AMSS International Conference on Modelling and Simulation, Cambridge University, UK, doi: 10.1109/UKSim.2015.33.

[8] G. Pearce, L. Mirtskhulava, J. Wong, S. Al-Majeed, K. Bakuria, N. Gulua, Artificial neural network and mobile applications in medical diagnosis, Conference: 2015 17th UKSIM-AMSS International Conference on Modelling and Simulation, Cambridge University, UK, doi: 10.1109/UKSim.2015.34.

[9] C. Yalung, S. Al Majeed, J. Karam, Analysis and Interpretation of Brain Wave Signals, in: International Conference on Internet of Things and Cloud Computing (ICC 2016) ACM International Conference Proceedings, ACM, Cambridge University, UK, March 22 23, 2016, ACM, 2016. https://dl.acm.org/doi/abs/10.1145/2896387.2900319?download=true.

[10] O.A. Padierna Sosa, Y. Quijano, M. Doniz, J.E. Chong Quero, Development of an EEG Signal Processing Program Based on EEGLAB, IEEE Research Paper, 2011, p. 199.

[11] K. Michalopolous, V. Iordanidou, G.A. Giannakakis, K.S. Nikita, M. Zervakis, Characterization of Evoked and Induced Activity in EEG and As-Sessment of Intertrail Variability, IEEE Research Paper, 2011, p. 978.

[12] A. Bhattacharya, N.G. Bawane Dr., S.M. Nirkhi Ms., Brain Computer Interface Using EEG Signals, 2011, p. 2. IEEE Research Paper.

[13] D. Al-Dabass, D.J. Evans, K. Sivayoganathan, Derivative abduction using a recurrent network architecture for parameter tracking algorithms, in: IEEE 2002 Joint Int. Conference on Neural Networks, World Congress on Computational Intelligence, (IJCNN02), 12-17 May 2002, Honolulu, Hawaii, 1570–74, 2002. ISBN 0-7803-7278-6. ISBN 0-7803-7279-4 and ISBN 0-7803-7281-6 (CD-Rom).

[14] D. Al-Dabass, D.J. Evans, K. Sivayoganathan, A recurrent network architecture for non-linear parameter tracking algorithms, in: 2nd Int. Conference on Neural, Parallel, and Scientific Computations, Morehouse College, Atlanta, U.S.A., August 07–10.167–170, 2002, ISBN 0-9640398-2-6 (vol. 2).

[15] D. Al-Dabass, D.J. Evans, K. Sivayoganathan, Signal parameter tracking algorithms using hybrid recurrent networks. I, J. Comput. Math. 80 (10) (2003) 1313–1322. ISSN 0020-7160 print, ISSN 1029-0265.

[16] D. Al-Dabass, Chapter 12: Nature-Inspired Knowledge Mining Algorithms for Emergent Behaviour Discovery in Economic Models, IGI Global, 2007.

[17] S. Bailey, R.L. Grossman, L. Gu, D. Hanley, A data intensive approach to path planning and mode management for hybrid systems, in: R. Alur, T.A. Henzigner, E. Sontag (Eds.), Hybrid Systems III, Proceedings of the DIMACS Workshop on Verification and Control of Hybrid Systems, Springer-Verlag, 1996. LNCS 1066.

[18] D.J. Berndt, J. Clifford, Finding patterns in time series: a dynamic programming approach. Adv. Knowl. Discov. Data Mining, in: U.M. Fayyad, G. Piatetsky-Shapiro, P. Smyth, R. Uthurusamy (Eds.), AAAI Press/MIT Press, 1996, pp. 229–248, 20.

[19] D.P. Bovet, P. Crescenzi, Introduction to the Theory of Complexity, Prentice Hall, 1994, ISBN 0-13-915380-2.

[20] P. Cawley, The reduction of bias error in transfer function estimates using FFT based analysers, J. Vib. Acoust. Stress Reliab. Des. (1984) 29–35.

[21] C.M. Close, D.K. Fredrick, Modelling and Analysis of Dynamic Systems, second ed., Wiley, 1994.

[22] D. Dewolf, D. Wiberg, An ordinary differential-equation technique for continuous time parameter estimation, IEEE Trans. Automat. Contr. 38 (4) (1993) 514–528.

[23] W. Gersch, Least squares estimates of structural system parameters using covariance function data, EEE Trans. Auto. Control. 19 (6) (1974).

[24] T. Kailath, Lectures on Linear Least-Squares Estimation. CISM Courses and Lectures No. 140, Spring-Verlag, New York, 1978.

[25] R. Kalman, A new approach to linear filtering and prediction problems, Tans. Same: J. Basic Eng. Ser. D 82 (1960) 35–45.

[26] H. Mannila, H. Toivonen, I. Verkamo, Discovery of frequent episodes in event sequences, Data Min. Knowl. Discov. 1 (1997) 259–289.

[27] Z. Man, Parameter-estimation of continuous linear systems using functional approximation, Comput. Electr. Eng. 21 (3) (1995) 183–187.

[28] R.C. Schank, K.M. Colby (Eds.), Computer Models of Thought and Language, Freeman & Co, 1973, ISBN 0-7167-0834-5.

[29] F.L. Seveance, Systems Modelling and Simulation, an Introduction, Wiley, 2001.

[30] L. Mirtskhulava, S. Al-Majeed, G. Pearce, T. Gogoladze, I. Javakhishvili, Blood Clotting Prediction Model Using Artificial Neural Networks and Sensor Networks, GESJ: Computer Science and Telecommunications 3 (43) (2014) 60–66. Reviewed Electronic Scientific Journal, http://gesj.internet-academy.org.ge/en/list_artic_en.php?b_sec=comp.

[31] K. Holewa, A. Nawrocka, Emotiv EPOC neuroheadset in brain computer inter-face, in: 15th International Carpathian Control Conference, ICCC, 2015.

[32] R. Wolpaw, D.J. McFarland, G.W. Neat, C.A. Forneris, An EEG-based brain-computer interface for cursor control, Electroencephalogr. Clin. Neurophysiol. 78 (1991) 252258.

[33] Yuan, Detection of Epileptic Seizure Based on Eeg Signals, IEEE Research Paper, 2010, p. 4209.

[34] I.M.V. Caminiti, F. Ferraioli, A. Formisano, R. Martone, Strategies for Brain Sources and Tissues Properties Identification from EEG/MEG and EIT Sig-Nals, IEEE Research Paper, 2010, p. 978.

[35] H. Nolan, R. Whelan, R.B. Reilly, H.H. Bülthoff, J. Butler, Acquisition of Human EEG Data during Linear Self-Motion on a Stewart Platform, IEEE research paper, 2009, p. 2343.

[36] S. Gurumurthy etal, Int. J. Eng. Technol. (IJET), ISSN 5(3) 0975–4024.

[37] https://www.nature.com/articles/nrneurol.2016.113, http://bnci-horizon-2020.eu/database/data-sets-stroke dataset.

[38] D. Al-Dabass, A. Zreiba, D.J. Evans, K. Sivayoganathan, Parameter estimation algorithms for hierarchical distributed systems. I, J. Comput. Math. 79 (1) (2002) 65–88. ISSN 0020-7160.

[39] https://dl.acm.org/doi/abs/10.1145/2896387.2900319?download=true.

CHAPTER

2

Computational intelligence in healthcare and biosignal processing

Nagaraj Balakrishnan[1], *Valentina E. Balas*[2], *Arunkumar Rajendran*[1]

[1]Department of Electronics and Communication Engineering, Rathinam Technical Campus, Coimbatore, Tamil Nadu, India; [2]Automation and Applied Informatics, Aurel Vlaicu University of Arad, Romania, Arad, Romania

1. Introduction

Data science systems are used to extract knowledge and information, which provides insight into the core of structured and unstructured data (which is gathered in massive quantities). Scientific methods and algorithms are used to analyze and provide profound insights into the data. Further development requires more such insights for the rapid growth of fields such as industry, business, and medical. To get such in-depth knowledge of the areas mentioned earlier, data science needs to be involved with techniques such as data mining, big data, and machine learning. In this research, the optimal use strategy of deep learning algorithms and the clustering methodologies are addressed for unattended data classification applications [1].

Data classification is the process of categorizing the given dataset or a feature for applications like computer vision, medical diagnosis, and so on. The data classification is analyzed, and the decisions are made based on the assessment of the data. The techniques such as Support Vector Machines (SVM), linear regressions, and feature vectors are used for the application of data classification. During the past decade, the machine learning algorithm plays a significant role in the development of data science [2]. The machine learning technique generates a nonlinear logic to suit real-time problems and applications by considering their uncertainty. The machine learning algorithm can be broadly classified into various types, such as (1) Supervised Learning, (2) Unsupervised Learning, (3) Reinforcement Learning. An Artificial Neural Network (ANN) follows the supervised learning methodology, which is widely used in many applications. ANN algorithms are able to learn and

Handbook of Computational Intelligence in Biomedical Engineering and Healthcare
https://doi.org/10.1016/B978-0-12-822260-7.00015-7

understand scientifically the situation with the iterative learning process organized using previously generated/collected data. On the other hand, the Clustering analysis of data mining technique follows the unsupervised learning methodology, which finds a natural grouping of instance, give unlabeled data. To directly reach the objective of data classification, the supervised learning method can be chosen with models such as SVM, Decision Trees, ANN, Linear Regression, Logistic Regression, Linear Discriminant Analysis, and more. Even though the machine learning for data classification produce better results, the new requirements and advancements for the modern trend of applications have a surge for more accuracy. The proposal of a deep learning algorithm (a subset of machine learning algorithm) leads to a new era of research that surpassed the abilities of traditional machine learning methods. The deep learning algorithm is organized with the massive number of layers of ANN in various levels of the network. Hence the data is analyzed in-depth, which extracts an enormous feature that passed to the next layer. The process progressively handles the high-level data abstraction that holistically represents the learned features from the previous layer into another form. Due to these facts the deep learning can also be used for the multiclass classification. A deep learning algorithm performs well at many datasets and applications, but it has limitations that provide for a new research opportunity [2,3]:

(1) Generally, deep learning algorithms operate in the mode of supervised learning. Supervised learning is the process of training the network with labeled or annotated datasets. But to train and classify the real-time parameters of a specific application, the labeled datasets are needed, which requires a massive manual effort, and on the other hand is costly. (2) Deep learning algorithms need a high computation facility to process the massive amount of features generated. Also, the deep learning algorithm learns the pattern of understanding and produces a better outcome by training through the large dataset. This is the reason why the Central Processing Unit (CPU) and graphical processing units (GPU) is in the picture while discussing deep learning algorithms [4].

As mentioned in the previous section, the clustering algorithm groups the data points or features with a similar property. The clustering algorithms do this in an unsupervised mode of operation. Unlike the deep learning algorithms, it does not need a dataset to train itself to process the grouping and to proceed toward classification. Soft and hard clustering algorithms are the types of clustering methods which involve the applications based on various aspects of data. As the clustering algorithm is bounded with limitations, it has considerable difficulty in applying it for classification problems [2,4].

1.1 Objectives of the research

The focus of this research is to enhance the behavior and nature of the deep learning method with the nature of the clustering algorithm so that the unsupervised learning methodology can be implemented in the deep learning algorithms for efficient data classification.

The deep learning and machine learning methods are employed in a variety of applications to solve various problems that occur due to uncertainty. But these problems were solved with the help of data collected from the history of occurrences of the event. Most of the machine learning and deep learning algorithms are trained to solve the supervised learning problem, where the algorithms know the prediction requirement. On the other hand, the potential measure of the unsupervised learning method is quite high. The ability to explore new possibilities of the outcome is high. In general, supervised learning methods are bounded with biases, in which the set of rules are determined with the DOs and DONTs, which prohibit the thinking of other possibilities. Also, high effort, manual work, and time are required to label the data for the supervised learning process, in case the labeling is not available. So, the primary objective of the research is to enhance the ability of unsupervised learning into the deep learning methodologies. This objective triggers a series of subobjectives to concentrate on, such as (a) the selection of suitable unsupervised learning methodologies (in this case, the clustering algorithms are utilized for the analysis); (b) the selection of suitable and efficient deep learning methodology; (c) the selection of diverse datasets and problems to test and validate the research outcomes; and (d) the exploration of the optimal deep learning methodology for data classifications. The objective of the research is described in detail in the following topics of this section.

1.2 Challenges of data classification

A data classification scheme is one of the significant features of an information security system that is employed to maintain and protect data. The data classification scheme provides an effective solution during the process of risk management and data security preferences. It provides a natural hierarchy to manage the various levels of data. Depending on the applications, the data classifications are utilized with the tags of segregations defined by context, content, and behavior. The data classification is implemented in many ways, which are given as follows:

(a) Manual intervals: This method works well with the tiny dataset, in which all the segmentation is done manually with the help of a human. (b) Equal intervals: This method segments the data into several groups (as per the user requirements) with the evenly distributed information. (c) Quantities: The segmentation through the quantity of data in each class. (d) Natural Breaks: Depends on the changes that happened in each data, a variation from its nature. (e) The geometric intervals: this method defines the segmentation based on the geometrical interval in each class of the data. (f) Standard deviation intervals: during the segmentation of data using standard deviation intervals, the attributes of the data are defined, and the variation from its normal is calculated, based on which the process is carried. (g) Custom Range: This method works with the help of input from the user and changed at any point in time, based on the additional requirements [2–5].

1.3 Data classification: implementation hurdles

As per the quote "Information is wealth," every entity in the world is organizing their version of data to understand the customers to know the depth of market penetration and to make the strategies for the future. This may apply to all sorts of entities around us.

(1) The data classification would be much easier if the classification is done manually by their employees (or the in-charges) during the process of storage. But this is not an easy task if in case the data are generated on a massive scale. Most of the entities today realized the importance of classification and insisted on their process managers to do the preprocessing (classification) before storing. However, their traditional data, which is generated in the past decades, seeks the requirement for the modern algorithms and process segment/classification [5].

(2) There are many traditional algorithms available for the researcher to enhance and use them for applications of classification but most of these algorithms are linear, which is not efficient for uncertain data. Additionally, the accuracy may vary with the size of the dataset.

(3) These difficulties urges the researchers to move toward nonlinear algorithms such as machine learning methods. But again the use of machine learning needs specific requirement such as labeled data, etc., Yet the accuracy is not promising [6].

(4) Though the machine learning has the advantage of working in uncertain data, lacks in accuracy paves the way for the study of deep learning method (a subset of machine learning). Indeed, a supervised learning method needs a massive volume of data to train to reach the accuracy goal. The deep learning algorithm plays a significant role in today's Artificial Intelligence–based applications. However, this platform needs many requirements such as (a) high computational power like GPU, (b) similar to machine learning methods a massive labeled dataset for supervised learning, and (c) adequate parameter selection to avoid overfitting or underfitting [7].

1.4 Effective data classification: implementation strategies

As described in the previous section, the primary objective of the research is to enhance the deep learning methodology to work in unsupervised modes with the help of a suitable clustering algorithm. The strategy of adopting the unsupervised learning behavior in deep learning is as follows.

The reason behind the usage of the deep learning method instead of traditional artificial neural networks is its ability to extract the deep abstraction of features. It makes the machine to work similar to the human brain in terms of data classification and other applications. The first step in this strategy is to transform the elements of the dataset $\boldsymbol{X}$ into a latent feature space $\boldsymbol{Z}$. A suitable network of the deep learning algorithm processed by the nonlinear mapping is used to achieve the transformation. In addition to the transformation, the dimension of the latent features $\boldsymbol{Z}$ is reduced much lesser than $\boldsymbol{X}$. The next phase of the strategy is to process $\boldsymbol{Z}$ to form a $\boldsymbol{k}$ number of clusters by initializing the random centroid and later converging it to the right position. During the process, the clustering loss is calculated along with the reconstruction loss occurred while processing the deep network. With the consolidated loss function ($\boldsymbol{L_R}$ **and** $\boldsymbol{L_C}$), the deep network can be trained in several iterations until the loss is minimized. This final trained network can be used as an unsupervised learning structure for the rest of the process. Thus the unsupervised classification is implemented with various combination to explore the best solution. The details of the implementation and its related study is noted in the following sections [8–10].

1.5 Preliminaries

To travel with this research, the concepts mentioned in the following sections play vital roles in the subsequent development of new possibilities of Artificial Intelligence.

1.5.1 Data clustering algorithms

As per the objectives of the research, the process is entirely involved with the unsupervised learning methodology. To adopt this behavior in the proposed work, the clustering algorithms, such as the versions of k-means, Fuzzy C Means (FCM), and more, are analyzed.

In this regard, the mathematical modeling to perform the clustering such as (a) calculation of the cluster centers of the data, (b) manipulation for the membership function to perform grouping, and finally, (c) the convergence of the iterative clustering assignment through the objective function, is involved. Clustering loss is the primary metric used to influence the performance of the deep learning network.

1.5.2 *Deep learning algorithms*

In addition to the clustering algorithm, deep learning algorithms play a significant role in the research. However, to understand the methodology of the deep learning algorithms, the basic understanding of the structure of the ANN is needed. The knowledge toward the mathematical modeling on the input, hidden, output layer, gradient descendant optimization, and thresholding is required. This research especially initiates with the Autoencoders algorithm, which can powerfully perform the training operation.

1.5.3 *Simulation tools for the operation*

This research is carried out with the help of two primary tools available, i.e., ***TensorFlow***, an open-source algorithm developed by the Google Brain team. It has precious library resources available for the many machine-learning algorithms. It is strongly recommended for research related to the deep learning method. The library functions such as ***numpy***, ***pandas***, and ***Scikit-learn*** are the precious resource for this research, which helps to build an efficient model. MATLAB is also an excellent resource for this research with operational libraries. However, in this research, MATLAB is used to test and validate modules (mathematical model) of the study.

1.5.4 *Dataset requirements*

Testing the research experiments of deep learning is a convenient task due to the availability of the suitable datasets. Many datasets are available for learning, testing, and validating functions of modules that are proposed and generated in this research. The datasets, such as *ImageNet*, is an image and vision research-based dataset, which is organized based on the WordNet hierarchy. Consisting of 14 million data under various categories, Modified National Institute of Standards and Technology (MNIST) is the most widely used dataset, especially for deep learning algorithms. Consists of handwritten digits, build from the NIST. US Postal Service (USPS), which contains about 9298 handwritten numerical numbered images (7291 training and 2007 testing samples) with a size of 16×16. This is a highly difficult dataset. They are used widely for testing the text classification algorithm. *CIFAR-10* is also the publicly available dataset which is the subsets of tiny eighty-million images. It consist of 32×32 sized images in 10 classes. Each class is filled with 6000 images. The dataset is separated with 50,000 datasets for training and 10,000 dataset for testing. STL-10 is an image dataset with 10 classes. There are 500 images for training, and 800 images for the testing process, size normalized to 96×96. Using these data without the label is a challenging task. These datasets are frequently used for testing the proposed modules since it has labeled data, which can be much helpful for validating the results obtained.

2. Investigation on various deep clustering algorithms

Today, many organizations and well-established industries have difficulties in handling their data (which has been generated every second in a massive quantity). Analyzing these data is becoming a challenging task. The machine-learning algorithms are employed to suit the situation. The algorithm like deep learning play a vital role in enhancing the modern technologies and its detailed analysis. Supervised learning methods have high potentials, which provide a solution for many problems. However, the unsupervised learning method has the potential to unlock many unknown scales of opportunities. On the other hand, the clustering methods of data mining algorithms can group unknown data into structures. Based on knowledge discovery, this method finds a way to cluster the data without supervision. Most of these clustering algorithms use distant metrics to complete the process. The process of combining the operation of deep learning algorithms and the clustering algorithms are known to be deep clustering algorithms [2,3].

Deep Embedded Clustering (DEC) methods outperform the process of unsupervised learning through its clustering ability. The ability to implement the cluster assignment and feature learning enables more possibilities and also fills the gaps traditionally maintained by the supervised learning methods. There are some algorithms, which evolved the process of the basic procedures of DEC. During the process of DEC, a feature space is created with the help of Autoencoders by processing the actual data transformed into features of latent space. Further, the clustering process influences the training phase of the Autoencoder through the clustering loss constraints. DEC performs in two-stage of operation (1) pretraining and (2) fine-tuning. The clustering algorithm gets its initial parameters such as cluster centers and convergence criteria from the pretraining stage, and these parameters are utilized in the further sequence known as fine-tuning. This is the stage in which both feature learning, and the clustering is processed. DEC prefers to use the Autoencoder because the framework is very simple, robust, and highly suitable for the data reconstructions. This section shows the details on DEC along with its versions in brief. Also this section addresses the perception of the research and analysis on the algorithms as per the objective specified in the previous section [4].

The deep learning networks are employed in the process to concentrate and learn low dimensional features of the data. Autoencoder algorithm is widely used in deep learning network for this particular operation. During the process of learning, the system is calibrated with parameters such as (1) Loss Function, (2) Network Loss, and (3) Clustering Loss. The performance metrics of each algorithm is measured with the following factors such as (a) Unsupervised Clustering Accuracy (ACC), (b) Normalized Mutual Information (NMI), and (c) Adjusted Rand Index (ARI). The details of these parameters and the performance metrics are as follows.

2.1 Loss function: calibration on deep clustering

As shown in Fig. 2.1, the process of deep clustering method is involved with both nonlinear learning and the clustering methodology, the parameters such as Network Loss L_R (measured with the attributes of the neural network) and the Clustering Loss L_C (measured with the characteristics of the clustering algorithm on the data) are calculated (per Eq. 2.1). The Loss Function L is defined as

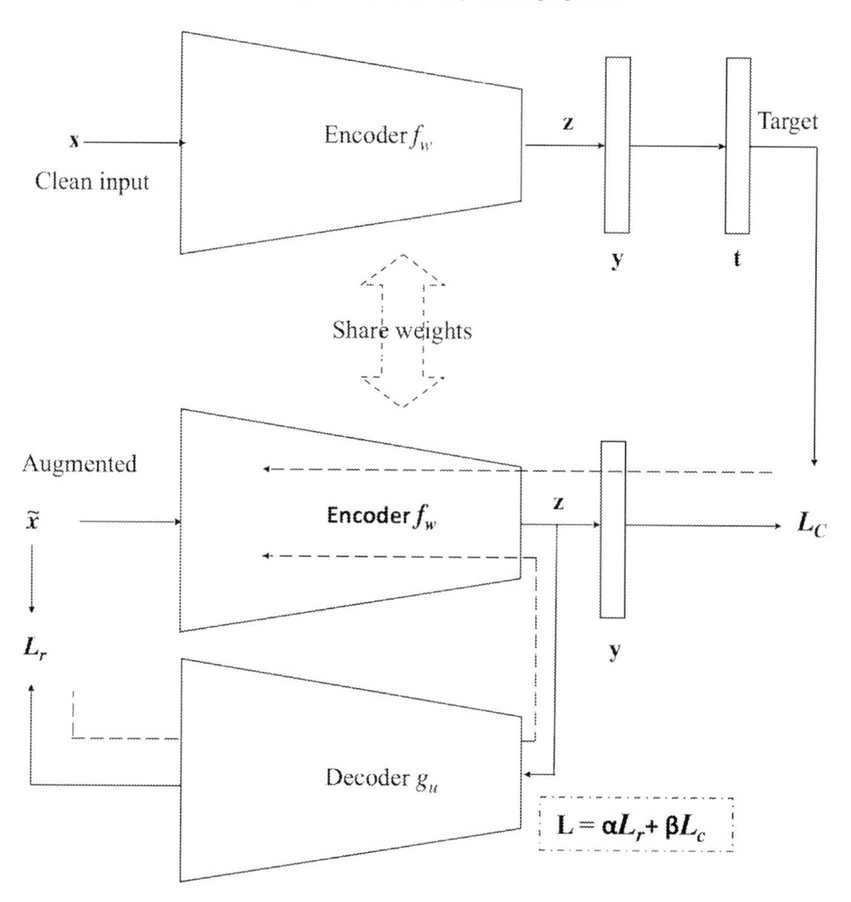

FIGURE 2.1 Basic process of DCN with stacked Autoencoder.

$$L = \lambda L_R + (1 - \lambda) L_C \tag{2.1}$$

Where λ is termed as hyperparameter, adjusted between 0 and 1 to balance the impact of network and the clustering loss.

2.1.1 Network loss (L_R)

During the process of deep learning strategy, the difference in the measurement toward the actual data and the reconstructed data is analyzed through reconstruction loss, only when the Autoencoder network is used. In the case of Variational Autoencoder (VAE) and Generative Adversarial Networks (GAN), the variational loss and an adversarial loss are considered. Networks loss is essential for training the deep learning network irrespective of its learning mode of operation (supervised or the unsupervised).

2.1.2 *Clustering loss (L_C)*

The clustering loss is a metric, which defines the ability of the clustering algorithm to make the clustering accurately. The method of this metric measurement may vary on the algorithm used for the process. Cluster Assignment loss and Cluster Regularization loss are the categories which are generally used in this research.

Cluster Assignment loss measures the loss occurred during the clustering process directly using the data points in the cluster. It is estimated by taking the distance between each data points and the cluster centers or through the student's t-distribution method. The methods such as cluster assignment hardening loss, k-means loss, and agglomerative clustering loss come under this category, which is briefly discussed in the below chapters with respective to the algorithm. Cluster Regularization loss is another category of measuring clustering loss. This measurement preserves the important discriminant information analyzed from the cluster data representation. The methods, such as group sparsity loss, locality loss, etc., are discussed in this section with respective to the algorithm.

2.2 Performance metrics

In this research, the performance of the existing systems and the research contributions are analyzed with the following metrics. The labeled data from the standard datasets were used to perform these measurements.

2.2.1 *Unsupervised clustering accuracy (ACC)*

Unlike general accuracy measurement methods, ACC performs a mapping through the mapping function (mp) between the ground truth(g) and the clustering assignment output(c) (as shown in Eq. 2.2). Since the final output is generated from the unsupervised learning procedures, the ground truth labels have no chance of matching with the output (even though the classification is the same).

$$ACC = max_{mp} \frac{\sum_{j=1}^{n} 1(g_j = mp(c_i))}{n} \tag{2.2}$$

where i and j are the loop variable to point out the particular datapoint.

Generally in ANN and Deep Learning methodologies the given input data will be mapped to the output or target or Ground Truth (g_i) during the training. n is the total number of input used for testing. So through ACC the percentage of input that appropriately mapped with the g_i is analyzed.

2.2.2 *Normalized mutual information (NMI)*

This measurement normalizes the average entropy (H) of the cluster assignments(c) and the ground truth labels(g) (as shown in Eq. 2.3). Through this analysis, the measure of mutual information present in the output compared to the ground truth is identified.

$$NMI(g, c) = \frac{I(g, c)}{\frac{1}{2}[H(g) + H(c)]} \tag{2.3}$$

2.2.3 Adjusted rand index (ARI)

ARI is also another form of accuracy measurement method that calculated the similarity between two data clusters.

As shown in Fig. 2.2, the measure and the evaluation of ARI is obtained in percentage of correct decisions made by the algorithm. The ARI is as shown below, using the permutation model (as shown in Eq. 2.4).

$$ARI = \frac{\sum_{ij}\binom{n_{ij}}{2} - \left[\sum_i\binom{a_i}{2}\sum_j\binom{b_j}{2}\right]/\binom{n}{2}}{\frac{1}{2}\left[\sum_i\binom{a_i}{2}\sum_j\binom{b_j}{2}\right] - \left[\sum_i\binom{a_i}{2}\sum_j\binom{b_j}{2}\right]/\binom{n}{2}} \tag{2.4}$$

where n_{ij}, a_i, and b_j are the values taken form the contingency table. Let the set S with n elements, and two clusters of the n elements be $X = X_1, X_2, \ldots X_r$ and $Y = Y_1, Y_2, \ldots Y_r$. The intersection or overlapping region of X and Y is framed to be the contingency table n_{ij}.

2.3 Deep embedded clustering (DEC) method [11]

DEC method utilizes the reconstruction loss produced by the Autoencoder along with the cluster assignment hardening loss. DEC is often considered to be a standard point of reference for the other algorithms to benchmark their performance (general DEC method is shown in Fig. 2.3). The process of DEC is as follows.

(1) Let the set $\{x_i \in X\}_{i=1}^{n}$ be the data points (which need to be learned in an unsupervised mode of operation) transformed into latent feature space Z through the nonlinear mapping $f_\theta(X \rightarrow Z)$ which is processed by the Deep Neural Networks (DNNs) (in this case an Autoencoder method). θ is the learnable parameters, although the dimension of the features in latent space is much less than the actual data.

(2) DEC extracts the k cluster centers from Z which is termed as $\{\mu_j \in Z\}_{j=1}^{k}$.

X \ Y	Y_1	Y_2	Y_3	Y_4	......	Y_S	Sums
X_1	n_{11}	n_{12}	n_{13}	n_{14}	......	n_{1s}	a_1
X_2	n_{21}	n_{22}	n_{23}	n_{24}	......	n_{2s}	a_2
⋮	⋮	⋮	⋮	⋮	⋱	⋮	⋮
X_r	n_{r1}	n_{r2}	n_{r3}	n_{r4}	......	n_{rs}	a_r
Sums	b_1	b_2	b_3	b_4	......	b_S	

FIGURE 2.2 Contingency matrix table.

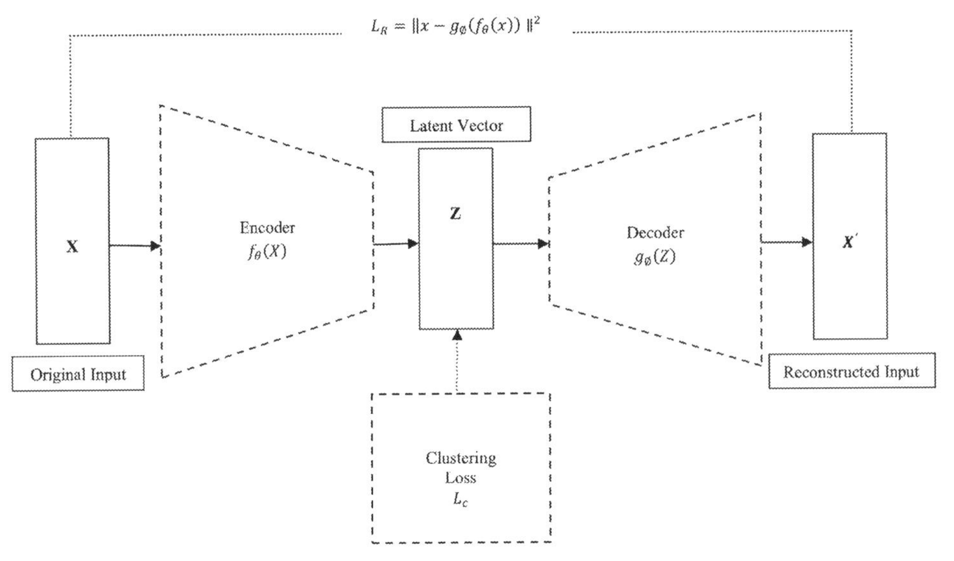

FIGURE 2.3 Autoencoder based deep clustering approach.

(3) DEC undergoes two-phase of operations such as (I) Initialization of Autoencoder, which deals with setting up the parameters and preprocess the data. (II) Parameter optimization, which is an iterative process that computes an auxiliary target distribution and also optimize the Kullback–Leibler (KL) divergence [11,12].

During each iteration, the optimization toward the autoencoder i.e., DEC algorithm, is handled as mentioned below. In general, most of the machine learning and deep learning algorithms use the Stochastic Gradient Descent (SGD) for minimizing the objective function of the network. In this case, the loss function using SGD is calculated concerning the feature space, which are embedded data points z_i and the centroid of each cluster μ_j as follows.

$$\frac{\delta L}{\delta z_i} = \frac{\alpha+1}{\alpha}\sum_j\left(1+\frac{\| z_i - \mu_j\|^2}{\alpha}\right)^{-1} \times (p_{ij}-q_{ij})(z_i-\mu_j) \tag{2.5}$$

$$\frac{\delta L}{\delta \mu_j} = -\frac{\alpha+1}{\alpha}\sum_i\left(1+\frac{\| z_i - \mu_j\|^2}{\alpha}\right)^{-1} \times (p_{ij}-q_{ij})(z_i-\mu_j) \tag{2.6}$$

The gradient loss function $\frac{\delta L}{\delta z_i}$ (Eq. 2.5) as per equation mentioned above is utilized for the computation of general backpropagation $\frac{\delta L}{\delta \theta}$ (Eq. 2.6), which is used for the network learning. α is the degree of freedom of student's t-distribution (which is assumed such that, $\alpha = 1$), and p_{ij} (Eq. 2.7) is the auxiliary target distribution given by

$$p_{ij} = \frac{\frac{q_{ij}^2}{f_j}}{\sum_{j'}\frac{q_{ij'}^2}{f_{j'}}} \tag{2.7}$$

$f_j = \sum_i q_{ij}$ are the soft cluster frequencies.

q_{ij} is termed as the soft assignment of the cluster as shown in Eq. 2.8, i.e., probability of assigning samples i to cluster j.

q_{ij} is given by

$$q_{ij} = \frac{\left(1 + \| z_i - \mu_j \|^{2/\alpha}\right)^{-\frac{\alpha+1}{2}}}{\sum_{j'}\left(1 + \| z_i - \mu_{j'} \|^{2/\alpha}\right)^{-\frac{\alpha+1}{2}}} \tag{2.8}$$

After the preliminary learning is attained by the DNN for the first time, the clustering algorithm processes the respective z_i features and also finds the cluster centers μ_j. In general, almost all the clustering algorithm iterates several times to find the convergence state. In this case, minimizing the KL loss is considered to be the convergence state.

KL is given by

$$L = KL(P||Q) = \sum_i \sum_j p_{ij} log \frac{p_{ij}}{q_{ij}} \tag{2.9}$$

Once the clustering algorithm attains the convergence state after several iterations, the current parameters such as z_i, μ_j, p_{ij}, and q_{ij} are used to generate SGD. This loss function is termed as clustering loss of L_C. Parallel to this, reconstruction loss L_R is also calculated with the DNN (as shown in Eq. 2.9). Finally, the learning parameter of the DNN is updated and prepares the network for the next training sessions. Thus the DEC algorithm operates as the Stacked Autoencoder (SAE) to simultaneously process both cluster and deep learning algorithms for the unsupervised learning process [2–4,12].

2.4 Discriminately Boosted Clustering (DBC) method [12]

DBC algorithm follows the same procedures as that of DEC, including the training scheme, clustering scheme, usage of KL divergence, and soft cluster assignment distribution. But the only difference is the utility of Convolutional Neural Network (CNN) based Autoencoder instead of general Autoencoder (feedforward based). This modification boosts the process of accuracy than the DEC, especially for the Image-based datasets.

This framework makes two state of operations, i.e., (1) Fully Convolutional Autoencoder (FCAE) and (2) the DBC, as shown in Fig. 2.4 (Part I is an FCAE (Fully Convolutional Autoencoder), and Part II is a DBC framework based on the FCAE). The training proceeded by the DBC part is considered as the preprocessing step, in which the soft k-means model categories the clusters. However, the data points will undergo varied training compared to the DEC method. Since CNN can learn the image features by extracting fine coarse image features, this method hardly used for the image datasets. Various scores on the features are generated from the clustering model depends on cluster assignment. Especially the highly scored assignments are highlighted with significants. The parameters such as r_{ij} (discriminative score/discriminative target distribution) and s_{ij} (soft k-means score) are evaluated from the clustering process [11,12].

The soft k-means score is defined as

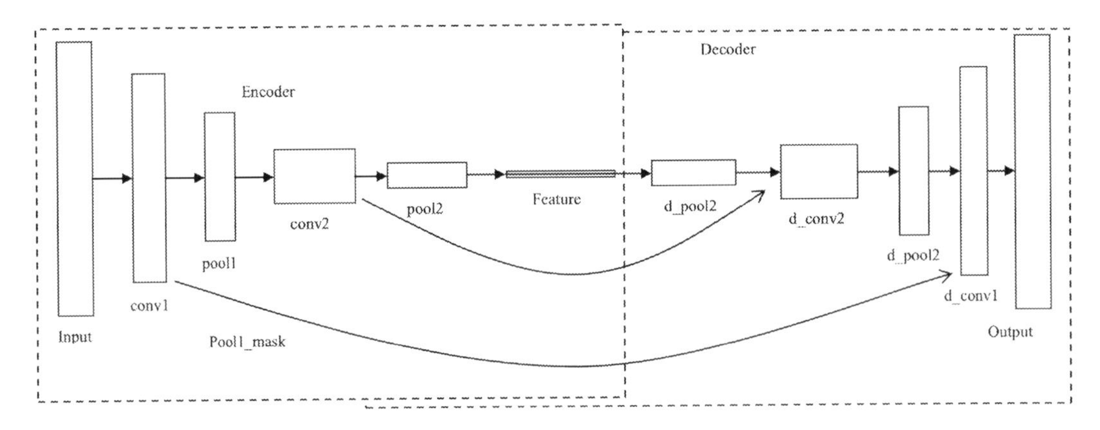

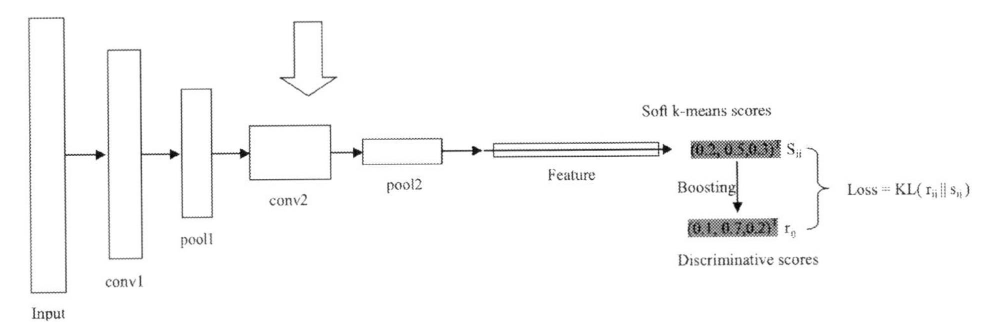

FIGURE 2.4 A unified image clustering framework.

$$s_{ij} = \frac{\left(1 + \| z_i - \mu_j \|^2\right)^{-1}}{\sum_j^k \left(1 + \| z_i - \mu_j \|^2\right)^{-1}} \tag{2.10}$$

where, $z_i = f_\theta(x_i)$

Eq. 2.10 is given by the t-distribution kernel through which the similarities or soft score between the cluster centers μ_j, where $(j \in 1 \text{ to } k)$ and the latent features z_i, where $(I \in 1 \text{ to } m)$ are analyzed.

On the other hand, the discriminative target distribution is given from s_{ij}, as shown in the equation below.

$$r_{ij} = \frac{\frac{s_{ij}^{\alpha}}{n_j}}{\sum_j^k \frac{s_{ij}^{\alpha}}{n_j}} \tag{2.11}$$

where, $n_j = \sum_{I=1}^{m} s_{ij}^{\alpha}$

Here, the α is a degree of freedom of the t-distribution.

As per Eq. 2.11, Soft k-means clustering method uses KL divergence as the objective function, which is given by

$$L = KL(R||S) = \sum_{i=1}^{m}\sum_{j=1}^{k} r_{ij}\log\frac{r_{ij}}{s_{ij}} \tag{2.12}$$

For each iteration, the clustering algorithm will try to convergence the KL loss and initiates the next session of neural network training, by calculating the gradient descent as follows formed with Eq. 2.12.

$$\frac{\delta L}{\delta z_i} = \frac{1+\alpha}{\alpha}\sum_{j=1}^{k}(r_{ij}-s_{ij}) \times \frac{z_i-\mu_j}{1+\parallel z_i-\mu_j\parallel^{\frac{2}{\alpha}}} \tag{2.13}$$

$$\frac{\delta L}{\delta \mu_j} = \frac{1+\alpha}{\alpha}\sum_{j=1}^{k}(r_{ij}-s_{ij}) \times \frac{\mu_j-z_i}{1+\parallel \mu_j-z_i\parallel^{\frac{2}{\alpha}}} \tag{2.14}$$

whereas Eqs. 2.13 and 2.14 is termed as L_C (clustering loss), which is combined with the reconstruction loss L_R to fine-tune the network training in each training sessions. Thus the DBC process as unsupervised learning, similar to the DEC.

2.5 Deep clustering network (DCN) method

DCN is a method that practically approaches the problem of adopting unsupervised learning methodology in deep learning by building a joint operation. According to this method, instead of processing the latent feature space (which is more complex) the latter genre such as weights w obtained after each training session can be used for the clustering process.

This method proposes a joint operation of DR and clustering method, i.e., k-means clustering assignment to accomplish the task. The optimization criteria used in the process consist of three practical actions such as (1) dimensionality reduction, (2) data reconstruction, and (3) cluster structure-based promoting regularization.

As mentioned earlier, instead of using the direct latent features, the weights that associated with each features is used, as discussed below.

$$h_i = f(x_i; w), f(;w): R^M \rightarrow R^R$$

where $f(;w)$ indicates the function to map the w i.e., the weights associated with the particular input x, in the respective hidden layer h_i. The joint DR and the clustering is formulated as

$$\min_{w,m,s_i} \widehat{L} = \sum_{i=1}^{N} \parallel f(x_i; w) - M_{s_i}\parallel_2^2 \tag{2.15}$$

$$s.t. S_{j,i} \in [0,1], 1^T s_i = 1 \forall i,j$$

where, $\widehat{L}$ of Eq. 2.15 is the clustering loss or distance in the Euclidean space. M is a cluster function which contains the clusters m_k, s_i is the assignment vector of data point i. Since k-means method is simple, yet powerful for the data with low-dimension and less effective for the data with high-dimension, it is assigned in such a way to operate with joint DR and clustering. As shown in equation above, the system calculates clustering loss which is insufficient to drive the DNN in right momentum, may lead to a trivial solution. Reconstruction loss is mandatory in this situation to influence the learning factor in DNN for more effectiveness. So the joint operation on DNN based DR and the k-means clustering is processed, where the weight w is directly used in the decoding part of DNN since the latent space features z are not considered for the operation. Based on this the cost function is given as

$$min_{w,m,s_i}\sum_{l=1}^{N}\left(\ell(g(f(x_i)),x_i)+\frac{\lambda}{2}\|f(x_i)-M_{s_i}\|_2^2\right) \tag{2.16}$$

$$s.t.S_{j,i}\in[0,1],1^Ts_i=1\forall i,j$$

where, $f(x_i;w)$ is simplified as $f(x_i)$ shown in Eq. 2.16.$g(h_i;z)$ is simplified as $g(h_i).\ell()$ is a loss function, which calculates the reconstruction loss and second part of this same equation extracts the clustering loss, through which the learning factor is influenced with the help of back propagation based SGD [3,12,13].

$$\nabla_\chi L^i=\frac{\delta\ell(g(f(x_i)),x_i)}{\delta\chi}+\lambda\frac{\delta f(x_i)}{\delta\chi}(f(x_i)-M_{s_i}) \tag{2.17}$$

where $\chi=(w,z)$ of Eq. 2.17 is a gradients and the parameters of the network is calculated by the back-propagation. During each training phase χ is updated as mentioned below.
$\chi\leftarrow\chi-\alpha\delta_\chi L^i$, α is predefined as >0.

2.6 Deep embedded regularized clustering (DEPICT) method

DEPICT operates way different than the algorithm which briefed before. Even though the previous algorithms (such as DEC, DBC, and DCN) solves the problems, it lacks scalability and efficiency (in such cases where the massive dataset is used). In this method, a multilayer convolutional autoencoder is used. Through which the relative entropy-based objective function is defined to regularize the cluster assignment. A data-dependent regularization strategy is employed to calculate the reconstruction loss, which prevents the network training to attain the overfitting. The joint learning framework introduced in this work optimally minimize the clustering loss L_C, reconstruction loss L_R, and simultaneously train the network.

The first step in the joint operation is the calculation of probabilistic cluster assignment as mentioned in the below equation.

$$p_{ik}=\frac{\exp(\theta_k^Tz_i)}{\sum_{k'=1}^{K}\exp(\theta_{k'}^Tz_i)} \tag{2.18}$$

which can also be defined as the $P(y_i = k|z_i, \Theta)$. Here, $\Theta = [\theta_1, \ldots \dot{\theta}_k]$ is the softmax function that indicate the current probability distribution of the ith data point. p_{ik} is the calculation of probability of ith data point in kth cluster as shown in Eq. 2.18.

In general, the KL divergence is utilized to impose the optimization of the clustering assignment such as $L = KL(Q \parallel P) = \frac{1}{N}\sum_{i=1}^{N}\sum_{k=1}^{K} q_{ik} log \frac{q_{ik}}{p_{ik}}$, through the approximation of the gradient $\frac{\delta L}{\delta q_{ik}}$ the q_{ik} is defined as follows

$$q_{ik} = \frac{p_{ik}/(\sum_{i'} p_{i'} k)^{\frac{1}{2}}}{\sum_{k'} p_{ik}/(\sum_{i'} p_{i'} k')^{\frac{1}{2}}} \tag{2.19}$$

Through the investigation of the p_{ik} and the q_{ik} (as shown in Eq. 2.19) the network parameter such as $\psi = [\Theta, w]$ (as shown in Eq. 2.20) is updated using the following objective function.

$$\min_{\psi} -\frac{1}{N}\sum_{i=1}^{N}\sum_{k=1}^{K} q_{ik} \log p_{ik} \tag{2.20}$$

For the applications such as classification, which updates the w (embedding function) and the Θ (softmax function) effectively with the help of backpropagation loss [11,12].

2.7 Variation deep embedding (VaDE) method [13]

VaDE is a novel approach which utilizes the Variational Autoencoder (VAE). VaDE does an unsupervised generative clustering approach with the combined operation on Gaussian Mixture Model (GMM) and the DNN (i.e., VAE). Since VaDE is a generative model based, the latent representation space can be interpolated and new samples from the data distribution can be created. The VaDE clusters data with GMM, prior to DNN. Then the latent embedding features Z were generated, which is let to the DNN for decoding into an observable. The joint action of VAE and the GMM is given by Evidence Lower Bound (ELBO) as shown below [12,13].

$$L_{ELBO}(x) = E_{q(z,c|x)}[\log p(x|z)] - D_{KL}(q(z, c|x)||p(z, c)) \tag{2.21}$$

where, the $E_{q(z,c|x)}[\log p(x|z)]$ (as shown in Eq. 2.21) is a term, representing the reconstruction of the DNN and $D_{KL}(q(z, c|x)||p(z, c))$ is representing the KL divergence from the Mixture of Gaussians (MoG).

$p(x|z)$ represents the probability of input data and the latent embedded space. It can be given as $Ber(x|\mu_x)$, when $x \rightarrow binary$ and $\mathcal{N}(x|\mu_x, \sigma_x 2I)$, when $x \rightarrow real - value$. μ_x, σ_x are the parameters of the gaussian distribution generated by MoG.

$q(z, c|x)$ is the calculation of mean-field distribution of the DNN, which is given by the neural network model g as shown in Fig. 2.5. $\left[\tilde{\mu}; \log\tilde{\sigma}^2\right] = g(x; \varphi)$, $q(z|x) = \mathcal{N}(z; \tilde{\mu}, \sigma^2 I)$.

Thus the L_{ELBO} related the clustering as well as DNN parameters to influence the system toward better accuracy. The cluster assignment, which is directly inferred from MoG prior, optimizes the entire process [13].

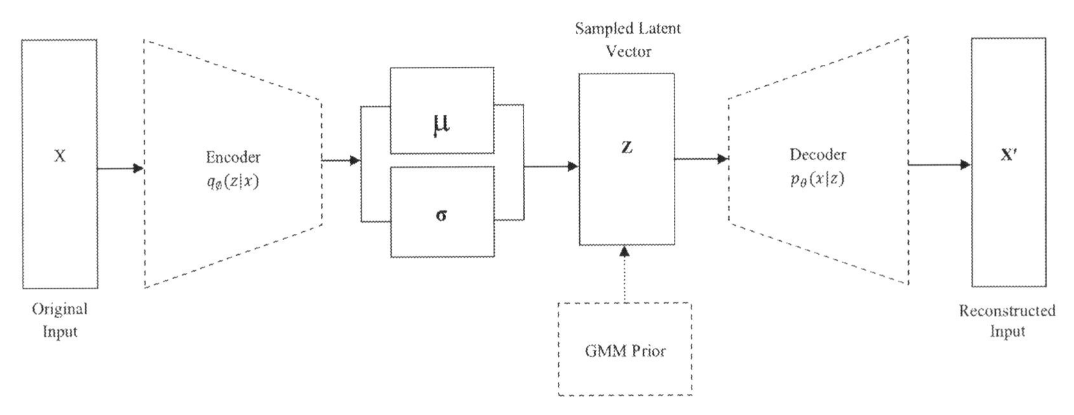

FIGURE 2.5 Process of variational deep embedding (VaDE).

2.8 Information maximizing generative adversarial network (InfoGAN) method [14]

The InfoGAN method is an advanced modified methodology of Generative Adversarial Network (GAN), which can learn with the disentangled representation methodology through unsupervised approach as shown in Fig. 2.6. GAN frameworks utilize a minimax game approach to train the deep generative models. The learning toward the mutual information between the generator distribution $P_G(x)$ and the real-data distribution $P_{data}(x)$ are achieved by this method. GAN learns the generator network G by playing against the ADN (Adversarial Discriminator Network) D. It transforms the P_G into a noise variable {z} into G (such as G(z)). With this process the optimal discriminator of ADN is given by $D(x) = P_{data}(x)/(P_{data}(x)+P_G(x))$.

InfoGAN formulated the variable mutual information maximization to address the lower bounding mutual information $I(c;G(z,c))$ which is termed to be $L_1(G,Q)$ i.e., $\leq I(c;G(z,c))$ is given by $\mathscr{E}_x\tilde{}G(z,c)\left[\mathscr{E}'_c\tilde{}p(c|x)[\log Q(c'|x)]\right]+H(c)$.where Q is the auxiliary distribution with posterior representation $\mathscr{E}_x[D_{KL}(P(|x))]\rightarrow 0$.

Now, the $L1(G< Q)$, a variational lower bound at maximum, is equal to $H(c)$ (which is the entropy of latent codes).

Thus the variable regularization (in minimax game) of InfoGAN is given by

$$\underset{G,Q}{min}\underset{D}{max}V_{InfoGAN}(D,G,Q) = V(D,G) - \lambda L_1(G,Q) \tag{2.22}$$

where λ (as shown in 2.22) is a hyperparameter that associated with the data points.

2.9 Joint unsupervised learning (JULE) method [13,14]

The JULE method process the data toward both deep learning as well as clustering. Since it uses CNN, which is arranged in the stack for the process, this method is best suitable for the image datasets. Here the recurrent framework is used which process the agglomerative clustering (expressed in forwarding pass) and the CNN learning process (shown in the backward pass). Unlike other methods, a single loss function is defined to optimize the recurrent framework (which includes the CNN as well as agglomerative clustering).

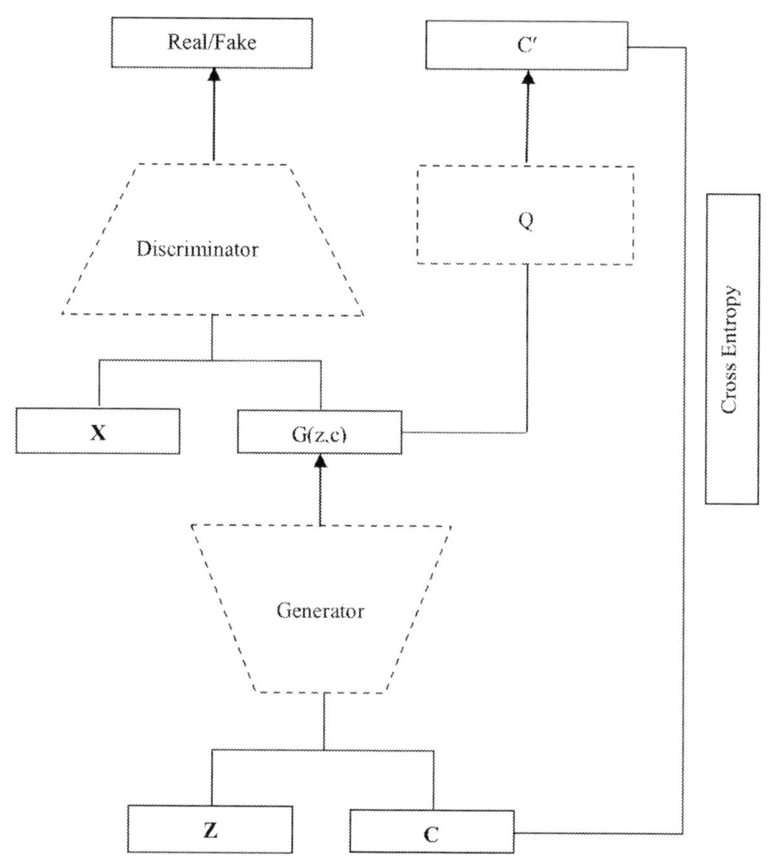

FIGURE 2.6 Process of information maximizing generative adversarial network (InfoGAN).

The process of the simplified recurrent framework is defined by $y^t = f_o(h^t) = h^t$, $h^t = f_m(X^t, h^{t-1})$, $X^t = f_r(I|\theta^t)$. where f_r is the transformation function to extract the features of the deep representation X^t on a particular input I. This function utilizes the θ^t upturned by the previous training session for the operation.

The agglomeration clustering has the property of merging two clusters in each iteration until the convergence is obtained. f_m is a merging function used in the process, which generates the h^t with the influence of X^t and a current hidden state h^{t-1} (also known as the label of the cluster at time step t).

The parameters of the agglomerative clustering are used to updates the CNN learning factors, through the backpropagation procedures.

The loss function or the objective function of the process is given by

$$L\left(\{y^1, \dots y^T\}, \left\{\theta^1, \dots \theta^T\right\}|I\right) = \sum_{t=1}^{T} L^t(y^t, \theta^t|y^{t-1}, I\Big) \tag{2.23}$$

Eq. 2.23 is derived into forward and backward pass.

During the forward pass this equation is used as

$$L^p\left(y^p|\theta^p, I) = \sum_{t=t_p^s}^{tpe} t_p^e L^t(y^t|\theta^p, y^{t-1}, I\right) \tag{2.24}$$

where, y^p is the image label sequence for a period p. The process of agglomerative clustering is processed with the time step $\left[t_p^s, t_p^e\right]$ in which the merging of two neighboring cluster (C_a, C_b) is implemented. $L^p(y^p|\theta^p, I)$ in the forward pass will influence the parameters of the conventional agglomerative clustering process.

During the backward pass this equation is given as

$$L\left(\theta|\{y_*^1 \ldots, y_*^p\}, I) = \sum_{k=1}^{p} L^k(\theta|y_*^k, I\right) \tag{2.25}$$

In the forward pass (as shown in Eq. 2.24) the process finds the clusters and merges it give an optimal sequence label with image clusters, i.e., $y_*^p = \{y_*^t\}$. The ultimate aim of the backward pass (as shown in Eq. 2.25) is to reduce and minimize the loss generated by the forwards pass, i.e., $L^p(y^p|\theta^p, I)$ and to extract the optimal θ, through which unsupervised learning is achieved.

2.10 Deep adaptive image clustering (DAC) method [15]

DAC is a direct cluster optimization process that recasts a binary pairwise classification framework for the clustering problem. In this algorithm, the images are taken in pairs, analyzed to find whether it belongs to the same cluster. *ConvNet* (Convolutional Network) is utilized as a single-stage along with the cosine distance method, which assesses the similarity among the features. The clustering of the images is processed based on the response for the learned labels. The singles stage *ConvNet* used in the systems is known as the Adaptive Learning method, which correlates and improves the clustering procedure.

Let χ be the image dataset given by $\{x_i\}_{i=1}^n$ and k be the number of clusters denoted in the binary Pairwise-classification process. The trained dataset D is given by the $\{(x_i, x_j, r_{I,j})\}_{i=1, j=1}^n$, where x_i and x_j are the unlabeled images from the dataset χ. r_{ij} represents the label for the comparison of similarities between x_i and x_j.

If x_i and x_j have similar features and belongs to same cluster then r_{ij} indicates *logical − one* else *logical − zero*.

According to this logic the objective function of DAC is defined by

$$\min_{w} E(w) = \sum_{i,j} L(r_{ij}, g(x_i, x_j; w)) \tag{2.26}$$

where this Eq. 2.26 indicates the loss function of estimated similarity $g(x_i, x_j; w)$ and r_{ij}, w is a model parameter in g.

The loss function formally represented as

$$L(r_{ij}, g(x_i, x_j; w)) = -r_{ij}log(g(x_i, x_j; w)) - (1 - r_{i,j})log(1 - g(x_i, x_j; w)) \quad (2.27)$$

where $g(x_i, x_j; w)$ is the process of measuring the similarities of two images x_i and x_j through cosine distance method.

i.e., $f(xi; w) \cdot f(x_j; w) = I_i \cdot I_j$. f_w is a function to map the labels and the input images, and I_i and I_j indicates the label features of x_i and x_j in k dimension.

The clustering criteria introduced in the DAC will remodel this equation into the following form

$$\underset{w}{min}E(w) = \sum_{i,j} L(r_{ij}, I_i \cdot I_j) \quad (2.28)$$

$$s.t. \forall i, \| I_i \|_2 = 1, \text{ and } l_{ih} \geq 0, h = 1, \ldots, k$$

where, l_{ih} is label feature of $h - th$ element.

On the other hand, the selection of the sample for training the ConvNets is handled with two observations, such as the untrained ConvNet can able to capture the Low-frequency features from the image and trained network can able to capture the high-frequency features (edges) from the image. With this, the model can be rewritten (ref. Eqs. 2.27 and 2.28) as

$$\underset{w,\lambda}{min}E(w, \lambda) = \sum_{i,j} v_{i,j} L(r_{ij}, I_i \cdot I_j) + u(\lambda) - l(\lambda) \quad (2.29)$$

where, v is an indicator coefficient, i.e.,

$$v_{i,j} = \begin{cases} 1, & \text{if } r_{i,j} \in \{0, 1\}, \\ 0, & \text{otherwise,} \end{cases}$$

$i, j = 1, \ldots, n$, $v_{i,j}$ enables and disables the loss function of the DAC based on the particular set i.e., $x_i, x_j, r_{i,j}$ is selected for the training. $u(\lambda)$ and $l(\lambda)$ defines the number of training samples that need to be processed known as a penalty term. DAC methods operates by increasing the same samples and by decreasing the penalty term.

2.11 Deep clustering framework based on orthogonal autoencoder (COAE)

Orthogonal AutoEncoder (OAE) constraints the process of the AE, which enhances the latent embedding space by discriminative representation. The fully connected layers use the OAE replaces the spectral clustering method that used with the DNN. SGD in the DCN influenced by the parameters of the OAE, which creates a natural regularization among the objective function to prevent overfitting. The proposal of the Clustering framework based on Orthogonal AutoEncoder (COAE) possess the components for dimension reduction and clustering. The OAE method have an ability to learn the features in low-dimensional embedding space. This method follows the SAE, in which the cross-entropy between the target and predicted assignments of loss function L along with the reconstruction error, L_R of OAE is computed [14,16].

As shown in Fig. 2.7, the dimension reduction part of the COAE generates the reconstruction loss is given as $\min_{W} \| X - \widehat{X} \|_F^2$ and the clustering loss is given by Eq. 2.30

$$\begin{aligned} KL(Q\|P) &= -H(Q) + H(Q;P) \\ &= \sum_{i=1}^{n}\sum_{j=1}^{k} q_{ij}\log q_{ij} - \sum_{i=1}^{n}\sum_{j=1}^{k} q_{ij}\log p_{ij} \\ &= \sum_{i=1}^{n}\sum_{j=1}^{k} q_{ij}\log\left(\frac{q_{ij}}{p_{ij}}\right) \end{aligned} \tag{2.30}$$

where the details of the parameters involved are already discussed in the previous section.

Let the data-matrix be X and k be the number of clusters, the hyperparameters involved in the process is λ, μ and the T be the maximum number of epochs [16].

Initial weights of the neural network (AE) is given by W i.e., $W = \{W_D, W_E\}$ (weights in the layers of encoder and the decoder) and Θ (fully connected layer).

For each Iteration (for $t = 1T$)

The process to compute the latent space features and the output via forward propagation method is as $z_i = \sigma(W_E * x_I)$ and $\widehat{X}_i = \sigma(W_D * z_I)$.

Simultaneously the computation of predicted assignment P (as shown in Eq. 2.31) and the target assignments Q (as shown in Eq. 2.32) are generated which is given by

$$p_{ij} = \frac{\exp(z_i^T \theta_j)}{\sum_{j'=1}^{k} \exp(z_i^T \theta_{j'})} \tag{2.31}$$

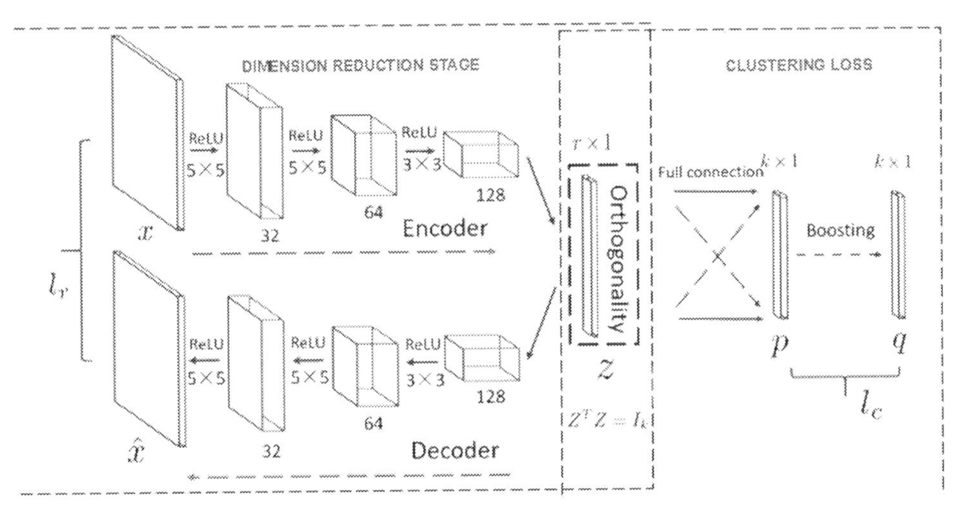

FIGURE 2.7 Process of clustering framework based on orthogonal autoEncoder.

$$q_{ij} = \frac{p_{ij}/\left(\sum_i p_{ij}\right)^{1/2}}{\sum_{j'=1}^{k} p_{ij'}/\left(\sum_i p_{ij'}\right)^{1/2}} \tag{2.32}$$

At this stage, the W and the Θ is updated, given by

$$\min_{\Theta.W} \parallel X - \widehat{X} \parallel_F^2 + \lambda \parallel Z^T Z - I_k \parallel_F^2 + \mu H(Q;P)$$

The iteration is repeated until the convergence state is attained.

2.12 Significance and challenges

The dimension of the data is the significant factor that defines the limitation and benefits of the clustering algorithm. Most of the clustering algorithms measure the distance between each data points to ensure the similarities, which does not work well when the complexity of the data dimension increases. The utility of deep learning algorithms alleviates the problem, but still, dimension reduction is required in the case of spectral clustering methods. Defining an absolute objective function to train the deep learning algorithm can improve the performance of the deep clustering algorithm in terms of high dimension data.

The process of deep clustering methods created the manifold to incorporate and perform the operations such as learning, data dimension reduction, clustering, and more. This methodology creates a complete end-to-end framework. Also, to process the massive data (high dimension), which includes the text, images, and videos, the DNN and DCA are incorporated. This also enhances the clustering algorithm to cluster the low dimensional space learned features.

2.12.1 *Challenges with direct autoencoder latent space manipulation*

The utility of the Autoencoder latent space features for the clustering process has a serious threat as the chances of corruption in latent space will affect the entire data and sometimes even lead to complete loss of data. Due to the fact that clustering loss does not always guarantee perfect outcome, the prevention of local structure formation is questionable which results in the incompetence classification.

2.12.2 *Challenges with implementation and training*

DNN model have a complex architecture, enhancing it with the clustering algorithm will increase more complexity in the phase of implementation and training. DNN and DCN have several hyperparameters involved in the process of learning, estimating loss, clustering and validating the performance, when used for the real data, which are not trivial enough to set, due the unavailability of the ground truth for the dataset(lack of interpretability). Optimizing these factors is highly complex to perform.

2.12.3 *Challenges with representation of clusters*

In DCN, clustering stage is considered to be very significant, since it enhances the unsupervised learning behavior in the DNN. The cluster obtained from the learning may sometime overlap the natural representation. The potential of this problem is high, since it affects the primary objectives of this research as supervision is needed to overcome.

3. Investigation on clustering algorithms for the unsupervised learning methodology

In the previous section, the detailed investigation of deep clustering algorithms is organized, which also includes the significance and challenges of implementing the enhancement of unsupervised learning methodology in the DNN. The difficulty of implementing DCN which is also discussed in the previous section is addressed in this section with the proposed solution for the same. The presection of this section contains the discussion on clustering algorithms and the evaluation that processed to address the various limitation of the predecessor. The postsection organizes the proposed work that contributed toward the clustering methodology [17].

One of the essential requirements of data science studies is the capability of analyzing and understanding the features extracted from the data. The applications such as computer vision utilize this capability of data science to understand the information available in the data that associated with the real-world. These real-world data is processed by inferring the statistical information, insight on spatial features, or sometimes divide and rule methodology (i.e., segmentation). The segmentation/clustering of data is the process of data mining techniques that obtains the particular region by analyzing the interconnected multivariate values of the area. The clustering process can be obtained through actions such as region extraction, thresholding, etc. The clustering process is highly important as the accuracy of the system solely depends on it. This process helps the system to segment the massive dataset to several groups through the measurements on the similarity between one element and another. The performance of the clustering algorithm depends on the distance measurement method and the objective function selected for the operation [18]. A multidimensional, unstructured data can be organized to detect the groups through the user-defined parameter. Once the dataset is clustered, these groups are used to make analytics, regression, classification, and anomaly detection. In this section, the focus given for the clustering algorithms as a continuity of the challenges that mentioned in the previous section.

In this research, data classification is the main application-focused, which is defined and attained by the deep learning algorithm. A data classification gained significance in the past decade as the field of information system is rapidly skyrocketing its growth. The data classification scheme provides an effective solution during the process of risk management and data security preferences. Besides, it gives a natural hierarchy to manage the various levels of data. According to the application need the data classification method segregates the tags with parameters such as context, tenure, content, and behavior. Clustering is categorized into hard and soft processing. In the case of this research, soft clustering methods are used in the existing circumstances, as well as the proposed research contributions [19–21].

3.1 Insight on various clustering algorithms

To induce more intelligence, the deep learning algorithms are incorporated with clustering to make the system classify like a human.

The *k*-means is one of the significant and famous algorithms among the various clustering methods, which defines the clusters using the distance measurements (Euclidean distance metrics). This method is simple yet powerful, among other techniques [22,23]. It classifies the data into k clusters as per the users requirement. k-means processes the data in spatial space, where the centroid is defined with the statistical method (*mean*) after each cluster. These clustering algorithms select random centroids and correct itself as per the conditions of the cluster and its divergence (objective function). This is the reason that these algorithms are called unsupervised. Once the process initiates, the iterations last until the convergence state is attained. This convergence state indicates the minimization of the distance between the data points and the cluster centers. The *k*-means method has many suboptimal states which vary based on the initial value chosen [24–26].

The objective function J of the k-means is given by Eq. 2.33

$$J = \sum_{i=1}^{m}\sum_{k=1}^{K} w_{ij} \parallel x^i - \mu_k \|^2 \tag{2.33}$$

where, m is number of data points, K is total number of clusters defined by the user, x^i is the data point in ith vector, μ_k is the kth cluster center, $w_{ij} = 1$ for data point x_i if it belongs to cluster k; otherwise, $w_{ij} = 0$.

k-means algorithm is a basis toward the development of the FCM algorithm [27,28].

The *FCM* method, on the other hand, follows the soft clustering strategy, which has an advantage over *k*-means for its property of finding mutual clusters (i.e., identifying data points that may belong to two or more clusters). Like the k-means method, FCM is an iterative process that progresses on minimizing the objective function. The action of convergence is analyzed through the means of square error. Since this method follows the nature of unsupervised learning, it also initializes and begins to cluster using random selections. The process of FCM is of three stages: (1) cluster center, (2) membership function, and (3) objective function, which is as follows.

The objective function of the FCM is given by Eq. 2.34

$$J_{iter} = \sum_{i=1}^{N}\sum_{j=1}^{nc} [M^f \parallel D_{x_i - c_j} \|^2] \tag{2.34}$$

where N indicates the number of data points in the data vector, $nc$is the number of clusters (this may be given by the user as per the requirement) [29,30].

M^f is a membership function given by Eq. 2.35

$$M^f\left(\frac{d_{ji}}{d_{ki}}\right) = M_{iter} = \frac{1}{\left[\sum_{k=1}^{nc}\left(\frac{d_{ji}}{d_{ki}}\right)^{\frac{2}{f-1}}\right]} \tag{2.35}$$

where d_{ji} and d_{ki} are the distance measurement (Euclidean distance method are used) between the data point D_{x_i} and the cluster center C_j which can also be given as $\| D_{x_i} - C_j \|$ and $\| D_{x_j} - C_k \|$.

f is a fuzzification factor.

C_j is a cluster center generally extracted from the M^f given by Eq. 2.36

$$C_j = \frac{\sum_{i=1}^{N} M^f D_{x_i}}{\sum_{i=1}^{N} M^f} \tag{2.36}$$

The process of FCM is as mentioned below.

(1) To initiate the process of *FCM* the parameters such nc, f, condition of the convergence state S_c are defined.
(2) Since the process is unsupervised the FCM initiates with selecting random membership function $C_{j,rand}$ or a random cluster center M^f_{rand}. In this case, M^f_{rand} is chosen for further processing.
(3) C_j is calculated using the equation from the current M^f.
(4) At this stage, the objective function J_{iter} is calculated using the above equation and checked for the state of convergence as mentioned below.

$J_{iter} - J_{iter-1} < S_c$, i.e., after each iteration, the difference between the value generated in the objective function of the current iteration and the previous iterations are taken. This difference is validated against theS_c.

(5) If the convergence state is attained, the iteration is stopped, and the current M^f is considered as the output, else the current M^f is updated with the new M^f (i.e., M^f_{new} calculated using the equation based on the current C_j). The iteration is continued until the objective is achieved [31,32].

Even though FCM produces better results, the limitation due to local optima increases the time complexity to reach the convergence state and sometimes never reach the state of convergence. The fuzzification factor f is introduced for this reason [33].

Based on FCM, many algorithms were introduced to overcome the limitations.

$$J_{iter} = \sum_{i=1}^{N} \sum_{j=1}^{nc} \left[M^f \| D_{x_i} - C_j \|^2\right] + \frac{a}{N_R} \sum_{i=1}^{N} \sum_{j=1}^{c} \left[M^f \| D_{x_r} - C_j \|^2\right], r \in N_i \tag{2.37}$$

(based on Eqs. 2.33–2.36) where N_R is the cardinality, a is a constant used as an adjustment or boosting parameter.

$$M^f = \frac{\left(\| D_{x_i} - C_j \|^2 \frac{a}{N_R} \sum_{r \in N_i} \| D_{x_r} - C_j \|^2 \right)^{\frac{1}{f-1}}}{\sum_{n=1}^{c} \left(\| D_{x_i} - C_j \|^2 \frac{a}{N_R} \sum_{r \in N_i} \| D_{x_r} - C_j \|^2 \right)^{\frac{1}{f-1}}} \tag{2.38}$$

$$C_j = \frac{\sum_{i=1}^{N} M^f \left(D_{x_i} + \frac{a}{N_R} \sum r \in N_i D_{x_r} \right)}{(1+a)\sum_{i=1}^{N} M^f} \tag{2.39}$$

It gives the objective function (as shown in Eq. 2.37), membership function (as shown in Eq. 2.38) and the cluster center (as shown in Eq. 2.39) of the version of FCM, i.e., FCM_S. The parameters in FCM_S is modified to make the algorithm into a new versions such as FCM_{S1} and FCM_{S2}.

$$\varepsilon_i = \frac{1}{1+a} \left(D_{x_i} + \frac{a}{N_R} \sum_{j \in N_i} x_j \right) \tag{2.40}$$

Further the D_{x_i} is replaced with ε_i to form a EnFCM (Enhanced FCM), where ε_i reduces the time complexity of the clustering process [34–36].

$$S_{ij} = \begin{cases} e^{-max\left(|p_i - p_j|, |q_i - q_j|\right)/\lambda_s - \frac{|D_{x_i} - D_{x_j}|^2}{\lambda_g \sigma_i^2}} & i \neq j \\ 0 & i = j \end{cases} \tag{2.41}$$

where, ith value of the data matrix is considered as the center of the local window (p_i, q_i) and jth value of the data matrix is considered for neighbor of center. In this scaling factor $(\lambda_s \text{ and } \lambda_g)$ are introduced for the efficient operation [22].

$$\sigma_i = \sqrt{\frac{\sum_{j \in N_i} \| x_i - x_j \|^2}{N_R}} \tag{2.42}$$

$$\varepsilon_i = \frac{\sum_{j \in N_i} S_{ij} x_j}{\sum_{j \in N_i} S_{ij}} \tag{2.43}$$

(based on Eq. 2.40)where S_{ij} (as shown in Eq. 2.41) gives the similarity between the data point and its neighbor to distinguish the gradient in it.

Using Eqs. 2.42 and 2.43 with the FCM is known as Fast Generalized Fuzzy C-means (FGFCM) method.

The generation of FCM evolved with some enhancements. One of the significant version of FCM algorithm is Fuzzy Local Information C-Means (FLICM), which is enhanced by the novel fuzzy factor ff_{ki} given by Eq. 2.44

$$ff_{ki} = \sum_{j \in N_{i,i \neq j}} \frac{1}{d_{ij} + 1} (1 - M)^f \| D_{x_i} - C_j \|^2 \tag{2.44}$$

where, ff_{ki} (which is substituted in the Eq. 2.45) is responsible for measuring the distance from the one data point and its neighbor, which preserves the local information from loss during the clustering process. Also ff_{ki} influences the membership function and the objective function of the FCM as shown below.

$$J_{iter} = \sum_{i=1}^{N}\sum_{j=1}^{nc}\left[M^{f} \parallel D_{x_i - C_j}\parallel^{2} + ff_{ki}\right] \tag{2.45}$$

$$M_{iter} = \frac{1}{\left[\sum_{k=1}^{nc}\left(\frac{d_{ji} + ff_{ki}}{d_{ki} + ff_{ki}}\right)^{\frac{1}{f-1}}\right]} \tag{2.46}$$

Eqs. 2.45 and 2.46 shows the Objective function and the Membership Function.

These characteristics of FLICM measures the similarity of the fuzzy local, which preserves the data from the noise and further information. ff_{ki} used in the algorithm is also known as the empirical adjustment parameter. The next section gives a detailed view on proposed algorithm, which is designed based on the FLICM. Thus the enhancement of error free clustering algorithm with deep learning methodologies will produce a new era of computational Intelligence in healthcare and biosignal processing [37–39].

3.2 Results and discussions on various clustering algorithms

The performance of the clustering methods is experimented by comparing the existing algorithms with the proposed. Here some standard image datasets are used to perform the analysis. The existing algorithms are compared by presenting numerical results on various standard and real-time images, incorporated with different types of noise levels. Here, the traditional algorithms such as $k - means, FCM_S, EnFCM, FGFCM,$ and *FLICM* are put to the test under noisy and original conditions to measure the proposed algorithm. The parameter for analysis is chosen based on the list of limitations mentioned and solutions given to overcome. During the test, a set of medical images is used, so that to test the boundaries of the proposed algorithm [40,41].

As per the result, it is found that the *FLICM* is capable of extracting maximum information from the smooth contour medical image. This test is made for about 250 medical images from various databases. Fig. 2.8 shows the sample input medical images (M1-M5) contains a copy of CT Scan, MRI scan, and X-ray images.

The test is carried out for original images as well as the noisy images to explore the information preserving ability of the algorithms. The algorithms such as FCM_{s1} and FCM_{s2} are specialized to cluster the noisy images. So that understanding toward the *FLICM* will be peculiar in this scenario. The parameters such as accuracy and the time-complexity are the main constraints that required for the decision toward DNN enhancement [42].

Table 2.1shows the comparison of various clustering algorithms for its accuracy. Among the 250 image samples from multiple datasets, seven samples are considered for the analysis, the result of some samples is listed in Table 2.1. Since k-means algorithm does the process of

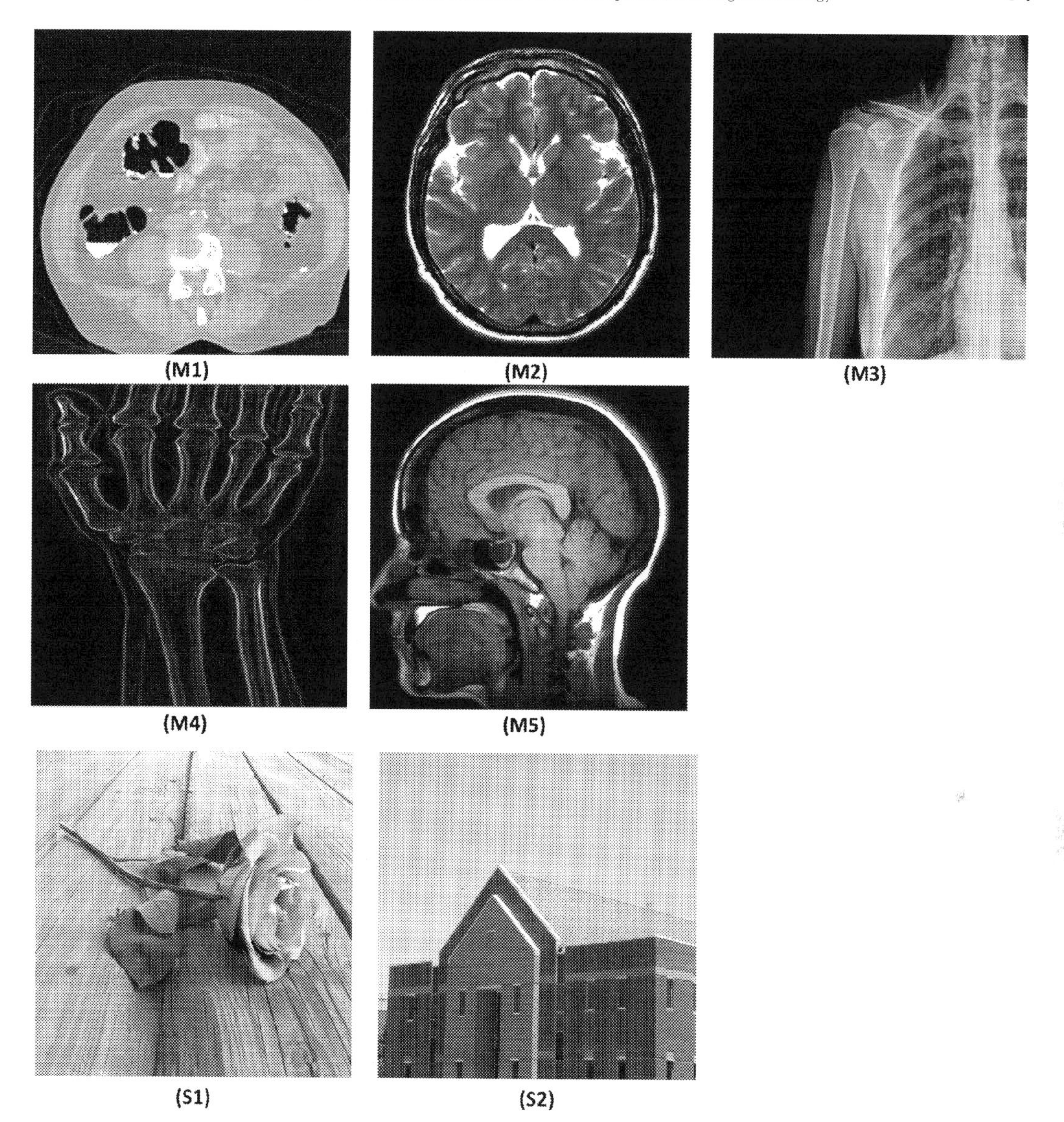

FIGURE 2.8 Samples of original medical images (M1–M5) and Standard original image (S1–S2).

the hard cluster, unlike the FCM version, it is excluded from the comparison. As the FLICM is incorporated with a fuzzy median strategy, the ability to take a median value from the window of neighboring pixels makes it unique to achieve better accuracy on average. Generally, the medical images were bounded with the more contour region(as the grayscale variations are not high). Still, the proposed algorithm is designed in such a way to extract even minor differences in the data neighborhood to form the clusters effectively. Table 2.2 shows the

TABLE 2.1 Comparison of clustering algorithms for accuracy under noiseless condition.

Images	*k-means*	*FCMs*	*EnFCM*	*FGFCM*	*FLICM*
M1	92.08	92.38	92.69	93.69	94.63
M2	88.96	89.09	89.23	90.04	96.009
M3	79.93	80.01	80.10	81.08	89.05
M4	82.88	83.12	83.38	84.51	93.46
M5	81.55	81.75	81.97	83.13	91.08
S1	92.82	93.13	93.46	94.55	98.53
S2	92.78	93.14	93.52	94.73	95.69

comparison of various clustering algorithms for its accuracy under noisy conditions. To study the nature of the algorithms, the noise level (variance) of the image is varied. The noise variance (σ) gives the noise intensity influencing the data. It infects the pixels of the image so that to make the classification or to cluster a challenging data. Noises may be of high or low frequencies that affecting the data in real-time. In this experiment, the high-frequency noise is used for analysis.

The noise in this experiment is varied in four levels (can be changed from 0 to 1) such as 0.2, i.e., 20% of the information is infected, 0.4, i.e., 40% of the information is infected, 0.6, i.e., 60% of the information is infected and 0.8, i.e., 80% of the information are infected. As a result of this validation, FLICM stands for a better accuracy ratio than other algorithms. But the significant parameter that needs to considered equal to accuracy is the time-complexity. As the deep learning algorithm requires GPUs or high computation facilities to simulate the learning process, the system cannot afford more time duration taken by the clustering algorithm in processing the data. Table 2.3 shows the average time taken by the algorithm in the simulation. Generally, the clustering algorithms choose their initial point in the first iteration as random, in which the time duration taken to complete each cycle may be different each time the situation is initiated for the same dataset so that the Table 2.3 indicates the best five simulation runs processed by the algorithms for analysis. This analysis shows that FLICM performing good at various datasets, as the average of the accuracy is taken from the medical as well as standard datasets. Time-complexity is another most important aspect

TABLE 2.2 Comparison of clustering algorithms for accuracy under noisy condition.

s	*k means*	*FCMS*	*EnFCM*	*FGFCM*	*FLICM*
0.2	90.56	91.12	91.09	92.21	93.13
0.4	84.26	84.85	85.28	86.65	88.11
0.6	76.31	80.17	80.31	81.52	84.93
0.8	61.78	66.74	64.91	66.17	71.37

TABLE 2.3 Comparison of clustering algorithms for time-complexity (in s).

Runno.	*k means*	*FCMS*	*EnFCM*	*FGFCM*	*FLICM*
1	14.6	20.45	19.96	16.76	16.25
2	14.24	20.11	20.31	16.33	16.18
3	15.61	19.91	19.45	17.02	16.32
4	14.76	20.22	19.29	17.11	17.06
5	14.78	19.86	19.82	17.22	16.92

of analyzing the clustering algorithm, as the primary objective of the research is to enhance this clustering algorithm in the DNN. These analyses are taken under the configuration—the Intel i7 third generation processor with 1 GB GPU and 4 GB Random Access Memory (RAM). During the investigation, it is clear that the proposed algorithm does not optimally work to complete the clustering task. Even though the algorithm produces excellent results in the phase of accuracy, the time consumed for the execution is more.

3.3 Results and discussions on various deep clustering networks

To make the comparison on DCN algorithms to conclude the research, standard datasets and performance measurement parameters were used. The primary objective of this research is to enhance the ability of unsupervised learning into deep learning methodologies to overcome the challenges mentioned in the introduction section, i.e., the corruption of features due to the clustering loss, the complexity of the model that let the process of training, the problem of tuning the hyperparameters, and the overlapping of clusters. So, it becomes mandatory to compare and analyze the following metrics on the algorithm, to standardize the measure.

(a) Unsupervised clustering Accuracy (ACC) (shown in Eq. 2.3).
(b) Normalized Mutual Information (NMI) (shown in Eq. 2.4).
(c) Adjusted Rand Index (ARI) (shown in Eq. 2.5).

To standardize the analysis, the dataset, such as MNIST, USPS, CIFAR-10, and STL-10, are used. The detailed view on these datasets is as follows.

The MNIST database, widely used for training, testing, and validation of machine learning or deep learning algorithms. It contains a massive number of handwritten numerical digits as images. This dataset has about 60,000 images for the training process and 10,000 datasets for the testing process. Due to image normalization size (of 28 × 28), it is most commonly preferred for the analysis of the USPS, which contains about 9298 handwritten numerical numbered images (7291 training and 2007 testing samples) with a size of 16 × 16. This is a highly difficult dataset. They are used widely for testing the text classification algorithm. CIFAR-10 is also the publicly available dataset which is the subsets of 80 million tiny images. It consist of 32 × 32 images in 10 classes. Each class is filled with 6000 images. The dataset is

TABLE 2.4 Clustering performance (ACC) of different algorithms on public datasets.

Datasets	MNIST	USPS	CIFAR-10	STL-10
k-means	0.5723	0.612	0.2289	0.1920
AE + k-means	0.801	0.682	0.287	0.204
DEC	0.834	0.772	0.253	0.211
VaDE	0.852	0.691	0.268	0.231
JULE	0.884	0.715	0.283	0.246
COAE	0.892	0.721	0.324	0.309

separated with 50,000 datasets for training and 10,000 dataset for testing. STL-10 is an image dataset with 10 classes. There are 500 images for training, and 800 images for the testing process, size normalized to 96 × 96. Using these data without the label is a challenging task.

This analysis further introduces the Deep Embedded Clustering algorithm that discussed in previous sections (i.e., DEC, DBC, DCN, DEPICT, VaDE, InfoGAN, JULE, DAC, and COAE). To simulate and analyze the process, the algorithms are programmed with the help of the Python programming language. The basic syntax on some of the algorithms mentioned above is gathered from the GITHUB, a repository that connects the most significant open-source community.

Tables 2.4–2.6 shows the details of the experiment based on ACC, NMI, and the ARI. Further the analysis is made based on dataset as follows. To make it convenient for the discussion, the below interference is specified in terms of percentage.

MNIST is the most popular publicly available dataset utilized for testing the deep learning–based architectures. Since this dataset consists of handwritten digits in a regularized size, the algorithms find comfortable in testing the parameters. The finding turnout to show that the COAE shows promising outcomes with ACC = 89%, NMI = 84%, and ARI = 77% since the algorithm provides a natural regularization by influencing the parameters of the SGD. This helps in preventing the algorithm process to reach overfitting. The

TABLE 2.5 NMI of different algorithms on public datasets.

Datasets	MNIST	USPS	CIFAR-10	STL-10
k-means	0.501	0.542	0.093	0.1471
AE + k-means	0.711	0.609	0.211	0.193
DEC	0.768	0.737	0.253	0.204
VaDE	0.854	0.512	0.268	0.221
JULE	0.823	0.710	0.271	0.238
COAE	0.846	0.713	0.314	0.299

TABLE 2.6 ARI of different algorithms on public datasets.

Datasets	MNIST	USPS	CIFAR-10	STL-10
k-means	0.912	0.627	0.299	0.254
AE + k-means	0.631	0.593	0.187	0.132
DEC	0.741	0.590	0.1607	0.1861
VaDE	0.761	0.505	0.253	0.218
JULE	0.739	0.675	0.244	0.187
COAE	0.771	0.698	0.296	0.281

performance of the algorithms in the datasets CIFAR-10 and STL-10 in terms of ACC, NMI, and ARI follows the same pattern as that of MNIST. Unlike the MNIST dataset, the USPS is a highly challenging dataset, even though it contains the numerical numbers as the dataset is developed initially from the US postal service and the regularization of the datasets is not performed well. For this dataset, it is seen that the performance metrics are normalized among all the algorithms and a significant fact the k-means algorithm can able to perform nominally well than most of the algorithms with ACC = 61%, NMI = 54%, and ARI = 63% (as the overfitting nature of other the algorithms limit the performance, due to the presence of DNN). Also, the DEC for this dataset outperforms than most of the algorithms, which gives ACC = 77%, and NMI = 74%. This result due to the reason that hyperparameters chosen by the algorithm is suitable for the dataset. At the same time, it provides ARI = 59%, which gives the inference that the data in the cluster have a multiple *Miss* and *False– hit* intersections. As the DEC is equipped with the basic autoencoder and DNN, it can able to perform better than other algorithms since it conducts the process through SAE.

As per the dataset CIFAR-10, the algorithms follow almost the same pattern as that of MNIST. But as the CIFAR-10 contains the tiny images with RGB planes, the features extracted and its gradients are much different from the MNIST dataset.

Even though the COAE shows less performance in the datasets such as MNIST and USPS, it gives better results in the CIFAR-10, since COAE inherits the convolution filters utilized in the CNN. Due to the difficulty level of the dataset, all the algorithms used, provides an average result with 25% accuracy. Like CIFAR-10, the STL-10 dataset has the class of data images with RGB plane. The dataset is organized after the regularization but still this dataset in the same category as of CIFAR-10 or sometimes even more complex than CIFAR-10. Since the COAE has the capability of the convolution filtering strategy of CNN, It computes the performance metrics with ACC = 33%, NMI = 32%, and ARI = 29%.

The significant difficulty faced during the study is the time-complexity, even though the algorithm can perform well compared to the existing models. The execution time is the major limitation, which has a huge research opportunity.

4. Conclusion

In this chapter, various deep clustering algorithms are investigated to understand the problem of enhancing the DNN with clustering algorithms to adopt the unsupervised learning behavior. This study also explored the strategies required to implement the unsupervised learning setup for the DNN optimally. The challenges encountered in the DCN models are briefly discussed. To understand these challenges and to design the optimal deep learning data classification model, the clustering algorithms are investigated in depth, and the insights are studied. To test the performance of the clustering algorithms and to study the nature, an experimentation is made with the standard datasets. As a result, it is found that the average performance of the FLICM algorithm is good compared to the other in noisy and noiseless condition. Based on these analyses, the suitable clustering algorithms are enhanced with the Autoencoder and CNN to perform unsupervised learning and data classification. The experiments are organized with standard datasets. The experimental outcome shows that the capability of the COAE algorithm and its performance are better than the other models. The inferences of these experiments provide a deep understanding of the study. Furthermore, this strategy and analysis give the scope for future research toward Artificial Intelligence.

References

[1] C. Birtolo, D. Ronca, Advances in clustering collaborative filtering by means of fuzzy C-means and trust, Expert Syst. Appl. 40 (17) (2013) 6997–7009.

[2] B. Fang, Y. Li, H. Zhang, J.C.W. Chan, Collaborative learning of lightweight convolutional neural network and deep clustering for hyperspectral image semi-supervised classification with limited training samples, ISPRS J. Photogrammetry Remote Sens. 161 (2020) 164–178.

[3] K.L. Lim, X. Jiang, C. Yi, Deep clustering with variational autoencoder, IEEE Signal Process. Lett. 27 (2020) 231–235.

[4] Y. Li, H. Zhang, J. Chen, P. Song, J. Ren, Q. Zhang, K. Jia, Non-reference image quality assessment based on deep clustering, Signal Process. Image Commun. (2020) 115781.

[5] X. Chen, Y. Duan, R. Houthooft, J. Schulman, I. Sutskever, P. Abbeel, Infogan: interpretable representation learning by information maximizing generative adversarial nets, Adv. Neural Inf. Process. Syst. 1 (1) (2016) 2172–2180.

[6] W. Zheng, H. Shu, H. Tang, H. Zhang, Spectra data classification with kernel extreme learning machine, Chemometr. Intell. Lab. Syst. 192 (2019) 103815.

[7] D.A. Naik, S. Seema, G. Singh, A. Singh, Exploration and implementation of classification algorithms for patent classification, in: Computing and Network Sustainability, Springer, Singapore, 2019, pp. 119–128.

[8] S. Venkataraman, R. Selvaraj, Optimal and novel hybrid feature selection framework for effective data classification, in: Advances in Systems, Control and Automation, Springer, Singapore, 2018, pp. 499–514.

[9] J. Zabalza, C. Qing, P. Yuen, G. Sun, H. Zhao, J. Ren, Fast implementation of two-dimensional singular spectrum analysis for effective data classification in hyperspectral imaging, J. Franklin Inst. 355 (4) (2018) 1733–1751.

[10] A.A. Yahya, A. Osman, M.S. El-Bashir, Rocchio algorithm-based particle initialisation mechanism for effective PSO classification of high dimensional data, Swarm and Evol. Computa. 34 (2017) 18–32.

[11] X. Guo, L. Gao, X. Liu, J. Yin, Improved deep embedded clustering with local structure preservation, in: IJCAI, June 2017, pp. 1753–1759.

[12] F. Li, H. Qiao, B. Zhang, Discriminatively boosted image clustering with fully convolutional auto-encoders, Pattern Recogn. 83 (2018) 161–173.

[13] S. Tian, Y. Zhang, J. Zhang, N. Su, A novel deep embedding network for building shape recognition, Geosci. Rem. Sens. Lett. IEEE 14 (11) (2017) 2127–2131.

[14] X. Chen, Y. Duan, R. Houthooft, J. Schulman, I. Sutskever, P. Abbeel, Infogan: interpretable representation learning by information maximizing generative adversarial nets, in: Advances in Neural Information Processing Systems, 2016, pp. 2172–2180.

[15] L. Wu, Y. Wang, J. Gao, X. Li, Deep adaptive feature embedding with local sample distributions for person re-identification, Pattern Recogn. 73 (2018) 275–288.

[16] W. Wang, D. Yang, F. Chen, Y. Pang, S. Huang, Y. Ge, Clustering with orthogonal AutoEncoder, IEEE Access 7 (2019) 62421–62432.

[17] Y. Ding, X. Fu, Kernel-based fuzzy c-means clustering algorithm based on genetic algorithm, Neurocomputing 188 (2016) 233–238.

[18] W. Gao, Forecasting of rockbursts in deep underground engineering based on abstraction ant colony clustering algorithm, Nat. Hazards 76 (3) (2015) 1625–1649.

[19] Z. Akkus, A. Galimzianova, A. Hoogi, D.L. Rubin, B.J. Erickson, Deep learning for brain MRI segmentation: state of the art and future directions, J. Digit. Imag. 30 (4) (2017) 449–459.

[20] X. Guo, L. Gao, X. Liu, J. Yin, Improved deep embedded clustering with local structure preservation, in: IJCAI, 2017, pp. 1753–1759.

[21] I.B. Aydilek, A. Arslan, A hybrid method for imputation of missing values using optimized fuzzy C-means with support vector regression and a genetic algorithm, Inf. Sci. 233 (2013) 25–35.

[22] C. Bai, D. Dhavale, J. Sarkis, Complex investment decisions using rough set and fuzzy C-means: an example of investment in green supply chains, Eur. J. Oper. Res. 248 (2) (2016) 507–521.

[23] G. Cheng, C. Yang, X. Yao, L. Guo, J. Han, When deep learning meets metric learning: remote sensing image scene classification via learning discriminative CNNs, IEEE Trans. Geosci. Remote Sens. 56 (5) (2018) 2811–2821.

[24] H. Chen, O. Engkvist, Y. Wang, M. Olivecrona, T. Blaschke, The rise of deep learning in drug discovery, Drug Discov. Today 23 (6) (2018) 1241–1250.

[25] L. Deng, A tutorial survey of architectures, algorithms, and applications for deep learning', APSIPA Trans. Signal Inform. Process. 3 (2014) 1–29.

[26] Y. Deng, Z. Ren, Y. Kong, F. Bao, Q. Dai, A hierarchical fused fuzzy deep neural network for data classification, IEEE Trans. Fuzzy Syst. 25 (4) (2016) 1006–1012.

[27] S. Bartunov, A. Santoro, B. Richards, L. Marris, G.E. Hinton, T. Lillicrap, Assessing the scalability of biologically-motivated deep learning algorithms and architectures, in: Advances in Neural Information Processing Systems, 2018, pp. 9368–9378.

[28] S. Ghosh, S.K. Dubey, Comparative analysis of k-means and fuzzy c-means algorithms, Int. J. Adv. Comput. Sci. Appl. 4 (4) (2013) 35–39.

[29] M. Hassan, A. Chaudhry, A. Khan, J.Y. Kim, Carotid artery image segmentation using modified spatial fuzzy c-means and ensemble clustering, Comput. Methods Progr. Biomed. 108 (3) (2012) 1261–1276.

[30] N. Hatipoglu, G. Bilgin, Cell segmentation in histopathological images with deep learning algorithms by utilizing spatial relationships, Med. Biol. Eng. Comput. 55 (10) (2017) 1829–1848.

[31] M. Hauser, M. Steinegger,¨J Soding, MMseqs software suite for fast and deep clustering and searching of large protein sequence sets, Bioinformatics 32 (9) (2016) 1323–1330.

[32] T.C. Havens, J.C. Bezdek, C. Leckie, L.O. Hall, M. Palaniswami, Fuzzy c-means algorithms for very large data, IEEE Trans. Fuzzy Syst. 20 (6) (2012) 1130–1146.

[33] D.C. Hoang, R. Kumar, S.K. Panda, Realisation of a cluster-based protocol using fuzzy C-means algorithm for wireless sensor networks, IET Wirel. Sens. Syst. 3 (3) (2013) 163–171.

[34] C. Hou, F. Nie, X. Li, D. Yi, Y. Wu, Joint embedding learning and sparse regression: a framework for unsupervised feature selection, IEEE Trans. Cybernetics 44 (6) (2013) 793–804.

[35] E. Dupont, Learning disentangled joint continuous and discrete representations, in: Advances in Neural Information Processing Systems vol. 1, 2018, pp. 710–720, 1.

[36] C. Fan, F. Xiao, Y. Zhao, A short-term building cooling load prediction method using deep learning algorithms, Appl. Energy 195 (2017) 222–233.

[37] J. Enguehard, P. O'Halloran, A. Gholipour, Semi-supervised learning with deep embedded clustering for image classification and segmentation, IEEE Access 7 (1) (2019) 11093–11104.

[38] S. Fogel, H. Averbuch-Elor, D. Cohen-Or, J. Goldberger, Clustering-driven deep embedding with pairwise constraints, IEEE Comput. Graph. Appl. 39 (4) (2019) 16–27.

[39] F. Ghasemi, A. Mehridehnavi, A. Perez-Garrido, H. Perez-Sanchez, Neural network and deep-learning algorithms used in QSAR studies: merits and drawbacks, Drug Discov. Today 23 (10) (2018) 1784–1790.
[40] C.W. Huang, K.P. Lin, M.C. Wu, K.C. Hung, G.S. Liu, C.H. Jen, Intuitionists fuzzy C-means clustering algorithm with neighborhood attraction in segmenting medial image, Soft Comput. 19 (2) (2015) 459–470.
[41] S.K. Adhikari, J.K. Sing, D.K. Basu, M. Nasipuri, Conditional spatial fuzzy C-means clustering algorithm for segmentation of MRI images, Appl. Soft Comput. 34 (2015) 758–769.
[42] S. Agarwal, G.N. Pandey, M.D. Tiwari, Data mining in education: data classification and decision tree approach, Int. J. e-Educ. e-Business, e-Manag. e-Learn. 2 (2) (2012) 140.

CHAPTER

3

A semi-supervised approach for automatic detection and segmentation of optic disc from retinal fundus image

Susovan Jana[1], *Ranjan Parekh*[1], *Bijan Sarkar*[2]
[1]School of Education Technology, Jadavpur University, Kolkata, West Bengal, India;
[2]Department of Production Engineering, Jadavpur University, Kolkata, West Bengal, India

1. Introduction

Eyes are the only organs responsible for human vision. Human visionary system works with the collective effort of different parts of the eye. The anatomy of the eye contains the cornea, iris, pupil, retina, lens, choroid, sclera, macula, aqueous humor, vitreous humor, optic nerve, etc. The cornea is the main entrance point of light in the eye. It also protects the lens, pupil, and iris. The adjustment of iris and pupil controls the amount of light to enter the eye. The lens, which is responsible for focusing the light on the retina, is a transparent component of the eye. The retina is a layer inside the eye that contains two types of light-sensitive cells i.e., rods and cones. The rods are mainly found in the outer edges of the retina and sensitive to low light conditions. The cones are mainly positioned near the fovea and sensitive to brighter light and color vision. The number of rods is more than cones. The outer coating of the eye is called the sclera. The middle layer between retina and sclera is choroid. The optic nerve is positioned into the backside of the eye. It carries visual signals from the retina to the brain. The macula, which is positioned at the center of the retina, is responsible for the straight, central, and fine vision in front of the eyes, like faces and written text. The macula is one of the most sensitive parts of the eye. Fig. 3.1A shows the anatomy of the human eye and Fig. 3.1B shows the color fundus image of the retina. The optic disc is the point where the optic nerves leave the retina. The starting region of major retinal blood vessels is also the optic disc. There are no light-sensitive cells in the optic disc region like other parts of the retina.

Handbook of Computational Intelligence in Biomedical Engineering and Healthcare
https://doi.org/10.1016/B978-0-12-822260-7.00012-1

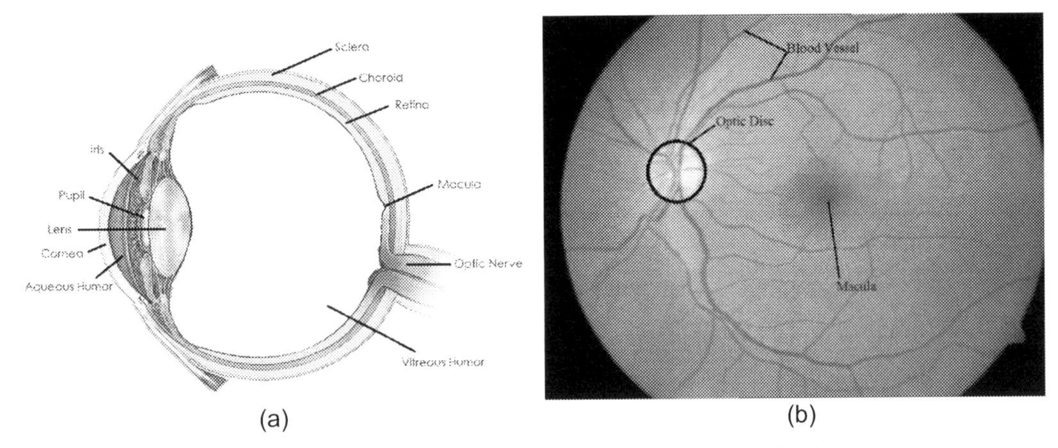

FIGURE 3.1 (A) Anatomy of the human eye; (B) retinal fundus image.

Visually it can be identified as the most bright region inside the retinal fundus image. Another assumption about the optic disc is that it is nearly circular in shape [1].

There are many eye diseases like diabetic retinopathy, diabetic maculopathy, macular edema, age-related macular degeneration, glaucoma, etc. Some of them may cause permanent vision loss. The earlier detection of those diseases may prevent vision loss. The disease diagnosis can be done from fundus photography and OCT image [2,3]. Among them, fundus photography analysis is more popular than OCT analysis. Some popular fundus image dataset is Drishti-GS [4], DIARETDB0 [5], DIARETDB1 [6], STARE [7], etc. Manual diagnosis of eye disease is very time-consuming. Sometimes the diagnosis varies over the expertise of the medical practitioner. This problem creates the need for automated diagnosis of eye disease. In the last decade, there has been a huge demand for automated retinal fundus image analysis for ophthalmologic disorder identification. Eye disease detection using retinal image analysis is still a thrust area for research. A huge amount of work has been carried out toward analysis of the digital fundus image of the retina for retinal diagnosis [8]. The works include retinal blood vessels extraction [9,10], arteriovenous nicking detection [11], optic disk segmentation [12], macula detection [13], etc.

There are also some disease detection works, which were performed based on features extracted from retinal fundus image, like diabetic retinopathy [14,15], diabetic maculopathy [16], macular edema [17], etc. Sometimes it is possible to detect diabetic retinopathy using color, texture, and geometric features from tongue image [18]. But the application of nonretinal image analysis is relatively low for eye disease detection because the nonretinal symptoms are not always sufficient. Diabetic retinopathy is at the top of the leading causes of worldwide vision loss. The symptoms are blurred vision, difficulty in color identification, and sometimes complete vision loss. This is happening due to leakage or damage to blood vessels in the retina. From an image analysis viewpoint, symptoms also include the appearance of exudates, which are liquefied proteins, lipids, cells, etc. Visually they appear as irregular shaped and brighter regions in retina except for the optic disc region. Another eye disease is age-related macular degeneration (AMD), which causes a severe vision problem for adults. Cells of the macula get damaged or stop functioning properly for AMD patients. Mostly it affects people aged 50 and above [19]. A blurry area appears at the center of vision in the early stage of AMD. It creates a problem in activities like reading, writing, recognizing small objects, etc. Over time this blurry area gets increased and may appear as a blank spot and gradually it leads toward blindness. It is the third leading cause of irreversible blindness in the whole world. The major visual symptom for AMD affected patients is the appearance of drusen in the macula. Drusens are also made with fatty protein and lipids. Visually it appears as little deposits of yellow color in the retina. Another eye disease is glaucoma, which gradually damages the optic nerve fibers due to high fluid pressure [20]. It is the second-highest leading cause of irreversible vision loss in the world. In most cases, the patients are unaware of the early stages. Therefore, diagnosis and treatment in the early stage may reduce the chance of blindness [21]. The optic cup and disc should be detected properly for the diagnosis of glaucoma. The cup-to-disc ratio (CDR) measurement is the most accepted test for glaucoma. The ratio is measured between the diameter of the optic cup and optic disc in vertical direction. The chance of glaucoma increases if this ratio is greater than 0.5. In case of glaucoma, the automatic detection of optic disc and optic cup becomes challenging due to the change of intensity in the fundus image of the retina.

Optic disc detection and segmentation are the initial stages for diagnosis of diabetic retinopathy, AMD, glaucoma, etc. There are many approaches to solve the automated detection of the optic disc. The computer vision-based solution is usually more efficient and accurate. The segmentation of optic disc is very challenging as vessels are concentrated there and there is no hard boundary between optic disc region and rest of the parts of retina in the fundus image. This challenge increases specifically when there are abnormalities due to disease in the eye. The abnormalities due to disease may be uneven illumination in fundus image, the appearance of exudate, drusen, etc. In this work, we have proposed a semi-supervised approach for the detection and segmentation of optic disc from the retinal fundus image. The proposed technique has been validated on the Drishti-GS dataset. The work has been structured as follows: Section 2 contains the state-of-the-art survey for segmentation of optic disc by analyzing a retinal image, Section 3 contains insight into the proposed method illustrated with images and equations. Experimentations and results are tabulated in Section 4. Section 5 makes the conclusion of the work with future scope.

2. State-of-the-art

There are different types of approaches for optic disc detection and segmentation from the fundus photographs of retina. The optic disc detection and segmentation approaches can be categorized into morphology-based processing, template matching, unsupervised machine learning, supervised machine learning, deep learning, and other approaches. A survey has been done on all kinds of approaches mentioned here.

2.1 Approaches based on mathematical morphology

Nugroho et al. [20] proposed a morphological reconstruction-based approach for optic disc segmentation. At first, they applied an average filter on the red channel of input color fundus image to soften the edge of the optic disc. Then vessels were removed using the bottom hat transformation. A morphological reconstruction was performed considering a point with maximum intensity detected as a center of the optic disc. Finally, Region of Interest (ROI) was segmented with the help of the active contour method. Another automatic optic disc segmentation was proposed based on multilevel thresholding [22]. Firstly, blood vessels were extracted by sequentially applying matched filtering and local entropy-based thresholding on the green channel of the input image. A range of threshold values was computed by analyzing the histogram valley of the red channel of the input image. After the ranged threshold computation, a morphological erosion and dilation were performed to smooth the disc contour. Bharkad proposed automatic optic disc segmentation from the retinal fundus image [23]. At first, some preprocessing steps were applied to eliminate blood vessels. The next few processing were performed to identify the circular region in the thresholded image. The surrounding region of the detected circle was cropped. The border pixels were removed from the cropped image. Then the cropped image was passed through some morphological and filtering operation to get a properly segmented optic disc. The databases, which were used in the experimentation, were DIRATEDB0, DIRATEDB1, DRIVE, and DRIONS. The parameters for measuring the segmentation performance were specificity, sensitivity, accuracy, average overlap, predictive value, computation time, and average absolute distance. In another method, they cropped the optic disc region and converted into lab color space. Then the density-based spatial clustering [24] was used for optic disc clustering. The thresholded image was passed through a median filter for removing noise. Finally, morphological operations were performed to segment the region properly. A two-stage method for optic disc segmentation has been proposed by Welfer et al. [25] using morphological operations. At first vascular tree was extracted to detect points inside the optic disc. Then, the optic disc boundary was generated from those points using the watershed transform. The datasets used in this work were DRIVE and DIARETDB1. Kulkarni et al. proposed another automated two-stage segmentation technique [26]. In the first stage, adaptive histogram equalization, and morphological operations were performed on the green channel of fundus image to enhance as well as extract the vessel tree. Vessel convergence was marked as the optic disc region. Then, a Markov random field (MRF) was used to remove the vessel from ROI. The method was tested on various public datasets.

2.2 Approaches based on template matching

A predefined template-based linear regression approach for optic disk detection was introduced by Aggarwal et al. [27]. Bright regions of the green channel were thresholded and labeled. Then previously defined fixed size template was used for comparison. The region with a maximum correlation coefficient was selected as the optic disc region. Processing time is the major shortcoming of this method. Aquino et al. [28] also proposed a template-based algorithm to detect optic disc boundary. They used edge detection and circular Hough transform over the green channel of RGB image for detecting the center of the optic disc. Most of the works considered images of normal eye for testing. The previous methods suffer when there is any disease in the eye. An efficient and automatic optic disc detection and segmentation technique were proposed here to address the issue. Reza proposed an automated optic disc detection technique from the color fundus image [29]. The author used a circle operator to detect the optic disc unlike the others using vascular tree or mask generation. The approach had been tested on six public datasets, i.e., DIARETDB1, DRIVE, STARE, DRIONS-DB, DIARETDB0, and Drishti-GS.

2.3 Approaches based on unsupervised machine learning

Mittal et al. proposed [13] an unsupervised approach. The candidates for optic disc center were identified based on the assumption that the intensity decreases if we move away from the center. The final optic disc center was selected using three rejection criteria, i.e., vessel direction, probability of vessels, and expected disc center. An innovative technique was proposed for optic disc and optic cup detection using image analysis [30]. The authors cropped ROI from the input image. They separated the green channel from the cropped image as the green channel contains a good amount of contrast and brightness among the three channels. A fuzzy C-means clustering technique was applied on the green channel of the cropped image after the morphological processing to get the improved segmentation result. Finally, a CDR was computed from the segmented optic disc and optic cup region to diagnose glaucoma disease. Thakur et al. [31] proposed a hybrid technique for the segmentation of optic disc and optic cup from the retinal fundus image. They combined "Adaptively Regularized Kernel-Based Intuitionistic Fuzzy C means (ARKIFCM)" with a level to segment the optic disc from the background. The proposed technique was named as "Level set based Adaptively Regularized Kernel-Based Intuitionistic Fuzzy C means (LARKIFCM)." They experimented on three different datasets i.e., Rim-ONE, Drishti-GS, and Messidor. The parameters for performance measurement were true positive, false positive, true negative, false negative, accuracy, Jaccard distance, dice similarity, and standard deviation. The proposed approach had been compared with nine other popular techniques of segmentation. Almotiri et al. proposed an automatic technique for the segmentation of optic disc [32]. The input fundus image had been passed through some preprocessing steps to enhance the contrast and removing noise. Then fuzzy c-means was applied to identify edges for Hough transform. The Hough transform technique was used to find the optic disc center. The technique was tested on three different datasets i.e., DiaRetDB1, DRIVE, and DRISHTI-GS.

2.4 Approaches based on supervised machine learning

A supervised technique was proposed for detection optic disc [33]. The minimum distance classifier was trained with LBP Histogram, energy, and correlation of Gray Level Co-occurrence Matrix (GLCM) from the cropped image of the optic disc. The gliding-box algorithm was used to decompose the image into multiple subimages. Then the subimages were tested with the trained minimum distance classifier to detect the subimage, which contains the optic disc. A four-stage optic disc region detection was proposed by Uribe-Valencia et al. [34]. In the first stage, candidate pixels were identified by using some pre-processing, ROI mask, contrast enhancement, and local thresholding steps. In the second stage, more promising regions were selected based on channel selection, size-based filtering. In the next stage, some distance-based features were extracted from the predefined set of possible optic disc regions. In the final stage, classification was done to select an accurate optic disc region. The experimentation was carried out on DIARETDB0, DIARETDB1, DRIVE, and e-ophtha-EX.

2.5 Approaches based on deep learning

A deep learning approach was proposed for optic disc and optic cup segmentation [35]. The region of interest was identified by the Hough transform technique. Some preprocessing technique was applied to enhance the image. The red channel was removed from image as the red channel does not carry much meaningful information. The images, which were used for experimentation, are collected from the REFUGE dataset and School of Medicine, Zhejiang University. Another deep learning–based optic disc and cup detection and segmentation approach was proposed for a glaucoma diagnosis [21] using deep learning. The authors proposed an advanced deep learning architecture by combining the popular ResNet-34 model as an encoder with a generic U-Net model as a decoder. The RIGA dataset was used to train the model. The system performance was measured by the outcome of the optic disc and optic cup segmentation, Jaccard index, dice coefficient, and CDR. The system was giving better performance than other state-of-the-art models even when it was tested with popular DRISHTI-GS and RIM-ONE dataset. A new deep learning framework was proposed for fine segmentation of optic disc from the retinal fundus photograph using a popular U-Net model [36]. At first, the grayscale vessel density map was extracted from the input color fundus image. Then the input fundus image and vessel density map were used separately to train the model. The two differently trained models gave two segmentation results. The results were merged to get a local image patch based on an overlap technique. This merged output again fed into the model for more fine segmentation of optic disc. The experimentation was done on six public datasets i.e., DRIONS-DB, DRIVE, DIARETDB0, DIARETDB1, MESSIDOR, and ORIGA. The authors proposed a new CNN architecture, which is named Fine-Net, to generate an optic disc segmentation system from high-resolution fundus image [37] for glaucoma diagnosis. The DRIONS-DB, MESSIDOR, and DRISHTI-GS datasets were used to train Fine-Net. The robustness of the system was tested by five-fold cross-validation. The segmentation performance had been measured by the parameters like Jaccard coefficient and Dice coefficient. An optic disc and optic cup segmentation from the retinal fundus photograph was proposed by Joshua et al. as a prerequisite of glaucoma diagnosis [38]. At first, the optic disc and cup region was cropped from the image. Then a spline interpolation applied for filling. The cropped images are resized. The quality of the image was improved by the application of histogram equalization. A CNN model was trained with those images to do the segmentation of optic disc and optic cup. DRISHTI-GS and RIM-ONE v.3 were used as image datasets for experimentation in this research.

2.6 Other approaches

A hybrid technique was proposed for optic disc segmentation from the fundus image of retina by Tuncer et al. [39]. The retinal fundus image was passed through some preprocessing steps. Then L_1 minimization technique was used for optic disc localization. Finally, a semiautomatic algorithm, which is based on region growing, was used to segment the region of interest. The experimentation was done on 30 color images from the Department of Ophthalmology, Medical Faculty at Firat University. The performance for both localization and segmentation was measured by Dice similarity index and Jaccard similarity index. Mei et al. proposed an optic disc detection technique using the low-rank matrix [40]. The author's extracted 53 features contain color, edge, and textures information. Those features were used to generate a recovery matrix with 1 for ROI and 0 for the background. The final segmentation of the ROI geometry was done by Hough transform. The authors experimented on the MESSIDOR dataset. The segmentation was done using simple image processing with a very small amount of processing. The approach is not data-based and does not require training or machine learning techniques. A novel blood vessel and optic disc segmentation method were proposed by Salazar-Gonzalez et al. [41]. At first, vessels were extracted from the vessel tree by using a graph cut technique. The vessel convergence point was also identified. Initially, the brightest pixel region was identified as the centroid of the optic disc. Then the centroid adjusted iteratively until it reaches the point of convergence of vessels. Two techniques were used to segment the optic are i.e., MRF and compensation factor. DRIVE, DIARETDB1, and STARE datasets were used to validate the proposed technique. An efficient and robust optic disc detection technique was proposed by Lalonde et al. [1]. The authors made three assumptions, i.e., (1) approximate location of the optic disc is known (located in right, left, or center); (2) the brighter region is the optic disc region; and (3) the shape of the optic disc is nearly circular. The pyramidal decomposition method was used for optic disc tracking. The contour of the disc was traced by Hausdorff distance. The proposed technique was tested on 40 low-resolution images. A new optic disc detection technique was proposed by combing different already published methods [42]. They had addressed the problem to detect the correct position of the optic disc where strong obstacles are present in the fundus image. They adopted a two-stage technique. In the first stage, the possible center of the optic disc was detected from the brightness information and vessel map. In the second stage, the best possible border around the center pixel was detected. The proposed technique was tested on 5052 images, which were collected from nine datasets. A "Modified Local Intensity Clustering" model had been applied to accurately segment the optic disc from the fundus image of the retina [43]. Initially, they applied some preprocessing techniques and Hough transform to detect a rough optic disc region. The proposed method was tested on the DIARETDB0 dataset. The automated diagnosis of diabetic retinopathy needs proper segmentation of optic disc. Kumar et al. used a watershed transform to segment optic disc as a prerequisite of diabetic retinopathy diagnosis [44]. Díaz-Pernil et al. proposed an automatic segmentation technique of optic disc from the fundus image of the retina [45]. The approach had been divided into six phrases to perform specific tasks. The phases are compression, color segmentation, binarization, mask detection, computing Hough maximums, and creation of optic disc. The proposed technique was tested on two publicly available datasets, i.e., DRIVE and DIARETDB1. A new technique [46] was given for optic disc segmentation based on the thresholding from color fundus image of the retina. They have selected and enhanced the contrast of green channel using Contrast Limited Adaptive Histogram Equalization (CLAHE). Then two Histogram-shaped thresholding methods, i.e., Triangle and Mean peak, were applied for optic disc detection.

3. Proposed method

The different types of approaches have advantages as well as future scopes for work. Refer to Table 3.1. The previous morphology-based approach requires lots of processing time and the latest deep learning approaches require lots of data for training. We have addressed those challenges in this work by proposing a semi-supervised approach of optic disc detection and segmentation from the fundus image of the retina.

TABLE 3.1 Summary of the selected papers from all type of previous approaches.

Sl no.	Paper title	Author, publication year	Methodology	Dataset	Limitations/future scope
1.	"Segmentation of optic disc on retinal fundus images using morphological reconstruction enhancement and active contour"	Nugroho et al., 2016 [20]	This paper proposes an application of mathematical morphology in optic disc segmentation from the retinal fundus image of a glaucomatous eye.	DRISHTI-GS	**(i)** The optic cup segmentation and glaucoma diagnosis can be done after the optic disc segmentation task. **(ii)** required processing time can be improved.
2.	"Retinal optic disk segmentation and analysis in fundus images using DBSCAN clustering algorithm"	Hamednejad et al., 2016 [24]	A density-based spatial clustering of applications with noise (DBSCAN) was applied to get more refined optic disc segmentation on LAB color space after the ROI masking.	DRIVE	**(i)** Lots of post-processing is needed. **(ii)** the result is very much sensitive to the processing parameters that can be improved.
3.	"Optic disc segmentation using graph cut technique"	Kulkarni et al., 2017 [26]	The graph cut technique was used to detect the optic disc region from the vessel map. Markov random field (MRF) was used to segment the optic disc properly.	Not mentioned	The approach can be tested with more images of diseased and healthy eyes from the different standard datasets.
4.	"A new method for optic disc localization in retinal images"	Aggarwal et al., 2016 [27]	The pixels with maximum correlation with predefined optic disc template are selected as optic disc regions. They used linear regression and blood vessel verification to identify maximum probable ROI.	DRIVE, and DIARETDB1	**(i)** The performance is highly dependent on the template. **(ii)** the regression also sufferers with a diseased eye image. **(iii)** high computation time is the main issue with blood vessel verification.
5.	"Automatic detection of the optic disc in color fundus retinal images using circle operator"	Reza, 2018 [29]	This approach gives balanced accuracy using circle operator in less processing time than the costly vessel-based or background segmentation approaches.	DIARETDB1, DRIVE, STARE, DRIONS-DB, DIARETDB0, and Drishti-GS	**(i)** The limitation of this approach is that the radius length of the optic disc should be known before detection. **(ii)** the approach fails when there is the appearance of exudates due to disease in the eyes.

6.	"Optic disc and optic cup segmentation from retinal images using hybrid approach"	Thakur et al., 2019 [31]	Level set based clustering technique (LARKIFCM) applied for the segmentation of optic disc and optic cup from retinal fundus image. They have compared their method with 9 other segmentation methods.	Drishti-GS, Rim-ONE, and Messidor	The processing time is very high.
7.	"An automated region-of-interest segmentation for optic disc extraction"	Almotiri et al., 2018 [32]	They overcame the challenge of quality enhancement of retinal fundus image. Then fuzzy c-means and Hough transform was applied for locating the optic disc region.	DiaRetDB1, DRISHTI-GS, and DRIVE	**(i)** Fine segmentation can be done from the ROI block detected by the proposed method. **(ii)** the approach can be compared with the other approach also.
8.	"Combining LBP and Co-occurrence matrix information to accurate recognition of the optic disc in retinal image"	Ichim et al., 2016 [33]	The gliding-box algorithm divided the image into multiple sub-images. Structural and statistical texture features were extracted to recognize the sub-image containing the optic disc region.	MESSIDOR	**(i)** The limitation of the work is the box size; hence it is more dependent on the dataset. **(ii)** more accurate segmentation can be done from the ROI.
9.	"Automated optic disc region location from fundus images: Using local multi-level thresholding, best channel selection, and an intensity Profile model"	Uribe-Valencia et al., 2019 [34]	ROI mask and local thresholding were used to detect optic disc candidates. The feature-based classification was done if the detected candidates were more than one.	DIARETDB0, DIARETDB1, DRIVE, and e-ophtha-EX.	The performance of this approach is highly dependent on the retinal vasculature and it degrades if- **(i)** The vessels are not distinguishable. **(ii)** the vessels that spread from the optic disc vertically are not present or appear in low contrast.
10.	"Optic disc and cup segmentation based on deep learning"	Qin et al., 2019 [35]	A convolutional neural network was applied to segment both the optic disc and optic cup at the same time from the preprocessed retinal fundus image.	REFUGE and the dataset from School of Medicine, Zhejiang University	**(i)** The optic cup to disc ratio can be measured for glaucoma diagnosis. **(ii)** there is also scope for improving segmentation performance.

(*Continued*)

TABLE 3.1 Summary of the selected papers from all type of previous approaches.—cont'd

Sl no.	Paper title	Author, publication year	Methodology	Dataset	Limitations/future scope
11.	"Robust optic disc and cup segmentation with deep learning for glaucoma detection"	Yu et al., 2019 [21]	The pretrained ResNet-34 model was combined with classical U-Net architecture to segment optic disc and optic cup. The pretrained network reduces the cost of training from scratch.	RIGA, DRISHTI-GS, and RIM-ONE	(i) The performance is highly dependent on fine-tuning with respect to the dataset. (ii) performance drop due to the appearance of disc atrophy in the case of myopia patient. (iii) image quality degradation causes a performance drop.
12.	"A coarse-to-fine deep learning framework for optic disc segmentation in fundus images"	Wang et al., 2019 [36]	A coarse to fine segmentation of optic disc was done using the U-Net deep learning model.	DRIONS-DB, DRIVE, DIARETDB0, DIARETDB1, MESSIDOR, and ORIGA	(i) The appearance of peripapillary atrophy and fuzzy rim causes wrong pixel estimation, which leads to a performance drop. (ii) Highly dependent on correct estimation of vessel density map. (iii) the approach can be experimented with high-resolution images than 256 × 256.
13.	"High-performance optic disc segmentation using Convolutional Neural Networks"	Mohan et al., 2018 [37]	Fine-net CNN model was used to segment the high resolution (1024 × 1024) retinal fundus image.	DRIONS-DB, MESSIDOR, and DRISHTI-GS	The reason for low performance on DRIONS-DB than other state-of-the-art approaches can be analyzed.
14.	"Segmentation of optic cup and disc for diagnosis of glaucoma on retinal fundus images"	Joshua et al., 2019 [38]	The popular U-net CNN model was improved to segment the optic cup and disc.	RIM-ONE v.3, and DRISHTI-GS	Performance can be improved irrespective of the dataset.
15.	"Optic disc segmentation method based on low rank matrix recovery theory"	Mei et al., 2018 [40]	The theory of low rank matrix recovery and circular Hough transform was used for the segmentation of optic disc. The approach reduces computation time by not using a data-based as well as a template-based technique.	MESSIDOR	(i) The approach can be compared with other approaches as well as in other datasets. (ii) the performance of optic disc detection on the glaucomatous eyes can be improved.

3.1 Complete system overview

The main goal of this work is to detect and segment the optic disc region from the input RGB color fundus image (I). Fig. 3.2 shows the detailed flow of the proposed system. Illumination and some image properties change over the dataset. Some preprocessing steps are needed to make this system compatible with every dataset. Here, preprocessing steps include RGB to grayscale (I_g) conversion, Contrast Enhancement, and Smoothing. I_{CLAHE} is the image after applying contrast-limited adaptive histogram equalization for contrast enhancement. Then Gaussian smoothing is applied on I_{CLAHE} to get I_{gs}. An edge detection technique is

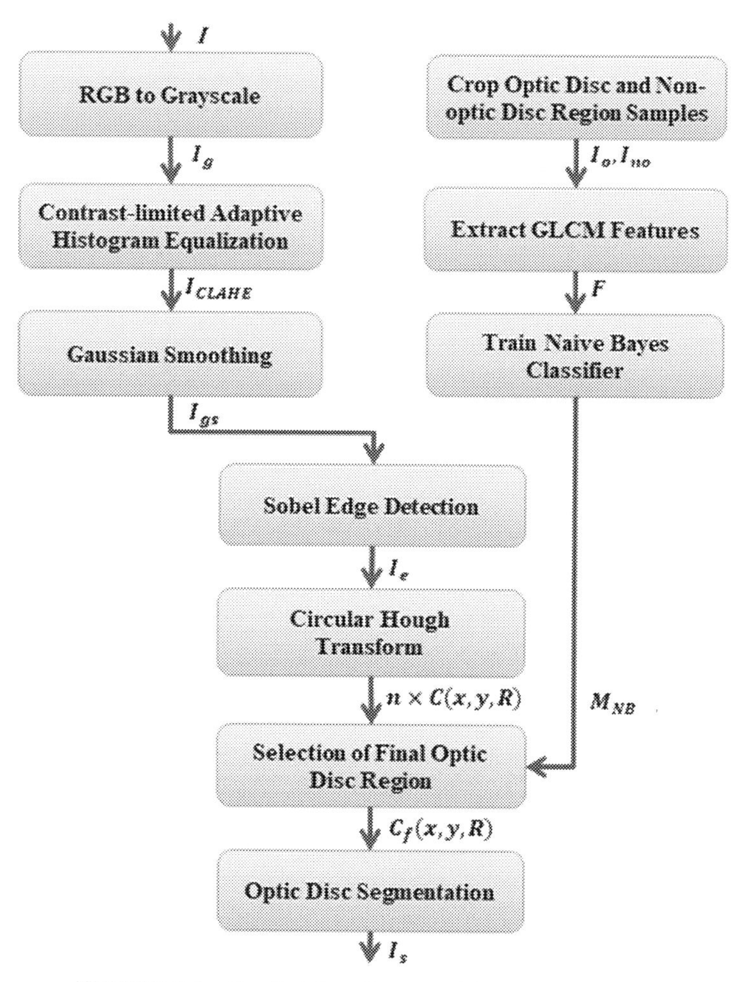

FIGURE 3.2 Detailed flow diagram of the proposed system.

applied on the preprocessed image to get the edge image (I_e). Then n number of possible centers ($C(x, y, R)$) for the optic disc are detected from the edged image using Circular Hough Transform. Some optic disc (I_o) and non-optic disc (I_{no}) samples are cropped from images of the dataset. Naïve Bayes classifier, a supervised classifier, is trained with the GLCM features (F) from those cropped regions. The trained Naïve Bayes model (M_{NB}) is applied to select the final optic disc region ($C_f(x, y, R)$) from the possible optic disc regions. Finally, the optic disc is segmented (I_s) from the input image using a region mask.

3.2 Preprocessing

The retinal fundus image is not always available with good quality. Most of the time it contains noise and uneven illumination. Even images within a dataset are not uniform with respect to quality. In that case, generally, the images are normalized for further processing. Some preprocessing steps are applied to normalize the input image across the dataset. Fig. 3.3A is the actual input image before preprocessing.

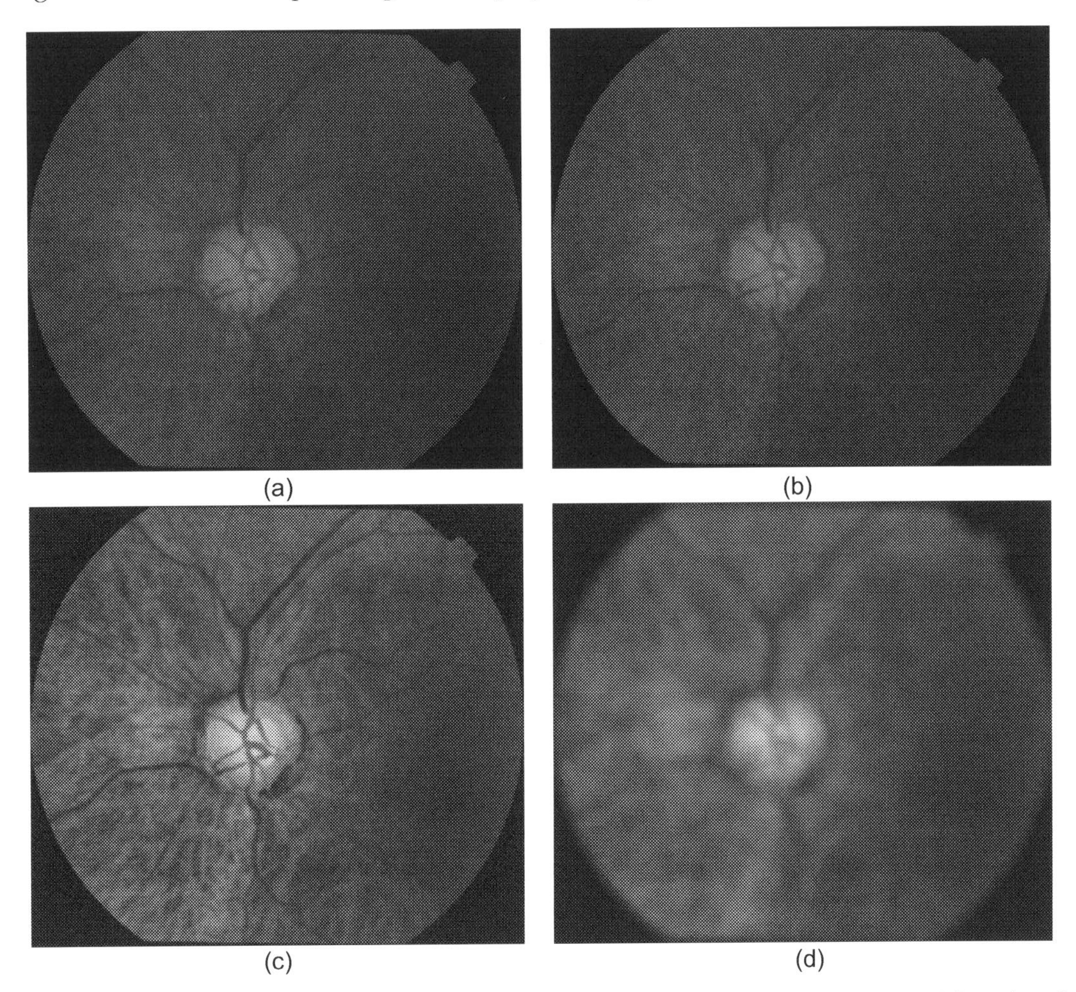

FIGURE 3.3 Preprocessing (A) input image(I); (B) grayscale image (I_g); (C) after applying CLAHE (I_{CLAHE}); and (D) after applying Gaussian smoothing (I_{gs}).

3.2.1 RGB to grayscale conversion

A grayscale image is very helpful for further processing of segmentation. Here, all the analyses will be performed on the grayscale image. So, the input RGB fundus image (I) must be converted to a grayscale image (I_g) using Eq. (3.1). Fig. 3.3B is the grayscale version of the input image.

$$I_g = 0.2989 \times IR + 0.5870 \times IG + 0.1140 \times IB \tag{3.1}$$

3.2.2 Contrast enhancement

Pixel distribution is not similar throughout the retinal fundus image. Especially, the optic disc region is significantly lighter than in other regions. Global histogram equalization is not effective for contrast improvement of the retinal fundus image. Adaptive histogram equalization enhances the image as well as enhances the noises of the region with a small intensity range. It causes appearances of unwanted artifacts in the image. But, we need to enhance the contrast of the image without enhancing the noise. Hence, CLAHE [47,48], a modified version adaptive histogram equalization, is applied to improve the image contrast. CLAHE divides images into multiple local histograms and enhances the local contrast. It limits the histograms of homogeneous regions. The limiting applied to the neighborhood pixel from where the distribution function is derived. It also improves the edges of the optic disc. At first, the image has been divided into multiple tiles. The number of tiles depends on the count of rows and columns for dividing the main image. The number of rows and columns should be at least 2. The contrast of each tile enhanced separately based on the specified clipping value so that output matches the histogram generated by the distribution function. The range of clipping value is between 0 and 1. Finally, all the tiles are merged using bilinear interpolation to reduce the unwanted boundary around the tiles. As a result, the overall contrast of the image will be improved through the enhancement performed blockwise. Contrast limiting guarantees that the noises will not be enhanced. Fig. 3.3C shows the output image (I_{CLAHE}) after applying CLAHE.

3.2.3 Smoothing

A Gaussian smoothing filter is applied here to the contrast-enhanced image using the Gaussian function. Fig. 3.3D shows the output (I_{gs}) after applying the Gaussian smoothing. 2D Gaussian function is mentioned in Eq. (3.2). The distance from the origin is x and y along the horizontal and vertical direction respectively. A scalar value for σ, which is the value of the standard deviation of the Gaussian distribution, is selected for generating the convolution kernel.

$$G(x, y) = \frac{1}{2\pi\sigma^2} e^{-\frac{x^2+y^2}{2\sigma^2}} \tag{3.2}$$

The convolution kernel is generally a square matrix. The selection of proper convolution kernel is very important. A sample of 3 × 3 convolution kernel for Gaussian blur is shown in Fig. 3.4. The convolution is performed by placing the center of the kernel at the pixel to be filtered. The filtered pixel gets a new value which is the weighted average of the neighbor

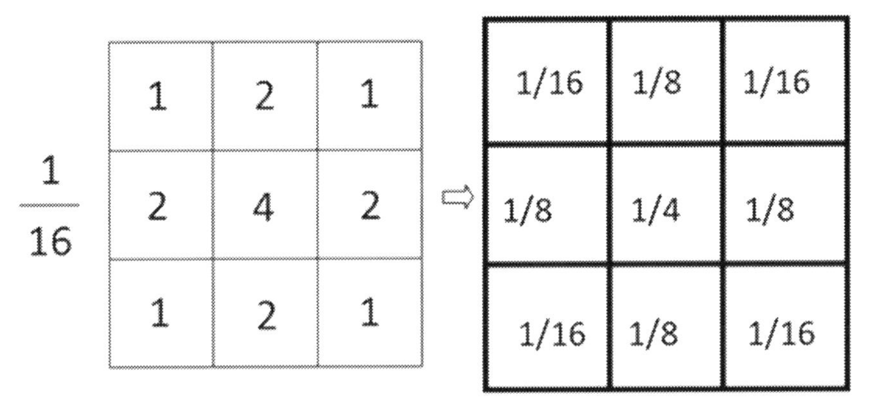

FIGURE 3.4 3 × 3 convolution kernel for Gaussian smoothing.

pixels. Center pixel gets more weight than the neighbor pixels. The convolution kernel does the smoothing as it progresses over the image. The degree of smoothing depends on the value of the standard deviation of the Gaussian distribution (σ). Increment of the value of σ means a more blurred image. It works best to reduce noise as well as blurs the outer boundary of the optic disc. It helps to give a smooth edge of the optic disc region in the next step.

3.3 Edge detection

An edge detection operation is performed on the blurred image. Sobel edge detection technique is used here. A special gradient magnitude operation has been done in both horizontal and vertical directions. Refer to Eq. (3.3) and Eq. (3.4) for the magnitude of the horizontal direction and the vertical direction respectively. An approximate magnitude of the gradients is measured by the sum of absolute values of the gradient in both directions. Refer to Eq. (3.5). The orientation angle of the edge (α) is calculated using Eq. (3.6). If the angle is 0, then the direction of the maximum contrast from black to white runs from left to right over the image. The measurements of the other angles are done in an anticlockwise direction from this reference. Fig. 3.5A depicts the edges (I_e) of the preprocessed input image.

$$g_x = \frac{\partial f}{\partial x} \tag{3.3}$$

$$g_y = \frac{\partial f}{\partial y} \tag{3.4}$$

$$M(x, y) = |g_x| + |g_y| \tag{3.5}$$

$$\alpha = \tan^{-1}\left[\frac{g_x}{g_y}\right] \tag{3.6}$$

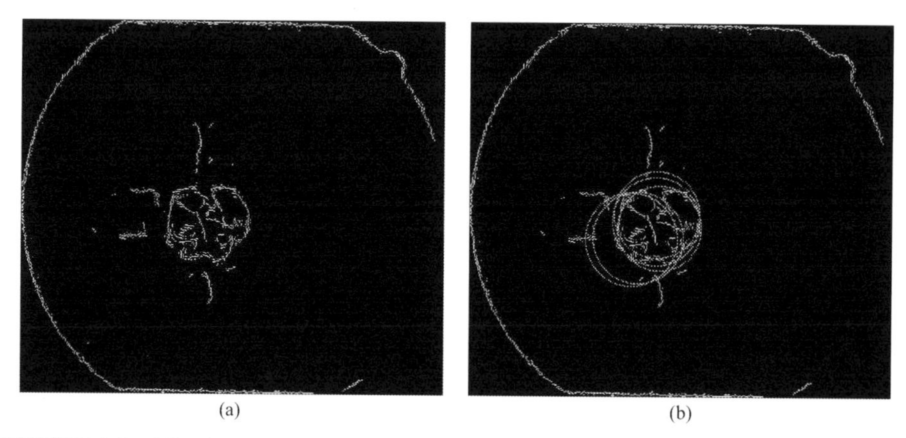

FIGURE 3.5 (A) After applying Sobel edge detection (I_e); (B) candidate regions to be optic disc.

Sobel operators give the edges where intensity changes faster. Another reason is that it gives the edges of the optic disk and nearby regions. The Canny operator is not suitable here because it gives edges of the complete image region.

3.4 Detection of possible centers for optic disc

It is assumed that the optic disc region will be always circular. Possible regions for optic can be detected by the circular Hough transform [49] of the edged image. Eqs. (3.7 and 3.8) are the equation of a circle where R is the radius and (a, b) is the center. Every point in xy has an equivalent position in ab space. Generate a circle in ab space for every edge pixel (x_1, y_1) in edged image and θ rotates from 0° to 360°. Refer to Eqs. (3.9) and (3.10). Every point on the circle of ab space votes in accumulator cell. The cells with a greater number of votes are the detected circle.

$$x = a + R\cos\theta \tag{3.7}$$

$$y = b + R\sin\theta \tag{3.8}$$

$$a = x_1 - R\cos\theta \tag{3.9}$$

$$b = y_1 - R\sin\theta \tag{3.10}$$

Using this iterative process the centers of all possible circles are detected. The possible centers and corresponding radii are stored in a $n \times 3$ matrix (C), where n is the number of possible circles. Refer to Eq. (3.11). Fig. 3.5B shows some of the circles, detected by this technique, which are the top candidates.

$$C_{n\times 3} = [(x_1, y_1, R_1); (x_2, y_2, R_2); \ldots; (x_n, y_n, R_n)] \tag{3.11}$$

3.5 Selection of optic disc region

A set of circles is detected from the previous process as candidates for the optic disc. The selection of actual optic disc is very challenging among all possible candidates. To select the correct optic disc center a supervised approach is proposed here. Some samples of the optic disk region and non-optic disc regions of the fundus image are cropped and saved. The optic disc and non-optic disc regions can be separated by texture features. We have seen that GLCM features are capable to discriminate the optic disc and non-optic disc.

3.5.1 Feature extraction

A GLCM defines the occurrence probability of a gray level i in the neighbor of another gray level j at a given distance d and angle θ, provided the total number of gray levels N are known. Refer to Eq. (3.12). Here, contrast (Ct), correlation (Cn), energy (Ey), and homogeneity (Hy) are calculated for four directions ($0°, 45°, 90°, 135°$) and a distance of 1 pixel. Refer to Eqs. (3.13)–(3.16). The final feature vector (F) for the classifier is formed as per Eq. (3.17). Fig. 3.6 shows the 16 features of the final feature vector. It represents how the statistical texture features are discriminating optic disc and non-optic disc region. The four rows represent contrast, correlation, energy, and homogeneity sequentially. The columns represent the four directions sequentially for each feature.

$$GLCM = \Pr(i,j)|d,\theta,N \tag{3.12}$$

$$Ct = \sum_{i,j} |i-j|^2 \, S(i,j) \tag{3.13}$$

$$Cn = \sum_{i,j} \frac{(i-\mu i)(j-\mu j)S(i,j)}{\sigma_i \sigma_j} \tag{3.14}$$

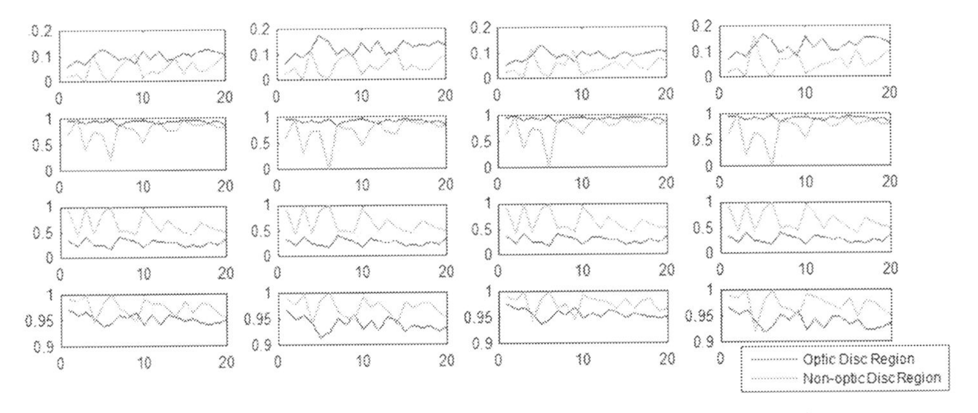

FIGURE 3.6 Features used for classification of the optic disc and non-optic disc region.

$$Ey = \sum_{i,j} S(i,j)^2 \tag{3.15}$$

$$Hy = \sum_{i,j} \frac{S(i,j)}{1+|i-j|} \tag{3.16}$$

$$F = \{Ct_{0^\circ}, Ct_{45^\circ}, Ct_{90^\circ}, Ct_{135^\circ}, Cn_{0^\circ}, Cn_{45^\circ}, Cn_{90^\circ}, Cn_{135^\circ}, Ey_{0^\circ}, Ey_{45^\circ}\ Ey_{90^\circ}, Ey_{135^\circ}, Hy_{0^\circ}, Hy_{45^\circ}, Hy_{90^\circ}, Hy_{135^\circ}\} \tag{3.17}$$

3.5.2 *Training of classifier*

A classifier is trained with texture features [50] extracted from GLCM for two classes. i.e., Optic Disc and Non-optic Disc. In this work, the Naïve Bayes classifier selected for training. The feature vector (F), which has 16 features, is fed into the classifier to train the classification model. It is very easy to understand and is a fast algorithm for binary as well as multiclass classification problems. It also performs well with less training data. It does probabilistic prediction based on the Bayesian theorem. Refer to Eq. (3.18). Bayes theorem calculates the probability of occurring an event where the probability of another event is given. Here, the probability of A will be calculated considering event B is true, which is evidence. $P(A)$ is the prior probability of A when evidence is not seen. $P(A \mid B)$ is the posterior probability of B where evidence was seen.

$$P(A|B) = \frac{P(B|A)P(A)}{P(B)} \tag{3.18}$$

$$P(c|F) = \frac{P(F|c)P(c)}{P(F)} \tag{3.19}$$

$$F = \{f_1, f_2, f_3, \ldots, f_n\} \tag{3.20}$$

$$P(c|f_1, f_2, f_3, \ldots, f_n) = \frac{P(f_1|c)P(f_2|c)P(f_3|c)\ldots P(f_n|c)P(c)}{P(f_1)P(f_2)P(f_3)\ldots P(f_n)} \tag{3.21}$$

The Bayes theorem can be explained in the context of this problem. Eq. (3.19) depicts that the c is class variable and F is the dependent feature vector of n dimension, refer Eq. (3.20). Finally, it can be expressed as in Eq. (3.21). Each of the features is contributing independently to the prediction of class.

3.5.3 *Classification of possible optic disc regions*

A subimage will be generated from each of the possible candidate regions to be optic disc. The subimage can be formed for the square region enclosing by the circle. The dimension of

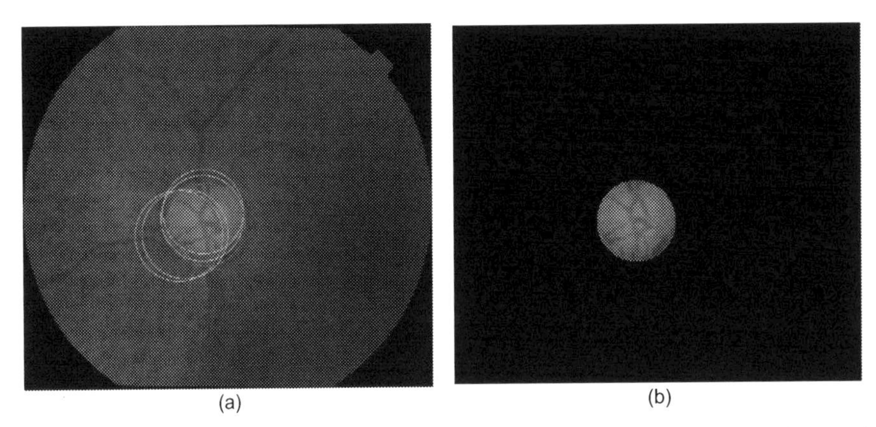

FIGURE 3.7 (A) Perfect location for the optic disc is enclosed by the red (light gray in printed version) circle (For interpretation of the references to color in this figure, the reader is referred to the web version of this chapter); and (B) segmented optic disc region.

the subimage will be $2R \times 2R$ where R is the radius of the circle. All the subimages are classified with the pretrained Naïve Bayes classification model. If only one subimage is predicted as optic disc class then the corresponding region is selected as the final optic disc region. There may have multiple subimages classified as optic disc regions. It creates a confliction that which one is more perfect among the subimages classified as an optic disc image. The corresponding prediction percentage is used to overcome this conflict. The subimage, which has the highest prediction percentage to be optic disc, is selected as the best candidate. The red (light gray in printed version) circle shown in Fig. 3.7A is selected as the best candidate to be the actual optic disc region.

3.6 Segmentation of optic disc

A binary mask is generated with the center (x, y) and radius (R) of the final optic disc region. The circular optic disc region is segmented from each channel of input color image using Eq. (3.22). Finally, red, green, and blue channels are combined to get the complete segmented RGB image. Fig. 3.7B represents the final segmented image.

$$I_s = I \times I_m \tag{3.22}$$

4. Experimentations and results

The experimentations are done to see the performance of the proposed approach over the existing approach on the same dataset. The system has been developed with the Matlab 2015a version. The machine has been configured with the Intel Core i5 processor and 4 GB RAM. The Drishti-GS [51] dataset is a very popular dataset of retinal fundus images. In our survey, most of the optic disc detection and segmentation work used this dataset for validation of their proposed techniques. Hence, this dataset has been used for the experimentations here also. The dataset [4] contains 101 retinal fundus images in PNG format, provided by Medical

Image Processing (MIP) group, IIIT Hyderabad. To test the effectiveness of the proposed approach both the diseased eye and normal eye are considered for experimentation. The proposed method has been executed on 70 images of glaucomatous and 31 images of a healthy retina. The dataset also contains a soft segmentation map in png format. Four experts marked the optic disc boundary manually and separately. The final soft segmentation map was created by fusing the segmentation from four experts. The soft segmentation map is used here as ground truth for experimentation and performance evaluation.

The size of the image in the dataset is around 2050 × 1750. Each image is scaled down to 25%. The preprocessing should be performed for all the images inside the dataset. The sigma specifies the standard deviation of the Gaussian smoothing kernel. The experimentation was done to find out the optimum value of sigma. The optimum value of σ is 5 in this work. Edge detection is done by the Sobel operator. Then Circular Hough transform was applied on the edges to generate five candidate regions to be optic disc. The range of the radius taken for circular Hough transform is 40–65 pixels. The radius range was determined by analyzing the dataset.

Those candidate regions need to be classified as optic disc and non-optic disc region. The circular Hough transform finds the possible optic disc centers with radius. Among the possible optic disc regions, there may have some non-optic disc regions. We need to discard those non-optic disc regions as well as find the best optic disc region for segmentation. The supervised machine learning algorithm does both with high accuracy. Forty images are cropped from the 40 samples of the database randomly. Out of those 40 images, 20 images cropped from the optic disc region and another 20 images from the non-optic disc region. In total 40 cropped images were used to train the Naïve Bayes classifier. The candidate regions are classified as the optic disc region and the non-optic disc region. The choice of the classifier is a tough call in the modern-day machine learning era. We have also experimented using few other supervised classifiers but the performance is better with Naïve Bayes than others. Another reason for the selection of the Naïve Bayes classifier is that it does probabilistic prediction with classification based on the Bayesian theorem. This probability helps to break the confusion if multiple regions are classified as the optic disc regions. The final optic disc region selection goes with the region with maximum prediction probability. Finally, the selected region is segmented from the input image by a region mask.

The most popular parameters to measure the performance of segmentation in medical imaging is the sensitivity, specificity, and accuracy. Generally, the pixels of the region of interest in the ground truth image is called as positive and the rests are negative pixels. The segmentation output has four types of pixels with respect to ground truth segmentation i.e., True Positive (TP), True Negative (TN), False Positive (FP), and False Negative (FN). The TP pixels are actually positive as well as predicted as positive. The TN pixels are actually negative as well as predicted as negative. The FP pixels are actually negative but predicted as positive. The FN pixels are actually positive but predicted as negative. Total pixels count should be equals to the summation of those four types of pixels. We have computed sensitivity, specificity, and accuracy here using Eqs. (3.23–3.25) respectively for all the images based on the ground truth soft segmentation map provided in the dataset.

$$\text{Sensitivity} = \frac{TP}{TP + FN} \times 100 \tag{3.23}$$

$$\text{Specificity} = \frac{TN}{TN + FP} \times 100 \tag{3.24}$$

$$\text{Accuracy} = \frac{TP + TN}{\text{Total Pixels}} \times 100 \tag{3.25}$$

Table 3.2 is showing the aggregate optic disc segmentation result using the proposed approach for both the glaucomatous retinal image and the healthy retinal image. Table 3.3 depicts the aggregate optic disc segmentation result of the Nugroho et al.'s for both glaucomatous retinal image and healthy retinal image. This is true that accuracy improvement is very nominal. Accuracy depends on the detection of both the TP and TN pixels. Here the number of negative pixels is approximately 30 times more than the number of positive pixels, which is a class imbalance problem. This is the reason that TP pixels contribute very little to the accuracy result. But, the segmentation performance of medical images is mainly measured by proper detection of the TP region. Here, the optic disc region is the TP region. The detection and segmentation performance of TP regions has been improved significantly than the Nugroho et al.'s approach. The improvement of TP detection is easily identifiable by the sensitivity result but not in accuracy and specificity result.

Figs. 3.8–3.10 shows the sensitivity, specificity, and accuracy percentage respectively for all the glaucomatous retinal samples using the proposed approach as well as Nugroho et al.'s approach. Figs. 3.11–3.13 depict the sensitivity, specificity, and accuracy, respectively, for the healthy retinal samples, using the proposed approach as well as Nugroho et al.'s approach. The red (gray in printed version) line indicates the result of the proposed approach and the green (light gray in printed version) line indicates the result of Nugroho et al.'s approach. The result is very consistent using the proposed approach for the entire glaucomatous eye samples as well as healthy eye samples. The percentage limit of the y-axis is 0%–100% in Figs. 3.8 and 3.11, and 90%–100% in Figs. 3.9, 3.10, 3.12, and 3.13. The difference of sensitivity between two approaches is interpretable with 0–100 range of y-axis. The specificity and accuracy are

TABLE 3.2 Optic disc segmentation results using the proposed approach on Drishti-GS dataset.

Type of eye	Total image	Average sensitivity (%)	Average specificity (%)	Average accuracy (%)
Glaucomatous	70	93.32	99.14	98.96
Healthy	31	91.76	99.01	98.79

TABLE 3.3 Optic disc segmentation results using the Nugroho et al.'s approach on Drishti-GS dataset.

Type of eye	Total image	Average sensitivity (%)	Average specificity (%)	Average accuracy (%)
Glaucomatous	70	82.14	98.57	98.06
Healthy	31	80.56	98.14	97.56

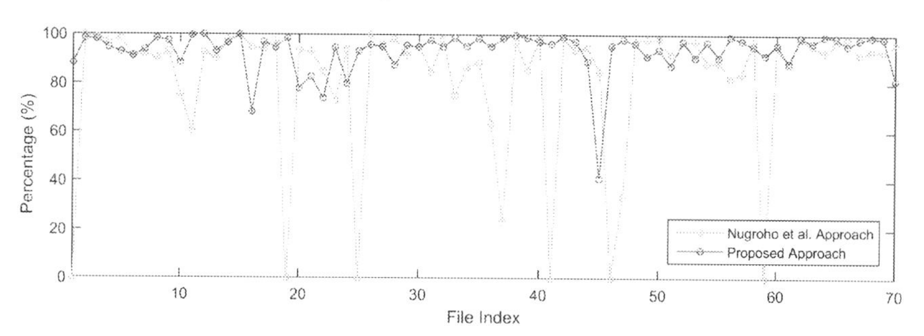

FIGURE 3.8 Comparison of sensitivity between the proposed approach and Nugroho et al.'s approach on retinal images of glaucomatous eye.

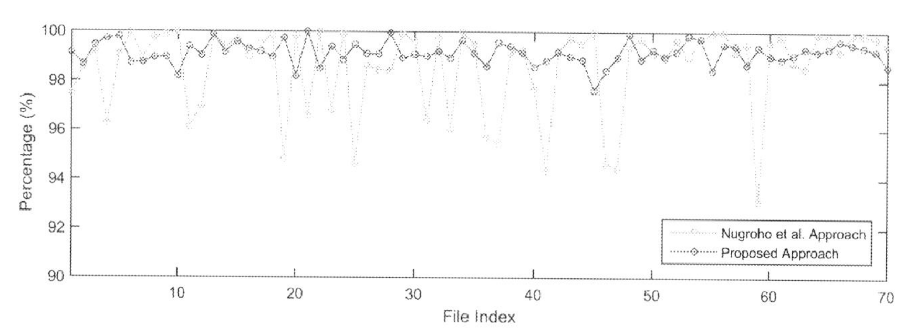

FIGURE 3.9 Comparison of specificity between the proposed approach and Nugroho et al.'s approach on retinal images of glaucomatous eye.

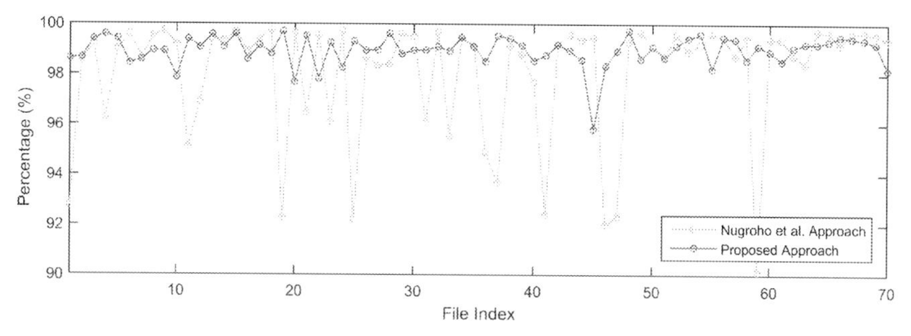

FIGURE 3.10 Comparison of accuracy between the proposed approach and Nugroho et al.'s approach on retinal images of glaucomatous eye.

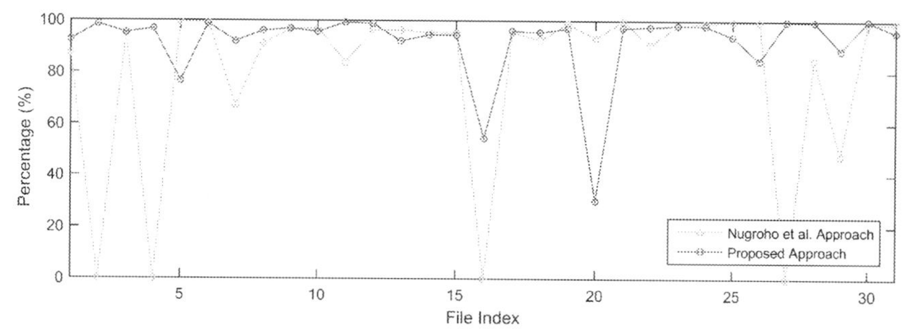

FIGURE 3.11 Comparison of sensitivity between the proposed approach and Nugroho et al.'s approach on retinal images of healthy eye.

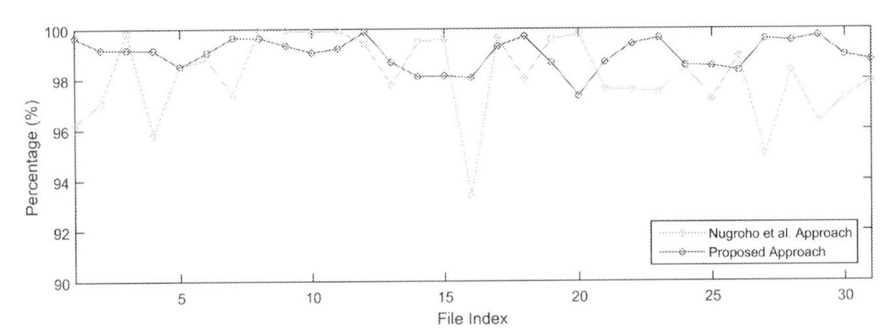

FIGURE 3.12 Comparison of specificity between the proposed approach and Nugroho et al.'s approach on retinal images of healthy eye.

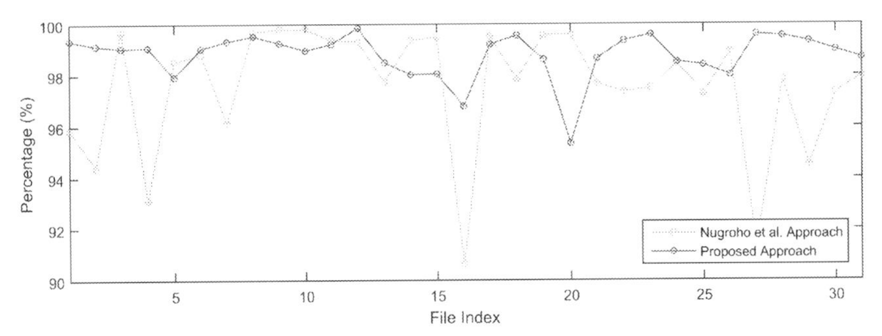

FIGURE 3.13 Comparison of accuracy between the proposed approach and Nugroho et al.'s approach on retinal images of healthy eye.

not good for interpretation with this same range in this work. This is the reason for taking the range from 90 to 100 for specificity and accuracy plot. Figs. 3.8 and 3.11 depict that few samples have zero sensitivity with the Nugroho et al. approach. The final segmentation output comparison has been shown later in Fig. 3.14 for those samples. The reason is that Nugroho et al.'s approach completely fails to segment the region of interest there. The proposed approach is performing far better there as well as for other samples. It is visible in Figs. 3.9 and 3.12 that the specificity of many samples is below 97% using the Nugroho et al. approach where has the specificity is consistently above 97% for most of the samples using the proposed approach. The same is observed in Figs. 3.10 and 3.13 for the comparison of accuracy between the proposed approach and Nugroho et al. approach. It is also observed that the samples which are showing very low sensitivity of segmentation also have their impact on the degradation of specificity and accuracy percentage.

Table 3.4 shows the required time for the segmentation of optic disc using the proposed approach as well as Nugroho et al.'s approach. The proposed approach requires less than one second for the segmentation of the optic disc from a retinal fundus image, where Nugroho et al.'s approach takes around 24 s to do the same from the same image.

Experimentation results show that the proposed approach is working with good accuracy and efficiency. Fig. 3.14 shows the input image and final segmentation outcome using Nugroho et al.'s approach as well as the proposed approach. It depicts the relatively poor

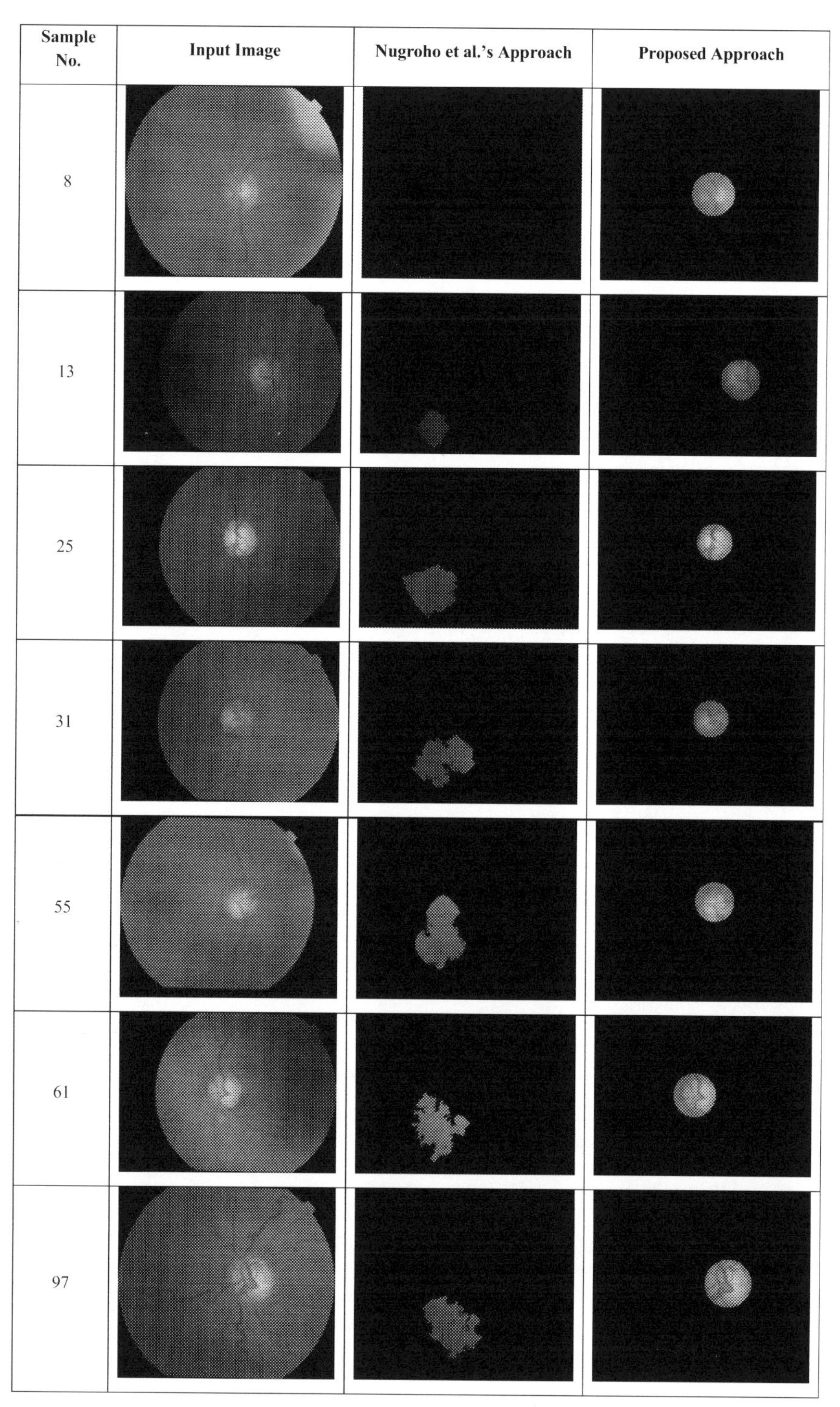

FIGURE 3.14 Segmentation result comparison between the Nugroho et al.'s approach and the proposed approach.

TABLE 3.4 Segmentation time using different approaches (sec).

Type of eye	Total image	Average segmentation time using proposed approach	Average segmentation time using Nugroho et al.'s approach [20]
Glaucomatous	70	0.78 s	24.25 s
Healthy	31	0.79 s	24.98 s

performance of Nugroho et al.'s approach vis-à-vis the proposed approach. The sample, which is demonstrated here, the sensitivity is zero using Nugroho et al.'s approach. In the case of a few images, Nugroho's approach completely fails to show anything after segmentation, the proposed approach is working very well there. These were the few challenging images where this approach gives a significant performance.

5. Conclusions

A semi-supervised approach is proposed here for the detection and segmentation of optic disc from the retinal fundus image. Input retinal fundus image is passed through some preprocessing steps to reduce noise and normalize contrast. A circular Hough transform is applied to the edged image to detect the possible regions for the optic disc. A supervised technique is applied to detect the final optic disc region among the possible regions. The overall sensitivity, specificity, and accuracy of the proposed approach are 92.54%, 99.08%, and 98.88% respectively. The previous technique was suffering badly with a few images, where the proposed approach provides very good segmentation results. The proposed approach is very efficient with respect to execution time compared with the previous approach. The main contributions of this paper are (1) a state of the art survey on different techniques of optic disc detection and segmentation; (2) a semi-supervised technique for detection and segmentation of optic disc from retinal fundus image; and (3) good performance in very less processing time. This approach can be very effective as an initial step for any kind of retinal disorder analysis in the future.

References

[1] M. Lalonde, M. Beaulieu, L. Gagnon, Fast and robust optic disc detection using pyramidal decomposition and Hausdorff-based template matching, IEEE Trans. Med. Imag. 20 (11) (2001) 1193–1200.

[2] V. Das, S. Dandapat, P.K. Bora, Multi-scale deep feature fusion for automated classification of macular pathologies from OCT images, Biomed. Signal Process. Contr. 54 (2019) 101605.

[3] C.S. Lee, D.M. Baughman, A.Y. Lee, Deep learning is effective for classifying normal versus age-related macular degeneration OCT images, Ophthalmol. Retina 1 (4) (2017) 322–327.

[4] J. Sivaswamy, S. Krishnadas, A. Chakravarty, G. Joshi, A.S. Tabish, A comprehensive retinal image dataset for the assessment of glaucoma from the optic nerve head analysis, JSM Biomed. Imag. Data Papers 2 (1) (2015) 1004.

[5] T. Kauppi, V. Kalesnykiene, J.K. Kamarainen, L. Lensu, I. Sorri, H. Uusitalo, H. Kälviäinen, J. Pietilä, DIARETDB0: Evaluation Database and Methodology for Diabetic Retinopathy Algorithms. Machine Vision and Pattern Recognition Research Group, vol. 73, Lappeenranta University of Technology, Finland, 2006, pp. 1–17.

[6] R.V.J.P.H. Kälviäinen, H. Uusitalo, DIARETDB1 diabetic retinopathy database and evaluation protocol, in: Medical Image Understanding and Analysis, vol. 2007, 2007, p. 61.
[7] A.D. Hoover, V. Kouznetsova, M. Goldbaum, Locating blood vessels in retinal images by piecewise threshold probing of a matched filter response, IEEE Trans. Med. Imag. 19 (3) (2000) 203–210.
[8] S. Morales, K. Engan, V. Naranjo, A. Colomer, Retinal disease screening through local binary patterns, IEEE J. Biomed. Health Inform. 21 (1) (2015) 184–192.
[9] J. Odstrcilik, R. Kolar, A. Budai, J. Hornegger, J. Jan, J. Gazarek, T. Kubena, P. Cernosek, O. Svoboda, E. Angelopoulou, Retinal vessel segmentation by improved matched filtering: evaluation on a new high-resolution fundus image database, IET Image Process. 7 (4) (2013) 373–383.
[10] R. GeethaRamani, L. Balasubramanian, Retinal blood vessel segmentation employing image processing and data mining techniques for computerized retinal image analysis, Biocybern. Biomed. Eng. 36 (1) (2016) 102–118.
[11] U.T. Nguyen, A. Bhuiyan, L.A. Park, R. Kawasaki, T.Y. Wong, J.J. Wang, P. Mitchell, K. Ramamohanarao, An automated method for retinal arteriovenous nicking quantification from color fundus images, IEEE Trans. Biomed. Eng. 60 (11) (2013) 3194–3203.
[12] S. Roychowdhury, D.D. Koozekanani, S.N. Kuchinka, K.K. Parhi, Optic disc boundary and vessel origin segmentation of fundus images, IEEE J. Biomed. Health Inform. 20 (6) (2015) 1562–1574.
[13] G. Mittal, J. Sivaswamy, Optic disk and macula detection from retinal images using Generalized Motion Pattern, in: 2015 Fifth National Conference on Computer Vision, Pattern Recognition, Image Processing and Graphics (NCVPRIPG), IEEE, December 2015, pp. 1–4.
[14] S. Roychowdhury, D.D. Koozekanani, K.K. Parhi, DREAM: diabetic retinopathy analysis using machine learning, IEEE J. Biomed. Health Inform. 18 (5) (2013) 1717–1728.
[15] S. Wang, H.L. Tang, Y. Hu, S. Sanei, G.M. Saleh, T. Peto, Localizing microaneurysms in fundus images through singular spectrum analysis, IEEE Trans. Biomed. Eng. 64 (5) (2016) 990–1002.
[16] J.P. Medhi, S. Dandapat, An effective fovea detection and automatic assessment of diabetic maculopathy in color fundus images, Comput. Biol. Med. 74 (2016) 30–44.
[17] A. Kunwar, S. Magotra, M.P. Sarathi, Detection of high-risk macular edema using texture features and classification using SVM classifier, in: 2015 International Conference on Advances in Computing, Communications and Informatics (ICACCI), IEEE, August 2015, pp. 2285–2289.
[18] B. Zhang, B.V. Kumar, D. Zhang, Detecting diabetes mellitus and nonproliferative diabetic retinopathy using tongue color, texture, and geometry features, IEEE Trans. Biomed. Eng. 61 (2) (2013) 491–501.
[19] V. Asokan, S.A. Jerome, Computer aided approach for detection of age related macular degeneration from retinal fundus images, in: 2016 International Conference on Circuit, Power and Computing Technologies (ICCPCT), IEEE, March 2016, pp. 1–7.
[20] H.A. Nugroho, A. Jalil, I. Ardiyanto, Segmentation of optic disc on retinal fundus images using morphological reconstruction enhancement and active contour, in: 2016 2nd International Conference on Science in Information Technology (ICSITech), IEEE, October 2016, pp. 362–366.
[21] S. Yu, D. Xiao, S. Frost, Y. Kanagasingam, Robust optic disc and cup segmentation with deep learning for glaucoma detection, Comput. Med. Imag. Graph. 74 (2019) 61–71.
[22] M. Kankanala, S. Kubakaddi, Automatic segmentation of optic disc using modified multi-level thresholding, in: 2014 IEEE International Symposium on Signal Processing and Information Technology (ISSPIT), IEEE, December 2014, pp. 000125–000130.
[23] S. Bharkad, Automatic segmentation of optic disk in retinal images, Biomed. Signal Process Contr. 31 (2017) 483–498.
[24] G. Hamednejad, H. Pourghassem, Retinal optic disk segmentation and analysis in fundus images using dbscan clustering algorithm, in: 2016 23rd Iranian Conference on Biomedical Engineering and 2016 1st International Iranian Conference on Biomedical Engineering (ICBME), IEEE, November 2016, pp. 122–127.
[25] D. Welfer, J. Scharcanski, C.M. Kitamura, M.M. Dal Pizzol, L.W. Ludwig, D.R. Marinho, Segmentation of the optic disk in color eye fundus images using an adaptive morphological approach, Comput. Biol. Med. 40 (2) (2010) 124–137.
[26] S. Kulkarni, S. Annadate, Optic disc segmentation using graph cut technique, in: 2017 Third International Conference on Sensing, Signal Processing and Security (ICSSS), IEEE, May 2017, pp. 124–127.
[27] M.K. Aggarwal, V. Khare, A new method for optic disc localization in retinal images, in: 2016 Ninth International Conference on Contemporary Computing (IC3, IEEE, August 2016, pp. 1–5.

[28] A. Aquino, M.E. Gegúndez-Arias, D. Marín, Detecting the optic disc boundary in digital fundus images using morphological, edge detection, and feature extraction techniques, IEEE Trans. Med. Imag. 29 (11) (2010) 1860–1869.
[29] M.N. Reza, Automatic detection of optic disc in color fundus retinal images using circle operator, Biomed. Signal Process Contr. 45 (2018) 274–283.
[30] N.E.A. Khalid, N.M. Noor, N.M. Ariff, Fuzzy c-means (FCM) for optic cup and disc segmentation with morphological operation, Procedia Comput. Sci. 42 (2014) 255–262.
[31] N. Thakur, M. Juneja, Optic disc and optic cup segmentation from retinal images using hybrid approach, Expert Syst. Appl. 127 (2019) 308–322.
[32] J. Almotiri, K. Elleithy, A. Elleithy, An automated region-of-interest segmentation for optic disc extraction, in: 2018 IEEE Long Island Systems, Applications and Technology Conference (LISAT), IEEE, May 2018, pp. 1–6.
[33] L. Ichim, D. Popescu, S. Cirneanu, Combining LBP and co-occurrence matrix information to accurate recognition of the optic disc in retinal image, in: 2016 International Conference on Development and Application Systems (DAS), IEEE, May 2016, pp. 254–259.
[34] L.J. Uribe-Valencia, J.F. Martínez-Carballido, Automated Optic Disc region location from fundus images: using local multi-level thresholding, best channel selection, and an Intensity Profile Model, Biomed. Signal Process Contr. 51 (2019) 148–161.
[35] P. Qin, L. Wang, H. Lv, Optic disc and cup segmentation based on deep learning, in: 2019 IEEE 3rd Information Technology, Networking, Electronic and Automation Control Conference (ITNEC), IEEE, March 2019, pp. 1835–1840.
[36] L. Wang, H. Liu, Y. Lu, H. Chen, J. Zhang, J. Pu, A coarse-to-fine deep learning framework for optic disc segmentation in fundus images, Biomed. Signal Process Contr. 51 (2019) 82–89.
[37] D. Mohan, J.H. Kumar, C.S. Seelamantula, High-performance optic disc segmentation using convolutional neural networks, in: 2018 25th IEEE International Conference on Image Processing (ICIP), IEEE, October 2018, pp. 4038–4042.
[38] A.O. Joshua, F.V. Nelwamondo, G. Mabuza-Hocquet, Segmentation of optic cup and disc for diagnosis of glaucoma on retinal fundus images, in: 2019 Southern African Universities Power Engineering Conference/Robotics and Mechatronics/Pattern Recognition Association of South Africa (SAUPEC/RobMech/PRASA), IEEE, January 2019, pp. 183–187.
[39] S.A. Tuncer, T. Selçuk, M. Parlak, A. Alkan, Hybrid approach optic disc segmentation for retinal images, in: 2017 International Artificial Intelligence and Data Processing Symposium (IDAP), IEEE, September 2017, pp. 1–5.
[40] D. Mei, D. Chen, Optic disc segmentation method based on low rank matrix recovery theory, in: 2018 Chinese Control and Decision Conference (CCDC), IEEE, June 2018, pp. 2626–2630.
[41] A. Salazar-Gonzalez, D. Kaba, Y. Li, X. Liu, Segmentation of the blood vessels and optic disk in retinal images, IEEE J. Biomed. Health Inform. 18 (6) (2014) 1874–1886.
[42] J. Dietter, W. Haq, I.V. Ivanov, L.A. Norrenberg, M. Völker, M. Dynowski, D. Röck, F. Ziemssen, M.A. Leitritz, M. Ueffing, Optic disc detection in the presence of strong technical artifacts, Biomed. Signal Process Contr. 53 (2019) 101535.
[43] Y. Gao, X. Yu, C. Wu, Y. Zhuang, S. Wang, Automated segmentation of the optic disc from retinal image using modified local intensity clustering model, in: 2019 Chinese Control and Decision Conference (CCDC), IEEE, June 2019, pp. 602–606.
[44] S. Kumar, A. Adarsh, B. Kumar, A.K. Singh, An automated early diabetic retinopathy detection through improved blood vessel and optic disc segmentation, Optic Laser. Technol. 121 (2020) 105815.
[45] D. Díaz-Pernil, I. Fondón, F. Peña-Cantillana, M.A. Gutiérrez-Naranjo, Fully automatized parallel segmentation of the optic disc in retinal fundus images, Pattern Recogn. Lett. 83 (2016) 99–107.
[46] L.J. Uribe-Valencia, J.F. Martínez-Carballido, Thesholding methods review for the location of the Optic disc in retinal fundus color images, in: 2016 13th International Conference on Electrical Engineering, Computing Science and Automatic Control (CCE), IEEE, September 2016, pp. 1–6.
[47] K. Zuiderveld, Contrast limited adaptive histogram equalization, in: Graphics Gems IV, Academic Press Professional, Inc., August 1994, pp. 474–485.
[48] G. Yadav, S. Maheshwari, A. Agarwal, Contrast limited adaptive histogram equalization based enhancement for real time video system, in: 2014 International Conference on Advances in Computing, Communications and Informatics (ICACCI), IEEE, September 2014, pp. 2392–2397.

[49] H.K. Yuen, J. Princen, J. Illingworth, J. Kittler, Comparative study of Hough transform methods for circle finding, Image Vis Comput. 8 (1) (1990) 71–77.

[50] S. Jana, S. Basak, R. Parekh, Automatic fruit recognition from natural images using color and texture features, in: 2017 Devices for Integrated Circuit (DevIC), IEEE, March 2017, pp. 620–624.

[51] J. Sivaswamy, S.R. Krishnadas, G.D. Joshi, M. Jain, A.U.S. Tabish, Drishti-GS: retinal image dataset for optic nerve head (ONH) segmentation, in: 2014 IEEE 11th International Symposium on Biomedical Imaging (ISBI), IEEE, April 2014, pp. 53–56.

CHAPTER

4

Medical decision support system using data mining: an intelligent health care monitoring system for guarded travel

L. Jegatha Deborah[1], *S.C. Rajkumar*[2], *P. Vijayakumar*[1]

[1]Department of Computer Science and Engineering, University College of Engineering, Tindivanam, Tamil Nadu, India; [2]Department of Computer Science and Engineering, University College of Engineering, Panruti, Tamil Nadu, India

1. Introduction

Remote health monitoring is a challenging issue for health care systems recently since inaccuracy of fitness records were collected using low efficient, limited recourse sensors. To improve the remote health monitoring system, we proposed the intelligent system, which takes the remote health monitoring into the next step. Recently, people prefer faster transportation to travel over places; for that vehicle drivers are influenced to drive the vehicle faster, which causes stress driving. In that case, a driver's health condition on driving is closely connected with a passenger's safety. While driving, a driver's sudden health complications lead to severe damages to the passengers, vehicle, alongside a vehicle, and in the environment surrounding a collision. In order to ensure guarded transport, continuous health monitoring status of drivers becomes inevitable.

Recently, a survey reports that 60% of vehicle accidents occur because of the driver's restless continual travel. In highways, vehicles are faster when it is compared with city side transportation and if any vehicle suddenly breaks down without notification will cause the severe damage, followed by alongside a vehicle. Moreover, the driver health condition is an important key factor and unexpectedly worsen health complications like brain stroke and cardiac arrest, which are unpredictable because the symptoms are very usual and happen within few minutes. This will cause collateral damages in vehicles, alongside vehicles, passengers,

https://doi.org/10.1016/B978-0-12-822260-7.00014-5

and surroundings. To avoid the stressful driving the intelligent assistant is required to monitor their health and suggests to do relaxation guidelines.

Rapid development of smart healthcare system combined with Internet of Things (IoT) technology provides promising solutions to monitor the driver's health conditions while driving [1]. The health readings, which are collected using various types of sensors and readings, are learned through intelligent learners who provide intelligent suggestions to avoid any unexpected health complication. Nowadays, wearable sensors provide valuable information of human health condition precisely. The wearable sensors collect the health readings from the body which means electrodes are continually interacts with human body and that leads to provoke various types of allergies on human skin and side effects. To overcome the side effects, in our proposed system the health sensors are placed inside the steering wheel and the sensors are highly configurable and able to transfer the readings using high-efficient, wireless-enabled medium continually. This proposed technique gives more reliable and higher accuracy results when it is compared with wearable sensors. The sensor readings are in the form of raw signals which records heart rate of the driver's health condition, from that we convert signals into optimized and easily understandable format of the record as Electronic Health Record (EHR) [2]. This EHR contains several sensitive information of the driver's health condition andinclusion of driver's vehicle location withparticular time stamp.This continual record of the driving driver's health condition isstored into the cloud server using effective communication protocol.

The health records are transmitted via Internet continuously, and when the vehicle goes offline the system transmits information using proximity-based data transfer [3], which ensures the reliable transmission until it receives the Internet connection. The low power and long range wireless communication uses wide area network technology is compactable to transmit health readings range more than 10 km. This work effectively discovers the vital health readings and transmits to the nearby wireless devices. It has the capability of wireless transmission and forwards the sensor readings using a hub-by-hub manner to reach the cloud server. This emergency system executes emergency procedures first to inform nearby emergency vehicle and store the record into the cloud server, second transfer the EHR to emergency healthcare system and third informs to the registered family members. This transmission is achieved using Internet and proximity services. Moreover, when the vital signals are detected the vehicle takes control from the driver and change into autonomous driving mode to park the vehicle safely without any collision. In the cloud server the driver's past medical histories and medication details are made available at every time, it also sends the emergency health care system.The sensor readings are handled with an intelligent learner system, Long-Short Term Memory (LSTM) [4], which predicts the driver vital reading before it captured. This intelligent memory-based model has the capacity of processing the health readings effectively using several gates. Finally, the proposed system achieves guarded transportation to the public and offers more attention to the driver's health condition.

2. Related works

In this section one can acquire the detailed study on various methods and techniques related to proposed system and it helps to understand the proposed system easily.

Nowadays, researchers proposed remote health monitoring system based on cloud services. The special type of clouds such as seed cloud and V-cloud are utilized to provide wireless healthcare solution for onboard state. These clouds collect the health readings, using a number of improved applications to ensure safety throughout the journey [5,6]. Any prolong activity of the driver health complication leads to worsen the condition to the environment, and this system provides the solution, which incorporates with long-term static health record and that is maintained and understands the driver behavior effectively. The primary objective of the system is to avoid the traffic on the busy road and provide instant medic services to the road-side accidents. This research study is identified as effectively safe depending on driver's driving behavior and road conditions which are captured from VANET. The remote health monitoring system helps to provide effective medical assistance to the passengers' and drivers' in a daily basis [7].The reliable wireless technologies transmitted the signals of the drivers health readings with regular interval to the health centers using Internet connection and the combined results are collected from multiple heterogeneous networks [4].In order to provide customized health service the author's proposed mobile healthcare platform called Health Drive (HD), which uses a multitier architecture to sense the readings and analyze the vital information from the vehicle environment. It stores the health record into the cloud environment [5,6] to improve the reliable communication on rural-sides and disconnected regions of the health centers. The authors proposed effective communication protocol for sharing the vital information instantly but it takes longer time to reach the destination [7]. Sharing of the vital information to the open network is easily induced vulnerability communication, so this model provides security on communication. However the successful received information is just a record which allows the system, to take decision-making processes called ubiquitous patient surveillance system using ad-hoc networks. It provides more reliable, faster and better power management including rural places [8–10].The connectivity among rural places to health centers is achieved through the VANET system which collects and transmits specific patient information from rural areas to health service providers even in long-range city to obtain timely advice from top health professionals and to reduce overall health costs [11]. This research study explains four subsections as follows:

- **Study on Electronic Health Record (EHR).**
- **Study on LSTM.**
- **Study on Emergency Braking system.**
- **Study on Proximity Area.**

2.1 Study on Electronic Health Record (EHR)

In this section, we collected the survey related to health records, and these records are created from various sensors to maintain in an effective manner. The health records are more sensitive and handled by medical professionals to determine the patient current health condition compare with past medical records. Before previous five decades the records are stored manually using paper folders, and after that the patient medical histories are maintained in electronic form, which is stored in the hospital servers and processed only by experts. Recently the records are globalized, and it can access from any registered doctors in an emergency situation. The digital records are previous health histories maintained as

Electronic Medical Record (EMR) that monitor the patient health information in long-term manner. The EMR have patient's past and present health conditions that include previous hospital visits, disease and illness information, and medications, scan reports, lab test records, etc. The effective use of EMR is applied to the predictive modeling, which helps research studies such as early disease detection [12], comparative disease efficacy [13], and risk assessments [14]. To achieve the remote health monitoring system the sensor readings are collected wirelessly and directly transmitted to the cloud server and such records are called EHR. Table 4.1 summarizes the recent research activity on EHR, and Table 4.2 shows the sample static health record.The remote EHR contains a lot of noise and irregularities because of the scarcity of the sensor readings, and that is difficult to use the original EHR directly. The records are preprocessed before storing it into the cloud server and this gives more accuracy of the patient's health readings in a better and more robust way.To remove the noise from the records, the author suggest to convert the readings into zero value of the sparse matrix in order to build vectors [15]. The author's proposed neural network representation of the EHR which is mapped into observed medical characteristics and that are low-dimensional medical concepts, such as space through matrix decomposition, which uses linear combination such as skip-gram model to learn the medical terms exactly. To capture the real value relationship between diseases, the EHR helps to train a large number of medical datasets that uses a multidimensional, dense matrix [16].For finding a universal cooccurrence matrix in a medical representation, various grammar and semantic models are proposed to achieve real value vectors [17].This successful representation of health records induced the researchers to predict the prolong disease at any instant in real-world with higher accuracy. Moreover the research study on EHR helps health insurance companies to access their client health condition and provide instant help without delay [18].

TABLE 4.1 Recent research on Electronic Health Record (EHR).

Method	Year	Summary
Present study: Deep learning for EHR analytics	2019	Several deep learning algorithms were implemented to optimize the enhanced EHR. This various approaches provide complete knowledge discovery of the electronic health readings in an efficient manner. It covers the sensitive clinical data and analyzes the performance of daily activity of the patient, based on the health readings to provide intelligent solution with the help of IoT.
EHR: Systematic view, gene regularities, deep analytics, and continual patient monitoring system.	2018	The digital record provides opportunities and challenges in developing deep learning model. Analyzed the pharmacogenomics and drug targets easily. Finally the systematic representation of clinical task is utilized in the cloud servers.
Health informatics	2017	Health records are shaped in different deep learning architectures effectively, which converts the patient medical information in a simple and efficient manner. The EHR utilizes several techniques and algorithms like neural networks, image processing, pervasive computing, etc.

TABLE 4.2 Static: Health Record (HR).

Date time	Patient demographic information	Doctor information	Clinical notes	Lab test	Drug	Diagnosis	Next appointment
DD/MM/ YYYY HH:MM:SS	Patient ID: Age: Sex:	Doc ID:	Disease code: Patient condition:	Test code: Test results:	Drug code: Usage interval: (hours)	Diagnosis code: Type of diagnosis:	DD/MM/ YYYY HH:MM:SS
#22/09/18 #14:35:22	#P4356 #36 #Male	#D033	#CD305 #Normal	#L077 #ECG: #BP: #HP:	#M39 $ AF (after food) - 4 #M45 $ BF (after food) - 6	#O70: Medication #045: Operation	#22/10/18 #14:35:22

2.2 Study on LSTM

Intelligent machine learning approaches are used to predict the vulnerable diseases based on the disease learned behavior and symptoms shown to find the effective medication with more accuracy. The remote health monitoring system utilizes the combined benefits of intelligent health care system and IoT technologies to activate the Emergency Medical Services (EMS) to avoid any critical situation. This system is based on previous sensor readings which predicts the vital signs from the collected readings and suggests the user to avoid stress.The Logistic Regression (LR) algorithm calculates the vital signals, which is most sensitive and effective from the normal health readings and that is called stored medical information. Based on the record the LR achieves a good compromise between the reliability and interpretability of the system. Recently the authors have predicted leukemia [19], Alzheimer's disease [20], and cancer with higher accuracy [21–23] using LR. The predicted simulation outputs are applied in extended time series complex sparse models for deep learning. The proposed LSTM [24,25] is to overcome the major problems on gradient disappearance and gradient explosion of the RNN, so that the gradient flows into the network significantly. The LSTM is applied to a number of variants to improve the prediction rate.

The advanced LSTM algorithms are emerged and applied in various fields such as medical context-sensitive keywords, Dynamic Bayesian Network (DBN) that uses bidirectional LSTM, which consists of units for learning representation [26,27]. In the medical field the phased LSTM [28] is an advanced RNN architecture which incorporates the sensor inputs at any sample rate and that can be processed even in asynchronous timing events [29] to predict the health complications [30]. In this model, the LSTM is used to diagnose the disease; the proposed combination of logistic regression classification with three-layer auto encoder is used to estimate different ICD9 and predicts the accurate diagnostic method for the 128 types of diseases [31]. To predict the patient's future disease, the authors propose continual health monitoring system that is simulated with intelligent doctor assistant which is trained by the doctor's behavior using GRU network [32]. In summary, the researchers proposed lot of prediction algorithms to achieve better disease diagnostic methods, still there are many challenging issues that are not considered, such as difficulty to handle the dynamic time series pattern of EHR's, low learn rate of semantic similarity of recent medical with captured readings, and lack of executing the EMS on critical situation.

2.3 Study on Emergency Braking system

The General Estimate System (GES) collects accidental survey on recent real-time events from the crash database, it reports nearly 309,000 rear-end collisions happened in 2000, 65% of the incidents are occurred because of driver's mental absence and 35% are occurred due to poor environmental conditions, such as road damages, poor light sources, and severe fog [33]. To detect the sudden environmental changes,the vehicle needs an intelligent assistance system to avoid any prolong activity and to alert the vehicle drivers [34]. Mostly, the researchers have developed many autonomous vehicle control system of Adaptive Cruise Control (ACC), which calculates the distances of leading vehicles and it informs the driver frequently. The intelligent decision making algorithms are implemented to improve the safe driving, sensors are utilized to collect vision perception data using stereo cameras and laser scanners to avoid vehicle crash, even in worse weather condition [35,36]. This system executes brakes based on the intelligent decision controller, which is installed under the vehicle braking system and it decides to apply either normal brake or emergency brake based on the driver's health condition [37,38]. To improve the braking system, the system prefers to remove the unwanted warnings of the assistant and safety margins are highly examined to achieve higher accuracy prediction [39].

2.4 Study on proximity area

The proximity type communication is defined as any wireless device which offers the wireless transmission also known as proximity device. These devices can transmit the data using long-range (LoRa) communication and that even works without Internet connection and transmits up-to 7 km [40]. The wireless devices are considered as wireless nodes such as Long Term Evolution (LTE) towers, mobile phones, laptops, Wireless Fidelity (Wi-Fi) access points, Worldwide Interoperability for Microwave Access (WiMAX) towers and also other such networking devices. To improve the efficiency of communication, various networking protocols are studied. The graph-based routing algorithm transmits the sensed data from the source to sink with high speed range using Destination Oriented Acyclic Graph (DODAG) [41]. To achieve the reliable transmission, graph-based regional energy-efficient RPL routing protocol (ER-RPL) is implemented. In this graph, every node chooses the parent node with higher energy consumption. The high-efficient communication model called cluster-based relay is proposed. In that, the nodes choose their immediate neighbor node either source to sink transmission to achieve the reliable and faster transmission. There is a lot of cluster-based routing protocols are implemented such as LEECH, Fuzzy LEECH, LEECH-S, LEACH-M, Hybrid Hierarchical Clustering (HHCA), and Enhanced Clustering Hierarchy (ECH) [42–44].Various procedures and algorithms have achieved their accuracy in better way even though some of the issues are not yet solved such as node failure, packet lost, and faster response during transmission.

3. Proposed system

The overall system architecture is shown in Fig. 4.1, which depicts the intelligent transportation system is combined with advance health care system to offer guarded transportation throughout the passenger's journey. This system monitors the driver's health condition during travel, by predicting the vital signals using steering wheel sensor and execute the fast and efficient emergency protocols without delay. When the system receives vital signals from the driver, the system executes emergency sound alarm system to indicate all the nearby vehicles by alarming that the vehicle is in danger condition and the vehicle enables autonomous driving mode with the help of intelligent assistant system to take control of the vehicle and reduces the speed by providing safe parking with the help of magnetic sensor to calculate the vacant places on the road to park the vehicle without any collision. The emergency system immediately executes the transmission of the vital readings to nearby emergency health centers, cloud servers, hospitals, and the registered family members.

The proposed system follows five modules

1. Steering wheel sensor,
2. Electronic Health Record (EHR),

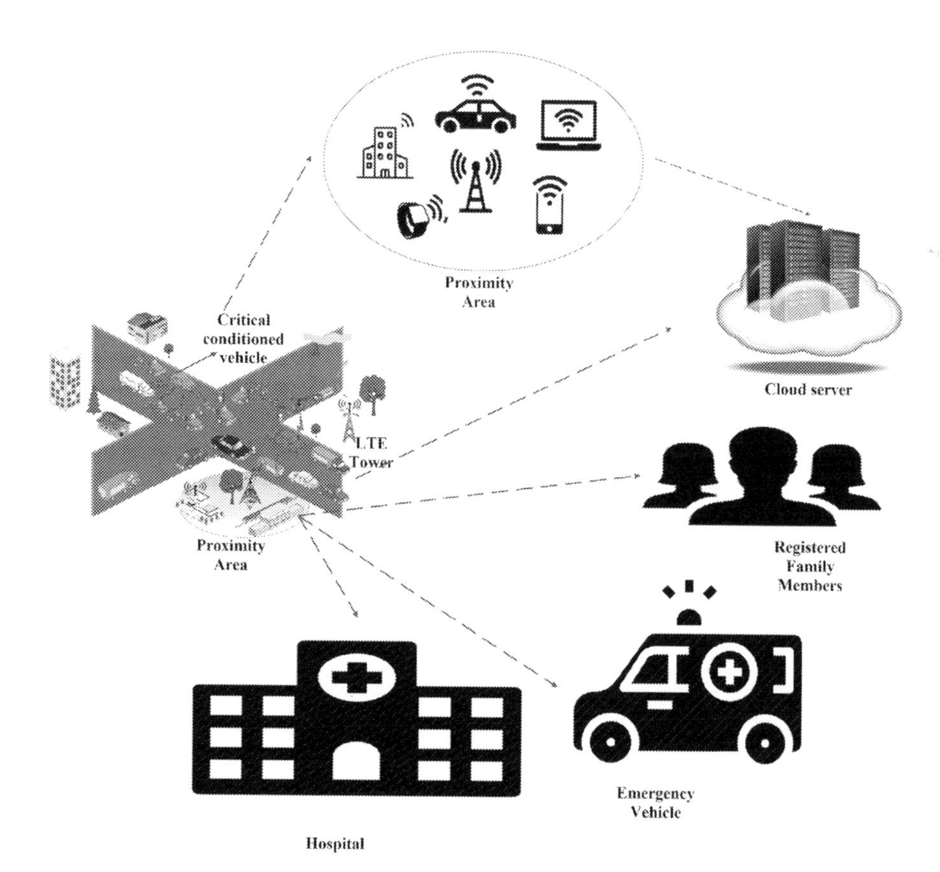

FIGURE 4.1 System architecture.

3. Proximity Area Network (PAN),
4. Automatic Emergency System, and
5. LSTM.

In this work, the system predicts the vital readings of drivers' critical health condition, and that is achieved using sensitive sensors. The sensors are deployed and tested under various states to improve the standard of the sensor quality. The sensor collects the ECG readings of the driver and is continually transmitted to the cloud server. The emergency procedure is activated in two ways such as 1. Execute emergency procedures from the vehicle itself is 2. Execute using cloud server which predicts the vital reading and executes the emergency system. The long-term monitoring of driver's behavior and the vital readings are easily predicted based on past records using cloud server decision. When a new driver handles the system it takes time to learn, in that case vehicle emergency system executes immediately without delay and no longer to wait for the cloud server response.

Both the types of the health readings is continually transmitted and stored in the cloud server. This work provides promising solution for guarded transportation for passengers, drivers, and nearby vehicles.

3.1 Steering wheel sensor

The steering wheel sensor collects the health readings of drivers and continually monitors their heart beats. It is used to predict the vital signals of onboard driver effectively and filter the readings into normal, fatigue, and abnormal which are based on the activity and fitness level of the driver and it is continually stored into the cloud server. The normal, fatigue signals indicates standard health condition but abnormal signals indicate the worst health condition of the driver and his life in danger state. We collected continual ECG readings of the driver and forwarded it to the intelligent system, which resides in the cloud server. It learns the heart beating pattern to predict the next heart rate accurately and helps to avoid the worst case of the health condition.

From the heart rate, the intelligent learner calculates the various vital signs and it helps to determine the range such as electrocardiogram, heart rate, blood pressure, respiration rate, blood oxygen saturation, blood glucose, skin perspiration, capnography, body temperature, and motion evaluation. In this research work, the sensors (A1, A2, A3, A4, A5, and A6) are placed in steering wheel and that is presented in Figs. 4.2 and 4.3. The sensor parameters are

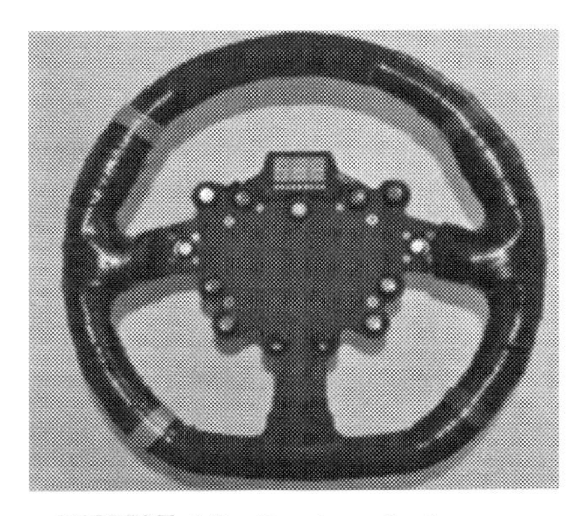

FIGURE 4.2 Steering wheel sensor.

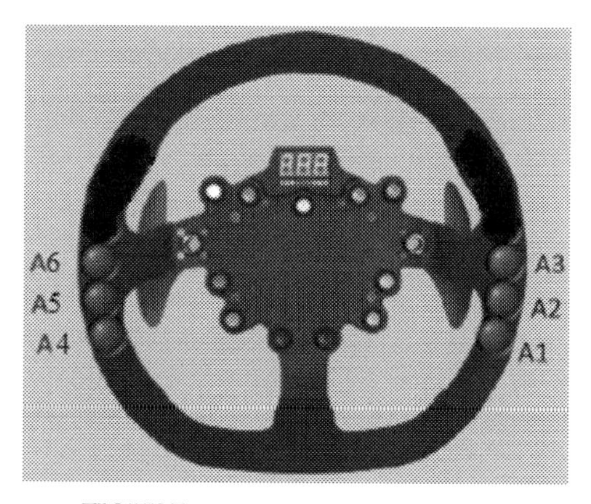

FIGURE 4.3 Positions of the sensor.

used to detect the physical state of the driver and it also detects the blood alcohol level. Sensors are deployed on the steering wheel, which remotely monitors the health readings into the cloud server with the help of proximity area communication. It achieves faster and reliable transmission to the cloud server. These types of optical sensors using photoplethysmographic technology which is used to find the blood flow to calculate the driver's blood pressure, pulse rate, and body temperature and the sensor characteristics are clearly mentioned in Table 4.3. This optical sensor uses light source in the transducer to transilluminate the tip of

TABLE 4.3 Sensor specification.

Steering wheel sensor	Physical characteristics	Type/Range
Sensor: ECG electrodes	Base material	Polyester
	Thickness	0.04–0.20 mm
	Surface	<0.04 Ω/sq
	Resistance	64–90 dB
	Gain effectiveness	500 (27 dB)
	Cut-off	0.5–34 Hz
A/D converter	Frequency resolution	12 bits
	Sampling rate	200 Hz
	Frequency band	2.4–2.485 GHz
RF transceiver	Sensitivity	95 dBm
	Transceiver rate	500 Mbps
	Mode	Rx: 18.8 mA, Tx: 17.4 mA
	Current drawn	Sleep mode 1μA
Power	Battery	3.3 v

the finger and a photoconductor identifies changes in light intensity within the finger due to pulsating changes in blood volume. One is made by the heart as it pumps blood into the arteries and through the vascular system. To calculate the peripheral blood flow of the body the efficient technique used is plethysmography [45], which is based on the differential light-absorbing characteristics of tissue and blood. Specifically, living tissue is relatively transparent to light in the infrared range, whereas blood is relatively opaque to light within this frequency range of blood pressure range from two forces. The increased blood flow from the force of the arteries resists the blood flow. This wireless physiological sensor is the unit of important observation and monitoring the human health using blood flow. The flow contains plethoric physiological information, which is collected from laboratories to determine the live blood flow through exercise and find the human typical blood flow meters by size, weight, and use of optic fibers. To resolve these disadvantages, proposed system uses latest technology that is easily available.

The proposed system uses microtechnology which integrates the laser Doppler blood flow meter that uses microelectromechanical technology and measures the skin blood flow in the forehead, fingertip, and ear lobe. The sensor readings are more accurate than other wearable sensors.Fig. 4.4A–D shows the real time sensor readings which is collected from two different age subjects test their normal, fatigue, and abnormal condition of heart readings in both the idle and driving states.

3.2 Electronic Health Record (EHR)

The long term maintenance of medical records for each drivers include their illness histories, medications, crucial signs, laboratory results, statistical significance, etc., are collected using previous medical reports and the lively health readings are collected from steering wheel sensor and that is updated to the cloud server with the help of LoRa communication. The driver's location information is appended with timely sensor, which are systematically collects through the raw ECG signal and that helps to monitor the driver's health condition remotely. Let the set D contains the collection of driver's health record represents as $D = \{d_1, d_2, d_3, \ldots d_n\}$ be a set of n drivers and $R = \{r_1, r_2, r_3, \ldots r_k\}$ be a set of k records. For each driver d in$D$, the sensor readings $S_i = \{r, t_j\}$, and r_i is the ith sensor reading recorded at time t_i. The readings are closely related to past recorded sensor readings with current readings, calculate the time interval between every sensor reading as Δt. Rewrite the sensor readings $S_i = \{r, \Delta t_j\}$ of ith recorded, Δt_i is the time interval between r_i and r_{i-1}. The prediction value P of the sensor readings uses multilevel list and that can be generated for l times of the driver d, as follow in Eq. (4.1).

$$P_d^l = \left\{ \begin{array}{l} \left((r_1^1, \Delta t_1^1), (r_2^1, \Delta t_2^1), \ldots, \left(r_q^1, \Delta t_q^1 \right) \right), \left((r_1^2, \Delta t_1^2), (r_2^2, \Delta t_2^2), \ldots, \left(r_q^2, \Delta t_q^2 \right) \right), \\ \ldots, \left((r_1^l, \Delta t_1^l), (r_2^l, \Delta t_2^l), \ldots, \left(r_q^l, \Delta t_q^l \right) \right) \end{array} \right\} \tag{4.1}$$

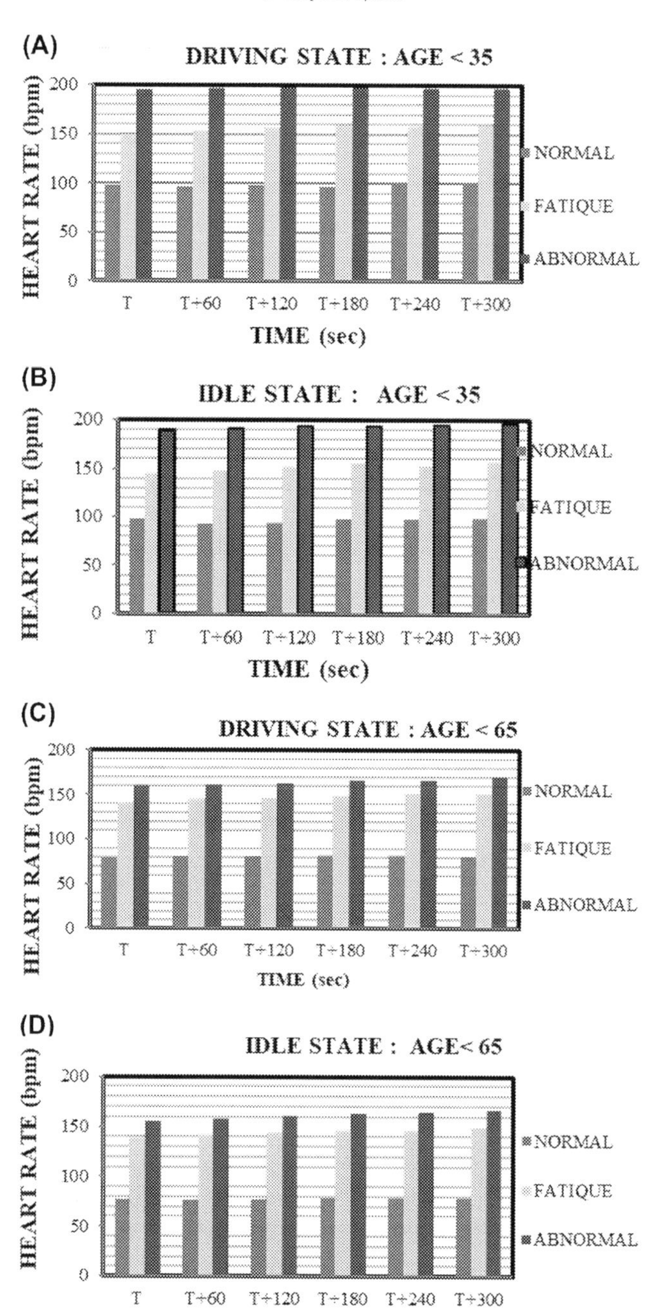

FIGURE 4.4 (A) Driving stage: age <35; (B) idle state:age <35; (C) driving stage: age <65; (D) idle state:age <65.

3.3 P-area

To improve the communication efficiency, the proposed system uses device-to-device (D2D) communications, because of the emergency conditions, the valuable sensor readings are transmitted to the destination without failure and the communication is based on time. Wireless proximity communication is one of the promising technologies which transmit the sensor readings in reliable manner. The sensor readings are collected from steering wheel and offers wireless transmission capability using the special type of standard, which is IEEE STD 11,073–20,601. This standard significantly broadcast sensor readings to nearby wireless devices until it receives the successful acknowledgment. However, the transmission initially checks the Internet connectivity if it is available, then it transmits the sensor readings to the cloud server directly. When the Internet connection goes offline, it uses the proximity type communication and broadcasts the data, which includes sensor readings, and it forwards them to the neighbor node and also has the ability to transmit the data in wireless manner. In this work, the proximity devices are considered, such as Wi-Fi, mobile phones, Internet connected vehicles, LTE-tower, and any wireless transmission enabled devices. This type of communication calculates the distance between the source and destination, if it is in long range it uses relays, otherwise chooses direct transmission. The relay is also a wireless device which acts as an intermediate member between the sender and the receiver, the successful transmission uses more number of relays based on the distance. In this paper, we utilized the joint relay selection and resource allocation scheme for relay-aided D2D communication networks. The objective is to maximize the total transmission data rate of the system while guarantee the Quality of Service (QoS).The transmission capability of the device is represented as $SE_T(n,t)$ and energy consumption for sending n bits of data is generated by sensor using Eq. (4.2).

$$SE_T(n,t) = nE_{elec} + n\rho_{amp}d^{\gamma} = \begin{cases} nE_{elec} + n\rho_{amp}d^2, & \forall d \leq d_0, \\ nE_{elec} + n\rho_{amp}d^4, & \forall d > d_0. \end{cases} \tag{4.2}$$

Where n-bit of packet at time stamp t, the energy consumption range for each bit transmission is E_{elec}. The power amplification factor ρ_{amp} is used to determine the communication range.

3.4 Automatic Emergency Braking (AEB) system

To avoid vehicle collision, the AEB system executes the emergency braking system when it receives the vital signals from the driver. The improved intelligent assistant system is taking control from the driver and enables autonomous driving mode to provide an emergency safe parking of the vehicle without any collisions. This work calculates the vehicle's current position and determines the number of vehicles moving along with the critical condition vehicle. This intelligent system uses magnetic sensor to calculate the vacant position on the road and computes the obstacles. The existing GPS system is used to find the vehicle's current location but cannot determine exact vehicle count. So, the proposed system shows the accuracy of the vehicle count. To compute exact vehicle count we use magnetic sensors which are implemented inside the vehicle and it senses the nearby ferrous obstacles up to 28 m (mostly all the vehicles are manufactured by ferrous material). AEB system engages the Dynamic Brake Support (DBS), which reduces the vehicle's speed rapidly by applying hydraulic system of brakes that are exposed. Calculating the vehicle's moment is represented in plane (x, y) and the Newton's second law of the center of the gravity of the vehicle's position using V_x, V_y is obtained from Eqs. (4.3) and (4.4).

$$\dot{V}_x = V_y r + \frac{1}{m}\sum_{i=1}^{4}(F_{xf}\delta_i + F_{xr}\delta_i) \tag{4.3}$$

$$\dot{V}_y = -V_x r + \frac{1}{m}\sum_{i=1}^{4}(F_{yf}\delta_i + F_{yr}\delta_i) \tag{4.4}$$

$$r = \sum_{i=1}^{4}\left(\frac{d_f F_{yf}\delta_i - d_r F_{yr}\delta_i}{I_z}\right) \tag{4.5}$$

Where $\dot{V}_x$ refers longitude and the front tire force defined as F_{xr}, rear tire force as F_{xf}. The lateral velocity is indicated $\dot{V}_y$, steering angle δ_i of vehicle wheel i =(front wheel1, front wheel2, rear wheel3, rear wheel4) front and rear end tires are represented as F_{yf} and F_{yr}. The rate of the yaw is defined as r, obtained using Eq. (4.5), and the mass of the vehicle is represented as m, the variable I_z refers the vehicle's moment of inertia to the Z axis. The distance between front axle d_f and rear axle d_r is calculated using center of gravity (CG) of the vehicle, respectively. Now, the magnetic sensor calculates the magnetic flux of nearby vehicles based on their respective flux. The wheel axle is controlled and turns the vehicle into predefined lane indication (left side) so, that the vehicle is parked without any congestion.

The magnetic flux is used to determine the exact vehicle count based on the ferrous material strength. Every vehicle has their respective magnetic flux signal, for example, bike has the ferrous material strength which is lower than bus. From that the magnetic flux accurately determines the vacant position of the road. The magnetic sensors implemented inside the vehicle's chassis can sense the flux up to 28 m surroundings. The sensor calculates x-axis which is the direction and it is parallel to the vehicle's moving direction, the y-axis represents perpendicular to the direction of the vehicle moving direction, and the z-axis is perpendicular to the road surface and it is upwards to the vehicle. The magnetic flux of a vehicle can be described as a magnetic dipole with a magnetic moment m center of the vehicle. The affection of the magnetic field depends strictly on the distance from the magnetic dipole to the vehicle. The Eqs. (4.6)–(4.8) are used to calculate the components of the magnetic field induced by m and that can be derived from Maxwell's equations [46].

$$B_x = \frac{\mu_0 \times (m_x(2x^2 - y^2 - z^2) + 3m_y xy + 3m_z xz)}{4\pi r^5}. \tag{4.6}$$

$$B_y = \frac{\mu_0 \cdot (m_y(2y^2 - x^2 - z^2) + 3m_x xy + 3m_z yz)}{4\pi r^5} \quad (4.7)$$

$$B_z = \frac{\mu_0 \cdot (m_z(2z^2 - y^2 - x^2) + 3m_x xz + 3m_y yz)}{4\pi r^5} \quad (4.8)$$

Where,μ_0 is the permeability air content and r is the distance between sensor position (x, y, z) and obstacles. This flux is used to analyze the vehicle's positions and obstacles accurately.

3.5 LSTM

The intelligent learner is residing in the cloud server to predict the abnormal state of the vehicle's driver. This learner collects the driver's past medical records and analyzes the health behavior. When the driver initiates driving, the lively steering wheel sensor starts sensing the ECG readings and transmits it to the cloud server continuously.

$$i_t = \sigma(w_{xi} \times x_t + w_{hi} \times h_{t-1} + w_{ci} \times h_{t-1} + b_i) \quad (4.9)$$

$$f_t = \sigma(w_{xf} \times x_t + w_{hf} \times h_{t-1} + w_{cf} \times h_{t-1} + b_f) \quad (4.10)$$

$$s_t = s_{t-1} \times f_t + i_t \tanh(w_{xc} \times x_t + w_{hc} \times x_{t-1} + b_c) \quad (4.11)$$

$$o_t = \sigma(w_{xo} \times x_t + w_{h0} \times h_{t-1} + w_{co} \times h_{t-1} + b_o) \quad (4.12)$$

$$h_t = o_t \tanh(s_{t-1} \times f_t + i_t \tanh(w_{xc} \times x_t + w_{hc} \times x_{t-1} + b_c)) \quad (4.13)$$

The collected ECG signals build knowledge base to convert the readings into vector representation to feed into the efficient learner system in a semantic manner. This system performs fine grained model to learn continuous time series pattern of sensor readings from that it predicts the next reading to avoid uncertain health condition. The LSTM uses the explicit model to find the variable time interval of the sensor readings and significantly takes the information accurately from the EHR. The memory based model which has already learnt the driver's behavior from the past medical records and it determines the current health status of the driver very accurately.

The sensor readings are controlled by three gates such as input gate, output gate, and forget gate. It has an independent memory cell which is capable of reading, writing and updating the long term continual dependencies. This model has strong processing ability to compute the health readings to find the inconsistent health readings.

Fig. 4.5 depicts the LSTM gate diagram, in that the variable i_t, f_t, and o_t refers input, forgetting, and output gates, the variable s_t refers memory unit, h_t indicates hidden state, input vector represent x_t, W refers corresponds weight to input and output reading and the bias value refers to b. The sigmoid function σ has value range from $0 \leq 1$. In general the EHR generation is strictly time series data, to store each driver's health record is more imbalance in the

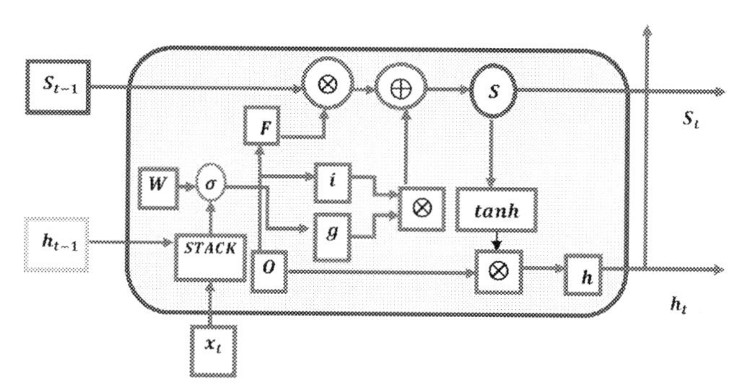

FIGURE 4.5 LSTM gate diagram.

occurrence of the time. All the sensor readings are recorded and stored into the memory cell without any loss to improve the sustainable readings. We combine the past health records and continual sensor readings with regular rich time and accurate location information is used to predict the outcome in reliable manner. We set the time control gate in LSTM architecture which records the sensor readings for every time interval to predict the next time interval reading. The time control gate Tg_t stores the time information and transmits it to the further gate to achieve the continual long term dependencies of the driver's health which is monitored using higher accuracy results and Tg_t is calculated in Eq. (4.14) using previous Eqs. from (4.9)–(4.13).

$$Tg_t = \sigma_s(w_{xs} \times x_t + \sigma_{\Delta s}(\Delta s_t \times W_{ss}) + b_s) \tag{4.14}$$

Where σ_s is the sigmoid time interval, s refers time interval of every sensor readings that is controlled by both the input and time information. The modified equation can be rewritten in LSTM as follows

$$s_t = s_{t-1} \times f_t + \sigma_s(w_{xs} \times x_t + \sigma_{\Delta s}(\Delta s_t \times W_{ss}) + b_s)$$
$$\times i_t \tanh(w_{xc} \times x_t + w_{hc} \times x_{t-1} + b_c) \tag{4.15}$$

Apply the value Tg_t in Eq. (4.15) we get new simplified value for s_t in Eq. (4.16) and o_t in Eq. (4.17).

$$s_t = s_{t-1} \times f_t + Tg_t \times i_t \tanh(w_{xc} \times x_t + w_{hc} \times x_{t-1} + b_c) \tag{4.16}$$

$$o_t = \sigma(w_{xo} \times x_t + \Delta s_t w_{h0} \times h_{t-1} + w_{co} \times h_{t-1} + b_o) \tag{4.17}$$

We use the sigmoid to predict the output value using variable y'_t and the real output value y_t.

$$y'_t = x'_{t+1} \tag{4.18}$$

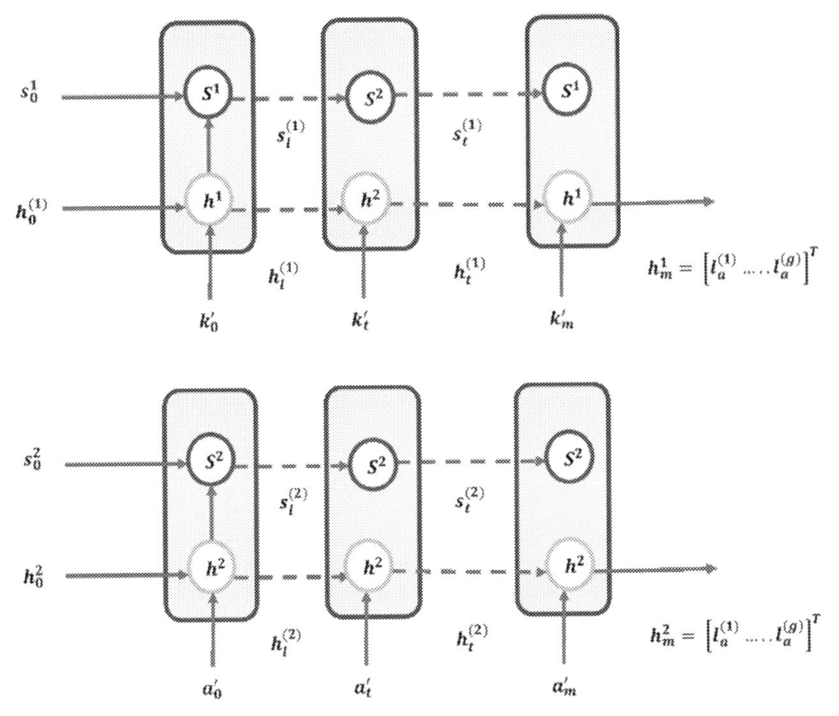

FIGURE 4.6 LSTM flow diagram.

The dropping value is predicted and the real-time output value is calculated using cross-entropy function. Where, N refers to the number of timestamps of the sensor readings, y_t' in Eq. (4.18) denotes predicted output value and y_t is the output value.

The flow of the predicted value is depicted in Fig. 4.6 and that is used to calculate using Eq. (4.19). The sensor signals are cumulated to determine the overall signal losses and take the average sum of the individual losses to improve the prediction accuracy.

$$L = \left(y_t, y_t'\right) = \frac{1}{N} \sum_{1 \le t \le N} \left(y_t \times \ln y_t' + (1 - y_t) \times \ln\left(1 - y_t'\right)\right) \tag{4.19}$$

This learner continually learns multiple driver health behavioral patterns and suggests the solution to the driver in order to avoid uncertainty health condition.

4. Performance analysis

Efficiency of the proposed system is examined with various comparative models to prove our system is more convenient to the drivers. Initially the sensor readings are collected from different human subjects to calculate the accuracy of the steering wheel sensor, ECG readings into different states such as idle, normal state, fatigue state, and abnormal state. The collected readings are converted into optimized EHR and that is forward to the cloud server to learn the pattern with the help of LSTM. The efficiency of the transmission is calculated using P-area and that is compared with different communication algorithms. The evaluated result provides prominence solution for continual health monitoring system for drivers. To find the accuracy of the vehicle positioning system, we use the higher accuracy of Differential Global Positioning System (DGPS), navigation system implemented by CHCB20 [47], which have a larger coverage area up to 750 m to monitor the experimental route. The sensor quality is determined by various factors such as reliability; durability, flexibility and configurability and are tested at different states such as driving and ideal states. The test vehicle implemented with speed controller of AEB system to provide Emergency Dynamic Brake Support (EDBS) system, which controls vehicle speed immediately by applying the hydraulic brakes. When the system receives the abnormal condition it automatically turns into autonomous driving mode and triggers the sound alarm to nearby vehicles to alert the nearby vehicles. The vehicle applies EDBS when it receives the vital signals and the vehicle count is calculated by the installed magnetic sensor inside the vehicle to detect the nearby ferrous material strength and to provide safe parking without collision. For evaluating the readings, the subject is asked to drive continuously for 4 h. Initially the subject's cardiac signals are checked with their full functional body actions to determine their normal health readings before attempting to drive. After that, subject's readings are collected and continually transferred it into cloud server with location and time stamp. When it receives the vital readings from the driver, it informs it to the nearby EMS, hospital, and sends message to the registered family members and cloud server at the same instance with the help of LoRa using different wireless devices to continually broadcast until it receives the receiver's acknowledgment. The duration of every single minute ECG signal report is collected and processed while driving state evaluates the accuracy. All the readings are successfully transmitted via P-area without any packet loss to the cloud server and other emergency system. This vehicle device transmits using beacon signals to nearby devices until it receives the acknowledgment from the receiver. Moreover, this proposed system achieves low power consumption during transmission of packet. From the received signals we determine the driver's state whether he is normal, fatigue or abnormal. All the three states readings are tested and compared with idle state reading of 60 s and it is achieved 100%, 99.2%, and 96.4% respectively. In order to perform reliability of the steering wheel sensors, the ECG readings are compared with standard medical ECG device and the results accuracy is achieved by 99.8%.

4.1 Measuring performance during driving

Measuring the performance of ECG readings while driving, the subjects are asked to drive the test track of 1 km in various formations such as straight, curved, lane changed and hairpin bend with temporary installed speed bumps on the test lane at the speed of 90 km/h. In that driving state, the driver's grip of steering wheel is stabled and the readings are collected and processed from different subjects and it achieved the accuracy of 98% on straight lane, 96.45% on curved, 95% on lane changed, and 94.56% on hairpin bend, respectively, using Eq. (4.25). These proposed method results compared with idle state readings has achieved the accuracy of 96.78% using Eq. (4.25). The proposed system achieves higher accuracy of the sensor readings on driving state.

4.2 Analysis of sensor stability

The stability of the sensor is determined at various driving state using various factors like varying the speed of the vehicles from time to time and also through a lot of obstacles like speed bumps, traffic signals, lane change, poor road and weather condition is analyzed. The stability of the dry electrode sensor and the driver's hand grip may be unstable in above scenarios. In that case, the sensor readings are not accurate and it is difficult to measure the ECG signal when the driver is in abnormal condition. In order to achieve the high performance of the proposed system, the received signals should be standardized and it should be able to recover the signal quality even in unstable contact between the driver's hand and steering wheel. The collected signals are processed using intelligent algorithm to improve the stability of the sensor readings with higher accuracy. The test vehicle achieves better performance at the speed of 65 km/h passing five speed bumps. The recorded reading shows higher accuracy after applying the signal recovery method and the quality of the ECG readings are disrupted while driving on the speed bumps.

4.3 Analysis of the ECG measuring performance for critical sensor reading

The strong relationship between the ECG readings with the driver's health activity is continually monitored and that is stored into the cloud server. This section concentrates when the vital signals are received from the sensors and execute the emergency procedures immediately to avoid critical stage. The critical state ECG readings compared with normal state ECG readings to determine the strength of the vital readings. For the experimental purpose, the ECG readings on normal state are collected for 1 min and that is compared with critical driving conditions. When the driver is in critical stage, the Heart Rate Variability (HRV) is calculated and it reflects the Autonomic Nervous System (ANS) because of that the ECG value slightly reduces from 67 to 54 which calculates the accuracy of vital circumstances at a rate of 99.42% and the normal state accuracy is achieved 100% respectively using Eq. (4.20). From the research study, we analyze the ANS system so that the accuracy of the vital readings is improved and that is compared with normal readings. While the driver is in critical condition the driver's hand grip of steering wheel may become unstable, in that case the accuracy of the readings evaluated and predicted using processed signal of LSTM. This intelligent learner already learns the driver's behavior from long period so that accuracy of the predication rate is high. When a new subject drives, the prediction rate is less compared with long term prediction even though it achieved 94.56% accuracy using Eq. (4.20).

4.4 Learning task

The sensor readings are collected and transmitted to the cloud server, in that health records are fed to learn the health behavior using LSTM. Different subjects health readings are collected and learned continuously. The intelligent learner analyzes each subject's normal, abnormal and fatigue readings. The results are collected using various states of driver's driving condition and the algorithm generates the emergency alerts based on the driver's health condition. The recorded readings are considered as episodes with particular time stamps, every 1 min of readings are recorded to proceed to the evaluation process. In that,

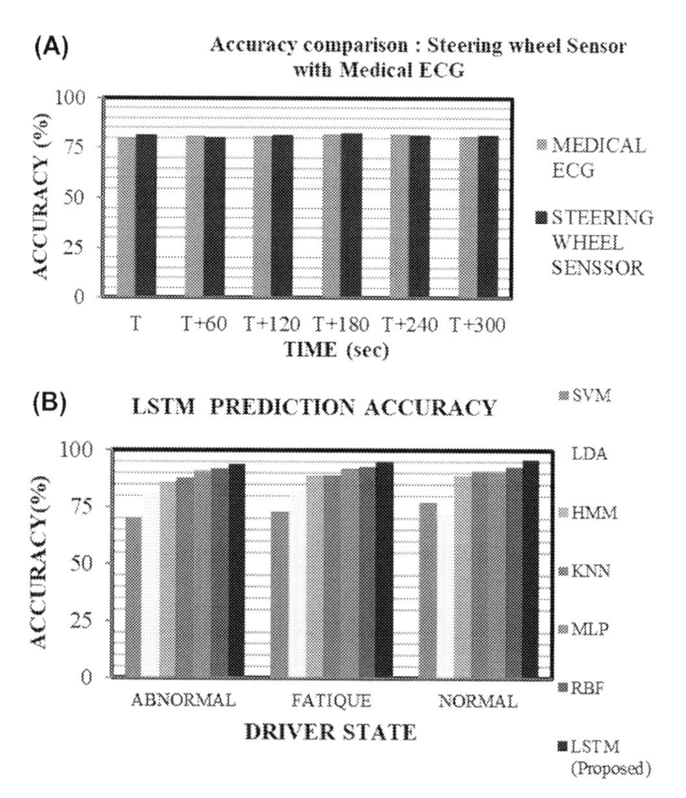

FIGURE 4.7 (A) Accuracy comparison: steering wheel sensor with medical ECG; (B) LSTM Prediction ACCURACY.

the reliability of the episodes is almost constant at the small intervals and the variations of sensor readings are captured and recorded into the episodes. Fig. 4.7A shows the reliability of the sensor readings which is compared with high-efficient medical ECG system.

The sensor reading gives more accurate results while driving when it is compared with normal ECG system and that is formulated in Eq. (4.20). Fig. 4.7B exhibits the proposed system prediction results which are compared with six different existing algorithms. In our system we achieved the higher prediction accuracy rate with the help of the sensor quality, effective transmission, and intelligent learner to predict abnormal state accurately using Eq. (4.25). Prediction values are determined using continuous real time ECG readings which are collected from steering wheel sensor and are fed to learn the cloud server. When the system predicts the vital readings, the predicted episodes are compared with normal and ideal readings and the efficiency is calculated using predictive parameters. Moreover, the transmission delay to reach the cloud server depends on the transmission capability of the every wireless node and that may not exceeded 20 s when we apply the greedy approach.The ECG readings are converted into episodes to achieve the optimized storage into the cloud server and also these episodes filter the unwanted signals and improve the quality value of the heart readings. Fig. 4.8A and B depict the episodes of three states with particular time stamp. From

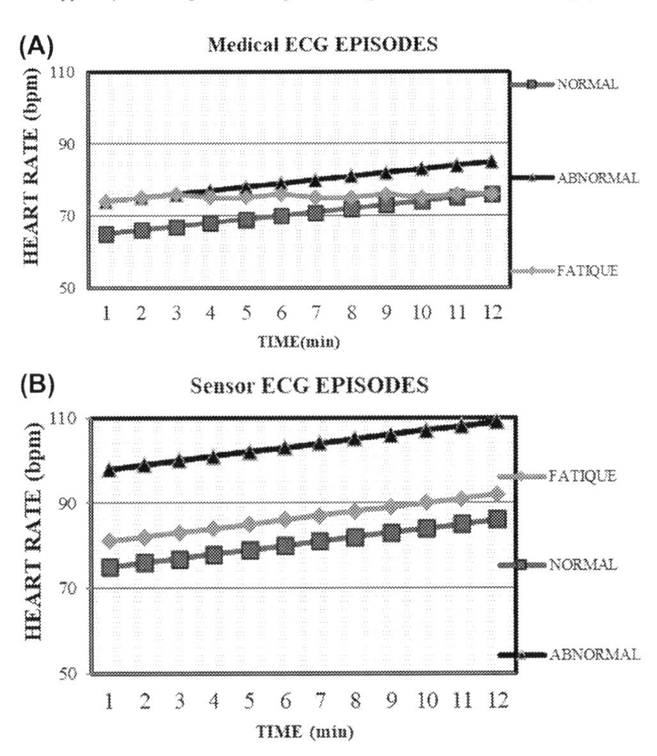

FIGURE 4.8 (A) Medical ECG episodes; (B) sensor ECG episodes.

that the sensor episodes show higher prediction rate when it is compared with medical ECG system episodes. Normally the medical ECG readings give higher accuracy on stable state but in driving state the readings achieves lower accuracy because of the movement. The sensor readings are learned and it predicts the next ECG reading of the subjects based on knowledge of long term evolution of the driver's heart beat pattern.

$$Accuracy = \frac{\text{Number of correctly classified samples}}{\text{Total number of test samples}} \tag{4.20}$$

The prediction results [45,48,49] are compared with actual readings using various methods such as Positive Predictive Rate (PPR), the Negative Predictive Rate (NPR), the False Positive Rate (FPR) and the False Negative Rate (FNR) are formulated in the following Eqs. (4.21)−(4.24).The predictive values are shown in Table 4.4 and the formulae are summarized as follows:

$$\text{PPR} = \frac{T_P}{T_P + F_P} \tag{4.21}$$

TABLE 4.4 Prediction accuracies using ECG data of three drivers.

Type	Definition
True positive (T_P)	Accurate abnormal ECG readingsusing steering wheel sensor
False positive (F_P)	Abnormal ECG readings divided into normal ECG readings
True negative (T_N)	The number of normal ECG beats correctly classified
False negative (F_N)	Number of normal ECG beats divided into number of abnormal beats

$$\mathrm{NPR} = \frac{T_N}{T_N + F_N} \tag{4.22}$$

$$\mathrm{FPAR} = \frac{F_P}{T_N + F_P} \tag{4.23}$$

$$\mathrm{FNAR} = \frac{F_N}{T_P + F_P} \tag{4.24}$$

$$\mathrm{Accuracy} = \frac{T_P + T_N}{T_P + F_N + T_N + F_P} \tag{4.25}$$

From the predictive values the system suggests the drivers to take medication and do some relaxing activity like listening to music, drinking water, and doing small exercise to improve the heart blood flow. The recorded episode values determine the vital readings when it is compared with normal readings. The smaller readings of the episodes are more sensible and lose at some period on critical condition when the emergency system also fails. And also the consideration of false emergency execution is avoided in our proposed system.

For example the sample episodes are taken in Table 4.5 from that the episodes two and four values are recoded with high variations when it is compared with other episodes. This proposed system achieves more than 99% of the precision prediction accuracy of vital readings from the cloud server. The experimental results are discussed with medical professionals and they recommend the system which predicts the vital signals effectively and the prediction time minimum of 7–10 min with higher accuracy.

The comparative vehicle location accuracy is plotted in Fig. 4.9A in which the real time readings were collected from six different areas such as A1, A2, A3, A4, A5, and A6 using Eqs. (4.6)–(4.8). The DGPS readings show higher accuracy of the exact vehicle location when it is compared with normal GPS readings. In Fig. 4.9B we achieved the efficient communication using greedy approach and the transmission time is faster than the brute force approach using Eq. (4.2).

TABLE 4.5 Prediction accuracies using ECG data of three drivers.

Episode size DD/MM/YYYY HH:MM:SS	Sensor reading parameters (%)														
	T = 1 s					T = 10 s					T = 60 s				
	PPA	NPA	FPAR	FNAR	Accuracy	PPA	NPA	FPAR	FNAR	Accuracy	PPA	NPA	FPAR	FNAR	Accuracy
(AN: T)	97.78	99.09	0.91	2.22	98.71	94.44	100	0.53	0	99.51	100	98.55	0	12.90	98.67
(AN: V)	100	97.26	0	4.25	98.31	83.33	100	2.12	0	98.08	100	99	0	6.89	99.12
(N + AN: V)	73.78	100	16.67	0	89.47	66.67	99.06	5.41	7.69	94.35	100	99.35	0	3.57	99.45
(N + AN: Ts)	48.89	100	47.92	0	67.12	55.56	100	10	0	91.11	100	100	0	0	100

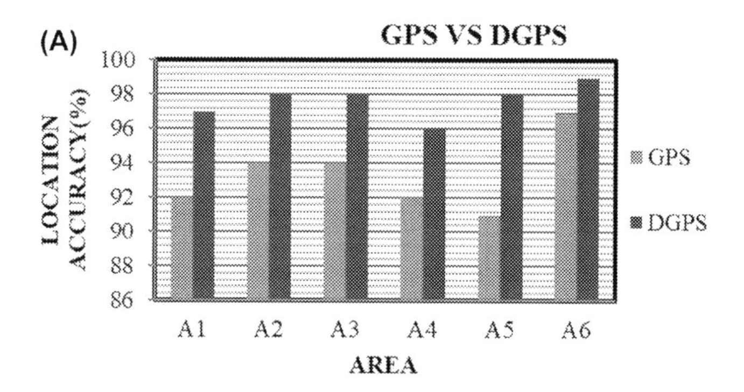

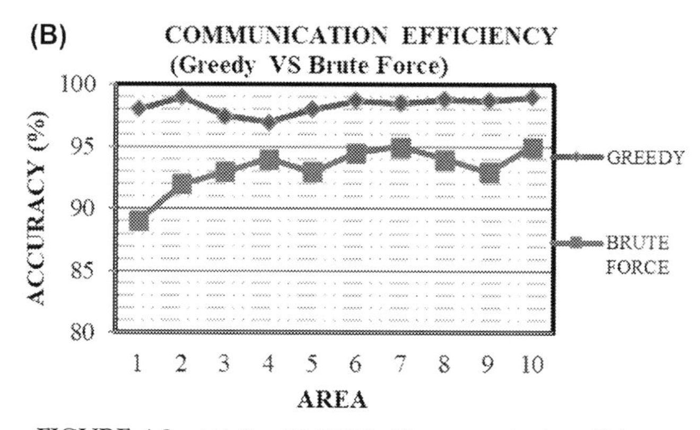

FIGURE 4.9 (A) Gps VS DGPS; (B) communication efficiency.

4.5 LSTM time series model

The proposed learning approach depends on the time series model which is obtained from the steering wheel sensor and the readings are stored as episodes. The episodes are univariant in time series $X(t) = \{x(1), x(2), \ldots, x(n)\}$, the average accuracy values of the time series based on the time stamp. From the episodes, the learner predicts the upcoming values using intelligent predictive model.

The training dataset that are collected from different subjects predicts the rate of ECG signal. Table 4.5 classify the readings into four types such as abnormal training set (AN: T), abnormal validation set (AN: V), combined data for both normal and abnormal validation set (N + AN: V), and testing set (N + AN: Ts). To predict the abnormal readings, the training set is learned using validation set to find the vital readings and the average prediction accuracy of three conditions are shown in Table 4.6. To calculate the rate of error prediction, the Multivariate Gaussian (MG) method is used for improving the maximum probability estimation. The stacked LSTM feeds the sensor readings into input layer with time stamp and makes prediction for the next time step. This multilayered recurrent network model is trained completely and produces the output layer. Table 4.7 shows the proposed model response time from the cloud server, which is compared with following models such as, Common Spatial Patterns (CSP), Multi-Channel Deep Belief Network (MCDBN), Fast Fourier

TABLE 4.6 Average accuracyprediction.

Time	Normal	Fatigue	Abnormal	Average
T = 1 s	99.01	96.03	97.02	98.76
T + 10 s	98.02	94.09	95.06	96.89
T + 20 s	99.81	96.44	96.08	98.78
T + 30 s	98.56	94.04	97.77	96.78
T + 40 s	99.24	92.05	96.67	95.07
T + 50 s	98.73	95.06	97.56	96.93
Average	98.72	95.01	94.08	

TABLE 4.7 Prediction accuracies using various methods versus proposed method (LSTM).

	Parameters collected from driving condition					
Episodes	CSP	MCDBN	FFTEM	GAFRN	GAFTCNN	LSTM
I	87.01	69.03	76.02	77.76	66.89	97.21
II	78.02	75.09	88.06	68.89	85.44	96.57
III	88.81	85.44	75.08	78.78	53.04	97.89
IV	89.56	57.04	88.77	89.78	67.05	98.07
V	77.24	65.05	76.67	77.07	85.81	97.76
VI	69.73	74.06	85.56	88.93	88.56	98.94
Average	80.27	70.24	84.67	76.08	83.90	94.06

TABLE 4.8 Response rate from the cloud server (Sec).

Method States	CSP	MCDBN	FFTEM	GAFRN	GAFTCNN	LSTM(Proposed)
Normal	150	167	190	181	129	64
Fatigue	160	171	200	192	141	69
Abnormal	178	190	198	197	174	67

Transform Energy Map (FFTEM), Gramian Angular Field using Residual Network (GAFRN), Gramian Angular Field using Titled Convolutional Neural Network (GAFTCNN).

Table 4.8 shows the comparison result of faster response time from the cloud server and the proposed system achieves the quicker response time when it is compared with other methods.

5. Conclusion

The guarded transportation system monitors the driver's heart readings continually and records them as episodes to the cloud servers. This work achieves faster and higher reliable services when the drivers are in critical condition. The real-time sensor data collected from different subjects are learned through the intelligent learner system. To show our proposed system has achieved higher prediction accuracy, it is compared with various existing methods to prove that our system offers higher efficiency than any other than the existing system. This model mainly concerns about prediction on vital signals while driving and reduces the collateral damages to the environment, nearby vehicles, and passengers. This system utilizes advanced level sensors, which can also sense the blood alcohol level from the sweat glands and monitors the stress level of both the physical and physiological signals. Moreover, this intelligent system learns the driving behavior of the driver and suggests useful information to avoid any prolong activity. This model guarantees the safety of the passengers as well as drivers to provide intelligent suggestions to the drivers based on their health readings. This system exactly identifies the nearby vehicle type and exact count, such valuable information helps to avoid collision. The intelligent system predicts severe health complication before it happens and informs it to the nearby health centers with their respective EHR. Finally, the overall guarded transportation system provides 94.5% of prediction accuracy of vital readings to assure the safety while driving and continual monitoring of the drivers and this model will save many lives.

References

[1] B. Xu, et al., Ubiquitous data accessing method in IoT-based information system for emergency medical services, IEEE Trans. Ind. Informat. 10 (2) (May 2014) 1578–1586.
[2] J. Zhou, J. Sun, Y. Liu, J. Hu, J. Ye, Patient risk prediction model via top-k stability selection, in: Proc. SDM, 2013, pp. 55–63.
[3] S.C. Rajkumar, L. Jegatha Deborah, P. Vijayakumar, Optimized traffic flow prediction based on cluster formation and reinforcement learning, Int. J. Commun. Syst. (Nov.2019), https://doi.org/10.1002/dac.4178.
[4] X. Shi, Z. Chen, H. Wang, D.-Y. Yeung, W.-K. Wong, W.-C. Woo, Convolutional LSTM network: a machine learning approach for precipitation nowcasting, in: Proc. Int. Conf. Neural Inf. Process. Syst., 2015, pp. 802–810.
[5] C. Baru, N. Botts, T. Horan, K. Patrick, S.S. Feldman, A seeded cloud approach to health cyberinfrastructure: preliminary architecture design and case applications, in: System Science (HICSS), 2012 45th Hawaii International Conference on, pp. 2727–2734, IEEE, 2012.
[6] H. Abid, L.T.T. Phuong, J. Wang, S. Lee, S. Qaisar, V-cloud: vehicular cyber-physical systems and cloud computing, in: Proceedings of the 4th International Symposium on Applied Sciences in Biomedical and Communication Technologies, ACM, 2011, p. 165.
[7] J. Barrachina, P. Garrido, M. Fogue, F.J. Martinez, J.-C. Cano, C.T. Calafate, P. Manzoni, VEACON: a vehicular accident ontology designed to improve safety on the roads, J. Netw. Comput. Appl. 35 (6) (2012) 1891–1900.
[8] S. Sneha, U. Varshney, Enabling ubiquitous patient monitoring: model, decision protocols, opportunities and challenges, Decis. Support Syst. 46 (3) (2009) 606–619.
[9] S. Sneha, U. Varshney, A framework for enabling patient monitoring via mobile ad hoc network, Decis. Support Syst. 55 (1) (2013) 218–234.
[10] A.Z. Doorenbos, A. Kundu, L.H. Eaton, G. Demiris, E.A. Haozous, C. Towle, D. Buchwald, Enhancing access to cancer education for rural healthcare providers via telehealth, J. Cancer Educ. 26 (4) (2011) 682–686.
[11] M. Barua, X. Liang, R. Lu, X.S. Shen, Rcare: extending secure health care to rural area using VANETS, Mobile Netw. Appl. 19 (3) (2014) 318–330.

[12] M.-K. Suh, et al., A remote patient monitoring system for congestive heart failure, J. Med. Syst. 35 (5) (2011) 1165–1179.
[13] W. Jionglin, R. Jason, W.F. Stewart, Prediction modeling using EHR data: challenges, strategies, and a comparison of machine learning approaches, Med. Care 48 (6) (2010) S106–S113.
[14] M. Markatou, P. Kuruppumullage Don, J. Hu, F. Wang, J. Sun, R. Sorrentino, S. Ebadollahi, Case-based reasoning in comparative effectiveness research, IBM J. Res. Dev. 56 (5) (2012) 468–479.
[15] J.J. Vanwormer, Methods of using electronic health records for population-level surveillance of coronary heart disease risk in the Heart of New Ulm Project, Diabetes Spectr. 23 (3) (2010) 161–165.
[16] J. Sun, W. Fei, J. Hu, S. Edabollahi, Supervised patient similarity measure of heterogeneous patient records, Acm Sigkdd Explor. Newslett. 14 (1) (2012) 16–24.
[17] J. Zhou, F. Wang, J. Hu, J. Ye, From micro to macro: data driven phenotyping by densification of longitudinal electronic medical records, in: Proc. ACM SIGKDD Int. Conf. Knowl. Discovery Data Mining, 2014, pp. 1–10.
[18] J.A. Minarro-Giménez, O. Marín-Alonso, M. Samwald, Exploring the application of deep learning techniques on medical text corpora, Stud. Health Technol. Inf. 205 (2014) 584–588.
[19] Y. Choi, C.Y.-I. Chiu, D. Sontag, Learning low-dimensional representations of medical concepts, AMIA Summits Transl. Sci. Proc. 2016 (Jul. 2016) 41–50.
[20] T. Manninen, H. Huttunen, P. Ruusuvuori, M. Nykter, Leukemia prediction using sparse logistic regression, PLoS One 8 (8) (2013). Art. no. e72932.
[21] A. Rao, Y. Lee, A. Gass, A. Monsch, Classification of Alzheimer's disease from structural MRI using sparse logistic regression with optional spatial regularization, in: Proc. Annu. Int. Conf. IEEE Eng. Med. Biol. Soc., Aug./Sep. 2011, pp. 4499–4502.
[22] Y. Kim, S. Kwon, S.H. Song, Multiclass sparse logistic regression for classification of multiple cancer types using gene expression data, Comput. Stat. Data Anal. 51 (3) (2006) 1643–1655.
[23] L. Zhenqiu, J. Feng, T. Guoliang, W. Suna, S. Fumiaki, S.J. Meltzer, T. Ming, Sparse logistic regression with Lp penalty for biomarker identification, Stat. Appl. Genet. Mol. Biol. 6 (1) (Feb. 2007) 1–22 [Online]. Available: https://ideas.repec.org/a/bpj/sagmbi/v6y2007i1n6.html.
[24] L. Meier, S. van de Geer, P. Bühlmann, The group LASSO for logistic regression, J. Roy. Statist. Soc. B 70 (1) (2008) 53–71 [Online]. Available: https://ideas.repec.org/a/bla/jorssb/v70y2008i1p53-71.html.
[25] R. McEliece, E. Posner, E. Rodemich, S. Venkatesh, The capacity of the Hopfield associative memory, IEEE Trans. Inf. Theor. IT-33 (4) (Jul. 1987) 461–482.
[26] S. Hochreiter, J. Schmidhuber, Long short-term memory, Neural Comput. 9 (8) (1997) 1735–1780.
[27] M. Wllmer, B. Schuller, F. Eyben, G. Rigoll, Combining long shortterm memory and dynamic Bayesian networks for incremental emotionsensitive artificial listening, IEEE J. Sel. Topics Signal Process. 4 (5) (Oct. 2010) 867–881.
[28] S. Ryali, K. Supekar, D.A. Abrams, andV. Menon, Sparse logistic regression for whole-brain classification of fMRI data, Neuroimage 51 (2) (2010) 752–764.
[29] D. Neil, M. Pfeiffer, S.-C. Liu, Phased LSTM: accelerating recurrent network training for long or event-based sequences, in: Proc. Adv. Neural Inf. Process. Syst., 2016, pp. 3882–3890.
[30] T. Bai, A.K. Chanda, B.L. Egleston, S. Vucetic, Joint learning of representations of medical concepts and words from EHR data, in: Proc.IEEE Int. Conf. Bioinf. Biomed., Nov. 2017, pp. 764–769.
[31] R. Miotto, L. Li, B.A. Kidd, J.T. Dudley, Deep patient: an unsupervised representation to predict the future of patients from the electronic health records, Sci. Rep. 6 (May 2016) 26094.
[32] Z.C. Lipton, D.C. Kale, R. Wetzel, Modeling Missing Data in Clinical Time Series with RNNs, Jun. 2016 arXiv:1606.04130. [Online]. Available: https://arxiv.org/abs/1606.04130.
[33] E. Choi, M.T. Bahadori, A. Schuetz, W.F. Stewart, J. Sun, Doctor AI: predicting clinical events via recurrent neural networks, Tech. Rep. 56 (2016) 301–308.
[34] B.N. Campbell, J.D. Smith, W.G. Najm, Examination of crash contributing factors using national crash databases, in: National Highway Traffic Safety Admin., Tech. Rep. DOT-VNTSC-NHTSA-02-07, Oct. 2003, pp. 02–07. Accession Number-00970427.
[35] A. Vahidi, A. Eskandarian, Research advances in intelligent collision avoidance and adaptive cruise control, IEEE Trans. Intell. Transport. Syst. 4 (3) (Sep. 2003) 143–153.
[36] T.-Y. Sun, S.-J. Tsai, J.-Y. Tseng, Y.-C. Tseng, The study on intelligent vehicle collision-avoidance system with vision perception and fuzzy decision making, in: Proc. IEEE Intell. Vehicles Symp., 2005, pp. 112–117.

[37] R. Labayrade, C. Royere, D. Aubert, A collision mitigation system using laser scanner and stereovision fusion and its assessment, in: Proc. IEEE Intell. Vehicles Symp., Las Vegas, NV, Jun. 2005, pp. 441–446.

[38] J. Hillenbrand, K. Kroschel, V. Schmid, Situation assessment algorithm for a collision prevention assistant, in: Proc. IEEE Intell. Vehicles Symp., 2005, pp. 459–465.

[39] F. Biral, M.D. Lio, E. Bertolazzi, Combining safety margins and user preferences into driving criterion for optimal control-based computation of reference maneuvers for an ADAS of the next generation, in: Proc. IEEE Intell. Vehicles Symp., 2005, pp. 36–41.

[40] M. Goodrich, E. Boer, Designing human-centered automation: trade-offs in collision avoidance system design, IEEE Trans. Intell. Transport. Syst. 1 (1) (Mar. 2000) 40–54.

[41] R. Sanchez-Iborra, J. Sanchez-Gomez, J. Ballesta-Vinas, M.-D. Cano, A.F. Skarmeta, Performance evaluation of LoRa considering scenario conditions, Sensors 18 (3) (2018) 772, https://doi.org/10.3390/s18030772.PMC 5876541. PMID 29510524.

[42] G. Pau, V.M. Salerno, Wireless sensor networks for smart homes: a fuzzy-based solution for an energy-effective duty cycle, Electronics 8 (2) (Jan. 2019) 131.

[43] A. Bendjeddou, H. Laoufi, S. Boudjit, LEACH-S: low energy adaptive clustering hierarchy for sensor network, in: Proc. Int. Symp. Netw., Comput. Commun., ISNCC, Jun. 2018, pp. 1–6.

[44] X. Long-long, Z. Jian-Jun, Improved LEACH cluster head multi-hops algorithm in wireless sensor networks, in: Proc. 9th Int. Symp. Distrib. Comput. Appl. Bus., Eng. Sci., Aug. 2010, pp. 263–267.

[45] N.B. Temesghen, T. Hani, S. Baker, M.A. Khandoker, M. Ismail, Low-power ECG-based processor for predicting ventricular arrhythmia, IEEE Trans. Very Large Scale Integr. Syst. 24 (5) (2016).

[46] N. Mittal, U. Singh, B.S. Sohi, A stable energy efficient clustering protocol for wireless sensor networks, Wireless Netw. 23 (6) (Aug. 2017) 1809–1821.

[47] R. Resnick, D. Halliday, J. Walker, Fundamentals of Physics, John Wiley, 1988.

[48] T. Mikolov, I. Sutskever, K. Chen, G. Corrado, J. Dean, Distributed representations of words and phrases and their compositionality, in: Proc.Adv. Neural Inf. Process. Syst., vol. 26, 2013, pp. 3111–3119.

[49] J.P. Phillips, M. Hickey, P.A. Kyriacou, Evaluation of electrical and optical Plethysmography sensors for noninvasive monitoring of hemoglobin concentration", J. PubMed Central Sens. (Basel) 12 (2) (2012) 1816–1826, https://doi.org/10.3390/s120201816. Published online 2012 Feb 9.

CHAPTER

5

Deep learning in gastroenterology: a brief review

Subhashree Mohapatra[1], *Tripti Swarnkar*[2], *Manohar Mishra*[3], *David Al-Dabass*[4], *Raffaele Mascella*[5]

[1]Department of Computer Science and Engineering, Institute of Technical Education and Research, Siksha 'O' Anusandhan (Deemed to be) University, Bhubaneswar, Odisha, India; [2]Department of Computer Application, Institute of Technical Education and Research, Siksha 'O' Anusandhan (Deemed to be) University, Bhubaneswar, Odisha, India; [3]Department of Electrical and Electronics Engineering, Institute of Technical Education and Research, Siksha 'O' Anusandhan (Deemed to be) University, Bhubaneswar, Odisha, India; [4]School of Science and Technology, Nottingham Trent University, Nottingham, United Kingdom; [5]Faculty of Communication Sciences, University of Teramo, Teramo, Italy

1. Introduction

Gastroenterology (GE) has gained enormous popularity during the last few decades. In the late 1960s, doctors had difficulty in getting a quick glimpse of bleeding varices and diagnose gastric ulcers [1]. But in modern days with the help of endoscopes, experts are able to get a distinct view of the entire digestive tract. It is also possible to treat early lesions and control bleeding from abnormal growths internally. From the late 1970s, students started gaining interest in GE largely due to enhanced vision and availability of different imaging modalities such as X-ray, positron emission tomography (PET), magnetic resonance imaging (MRI), ultrasonography and computerized tomography (CT) along with endoscopy. In the current scenario, almost every healthcare institute has a gastrology department providing the newest equipment and qualified professionals. Fig. 5.1 shows the rise in market value due to the endoscopic procedure [2].

Handbook of Computational Intelligence in Biomedical Engineering and Healthcare
https://doi.org/10.1016/B978-0-12-822260-7.00001-7

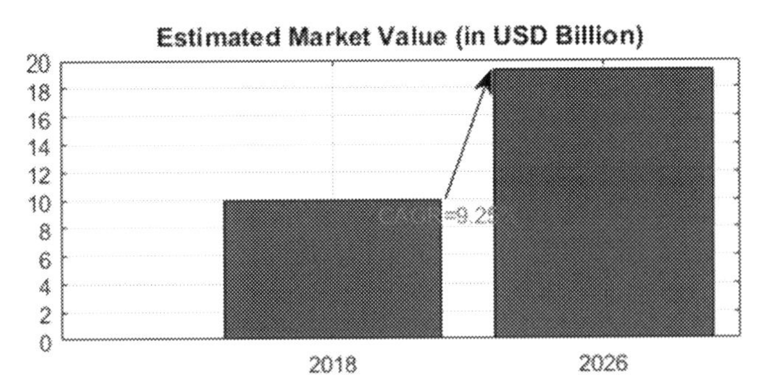

FIGURE 5.1 Estimated rising trend of the market value attribute to minimally invasive surgeries and need of endoscope for this procedure [2].

Healthcare is advancing to a new generation of technology where the rich biomedical facts and figures are playing a very significant role [3]. As the volume of medical statistics in public repositories is increasing speedily [4] it is quite difficult to render integrative investigation due to the heterogeneous nature of these data. The huge availability of medical data is bringing incredible prospects and challenges to biomedical research. Computational biology techniques are crucial and regularly used in different areas of medicine from drug discovery to biomarker development [5]. Investigating the relations among the various fragments of data in these datasets is an essential problem to design reliable healthcare tools created on data-driven methods and machine learning. Machine learning (ML) techniques are widely applied in the field of healthcare. Even though present models have great potential, but the predictive tools founded on ML methods have not been extensively practiced in healthcare [6]. Exploring and using the huge amount of medical data is still a challenge, due to their heterogeneity, irregularity, sparsity, and high-dimensionality.

Deep learning (DL) is a subset of ML which has potential in mining high-level concepts from the raw data of huge, varied, and large dimensional datasets. DL techniques are representation-learning methods with several stages of representation, achieved by combining simple but nonlinear units which convert the representation at one level into a representation at a higher, slightly more abstract level [7]. DL techniques have achieved good performance and have great potential in natural language processing, computer vision, and speech recognition tasks [8]. Based on its achieved performance in various fields, DL models have great potential in biomedical informatics. Researchers have started working on the application of DL paradigms to various departments of healthcare. Various features of DL like its ability to manage multidimensional and unstructured data, in-built feature extraction, and high performance are very helpful in the field of medicine. Enlitic is finding health issues on CT images and X-rays using DL techniques [9] while Google DeepMind has declared strategies to provide its skill and knowledge in healthcare [10]. Fig. 5.2 shows the basic difference between ML and DL.

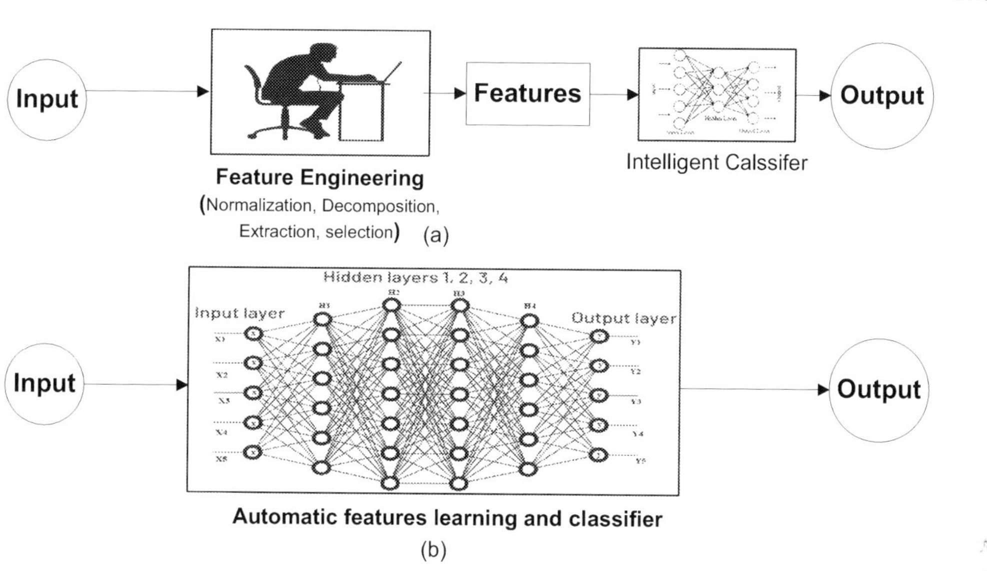

FIGURE 5.2 Basic architecture of (A) ML and (B) DL.

In GE-based research areas, general practitioners handle huge amounts of medical data and numerous varieties of imaging instruments. For example, esophagogastroduodenoscopy (EGD), colonoscopy, capsule endoscopy, and ultrasound equipment. The artificial intelligence (AI) has been successfully applied in the field of GE for effective diagnosis and analyzing gastrointestinal images. Several studies reported the success stories related to the AI-based applications in GE. The AI-based computer-aided methods are usually applied to detect, track, identify, and characterize the gastrointestinal (GI) lesions. However, the rapid development of AI-based ML and DL techniques demands that gastroenterologists should acquire the updated knowledge about the strength and challenges of each technique applied to medical diagnosis domain. Therefore, in this survey, the authors aim to provide a clear idea of state-of-the-art advancement in AI applications in GE-based medical domain. The initial phase of the report states about several medical image modalities for GE. Afterward, the authors summarize the AI applications in GE, which is divided into two categories. Lastly, several key issues/challenges are discussed, which will aid future research in the medical domain.

The remaining part of the chapter is divided into multiple sections as mentioned. In the following section different medical image modalities are discussed along with a comparative study. In Section 2, the application of conventional-ML based tools for detection of diagnosis of GI lesions are discussed. In Section 4, the different types of DL techniques are discussed along with related works done under each of them. In Section 5 different challenges are discussed briefly that need to be addressed in the near future, and finally concluding the chapter in Section 6.

2. Anomalies in GI-tract and medical image modalities for GE

This section provides a clearer idea of several anomalies in GI-tract and modalities available to detect these anomalies.

2.1 Anomalies in GI-tract

The gastrointestinal region is one of the major sections of the human body; it consists of the stomach, intestine, and entire digestive system. The GI-tract can be split into the esophagus, stomach (subdivided into the lower, middle, and upper stomach), small-bowel, and colon (subdivided into descending colon, transverse colon, descending colon, and rectum) [11,12]. The esophagus is a tube-like organ. It is used to carry foodstuffs to the stomach. The stomach is a towel-like organ that is used to grasp the foodstuff. Afterward, the digestive juices of the stomach and foodstuff are mixed which further moves into the entry of small-intestine called the duodenum. The small-intestine is identical to a contracted-tube of about 6 m in length. Finally, the colon is one of the important organs of GI-tract, which is more prone to tumors and polyps. There are numerous medical disorders, including simple warning signs and advanced diseases located in the GI-tract. Following are few major abnormalities that are popularly addressed by physician: (1) cancer (cancers start from an unusual growth of cells where the older cells do not die and unusual growth of cells form cancer); (2) polyps (polyps are similar to cancer; it is also an unusual mucosal growth and typically these are benign); (3) ulcer (ulcer is also referred to a disease caused by the acid that is produced by the stomach itself); (4) *Helicobacter pylori* (*H. pylori* causes inflammation in the mucosal wall, and therefore, an infection caused by this bacterium leads to various chronic abnormalities); (5) inflammation (inflammation refers to the condition of gastric abnormalities, involving dyspepsia, chronic gastritis, and acid reflex, are normally associated with the inflammation of gastric lining); (6) celiac disease (this is an autoimmune disorder in the small intestine, the intolerance of gluten found in wheat); (7) Crohn's disease (this is an intestinal inflammatory disease, but it may also cause severe abdominal pain); (8) bleeding (it may be caused by different other pathological conditions such as cancer); and (9) Barrett's esophagus (Barrett's esophagus is a disease, specifically, associated with the esophagus).

2.2 Medical image modalities for GE

In the last few decades, the minimally invasive medical imaging devices having more accuracy [13] have been going through a revolution. The fundamental medical imaging system comprises of sensors and sources of energy, which has the ability to enter the human system [14]. The energy that penetrates the body is attenuated at multiple levels depending on the atomicity of the various cells and tissues which generates signals. Special detectors help in detecting those signals and then they are mathematically altered to generate an image. Fig. 5.3 shows the general working of the medical imaging system [14].

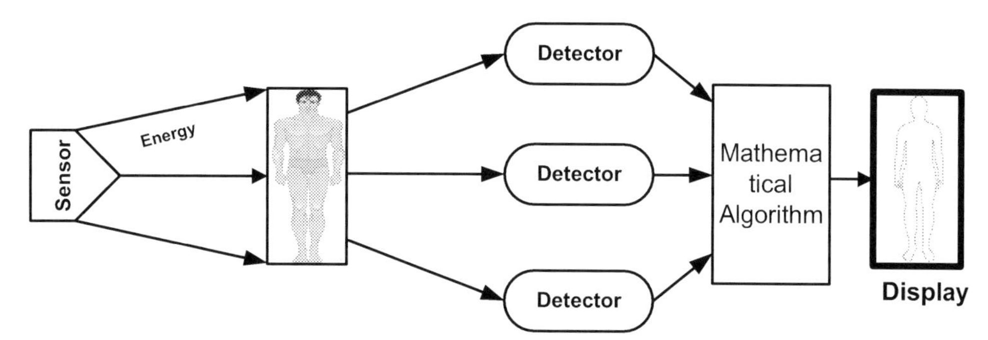

FIGURE 5.3 General working of the medical imaging system.

Detection of diseases in the GI tract requires an assessment of the internal structure of the GI tract along with its visualization [15]. Traditionally, standard radiological methods have played an important role in the imaging of the GI tract. But in recent times direct visualization and examination of the GI tract have become possible due to the introduction of endoscopic tests. From the past until the present, the quality of medical imaging is being enhanced and different advanced technologies are added into them for better investigation of the GI diseases [16]. The various techniques studied in this section are computed tomography, ultrasound, MRI, X-ray, PET, and endoscopy which are categorized according to the energy sources essential for the imaging purpose. Each of the techniques has its pros and cons. Table 5.1 gives a detailed comparison of each of these modalities on various parameters.

2.2.1 *Endoscopy*

In the present day, endoscopy is the most commonly used medical imaging method used for examining the GI tract. The concept of endoscope had started way back from 1850. But in 1932 the whole stomach was examined by the endoscopy procedure using a then-called instrument gastroscope [17]. The examination was carried out by Dr. Rudolph Schindler. Along with the GI tract, endoscopy procedure is also used for various other organs like the respiratory tract, urinary tract, female reproductive tract. According to the referring organ, the name of the endoscopy is different. Endoscopy is a minimally invasive procedure. In this test, a professional put in an endoscope, which is a long thin flexible tube having a camera and a light source, into an opening such as mouth or rectum or a minor incisor in the skin. It uses a specific blue and green wavelength which helps the professionals to detect some critical conditions more exactly [18]. In this procedure the patient is awake; however, in some cases, a mild sedative is given considering the patient's comfort. The internal of the body is examined by the doctor with the help of the images captured by the camera in the endoscope. Endoscopy is helpful in three important ways. First of all, it is used for the investigation of various critical symptoms related to different diseases. Secondly, it is used to confirm any findings through the process of biopsy. Lastly, it is useful in treating some of the medical conditions such as removing a polyp or stopping internal bleeding [18]. Some of the possible side effects can be like minor internal injury, pain, cramps, and oversedation.

TABLE 5.1 Comparison of different medical image modalities based on cost, pain, and accuracy advantage and disadvantages.

Name	Cost (approx. ... in INR)	Pain	Exactness	Advantages	Disadvantages
Endoscopy	1,500–3,500	Semiinvasive, insertion of the endoscope may cause pain	Observes internal organ and tissue in detail	Can be used for investigation, confirmation, and treatment	Mild cramping, over-sedation, tear of the inner lining
X-ray	250–2,000	Noninvasive, no pain	As it is just a black and white image, very exact details could be missed	Very short time and cost-effective procedure	If a dye is administered for an X-ray, some side effects may occur and the radiation is harmful to a developing fetus
Ultrasound	400–600	Noninvasive, no pain	As image resolution is low sometimes the exact problem could be missed	No radiations are used and it does not cause any discomfort to the patient	The image resolution is low interpretation is difficult at times, penetration is low through bone and air as well as in obese patients
CT	2,000–10,000	Noninvasive, no pain but can cause discomfort to the patient	Good visualization of bone and calcified lesions	Sedation not required, can be used in emergency conditions and more scanning speed	There is exposure to radiation, insertion of contrast medium can cause health problem
MRI	6,000–9,000	Noninvasive, no pain but can cause discomfort to the patient	A cross-sectional view of an organ is possible which is quite effective in many cases	Effective for soft-tissue imaging and sedation is not required	Study times are more along with loud noise and confined space
PET	5,000–30,000	Noninvasive except for the injection of tracer	Gives a detailed functioning of internal organs in the cellular level	When fused with other modalities like CT and MRI can give even better results	Much costly as compared to other modalities and complex procedure

2.2.2 *X-ray*

Physicist Welhelm Conrad Rontgen was the first to detect X-rays in 1895 [19], but it was first used under the medical condition in 1896. This finding was tagged as a miracle in the medical field and it became a very significant diagnosis device which helped doctors to visualize inside the body without any insertion. X-ray causes ionization and has a short wavelength in the electromagnetic spectrum. The properties of X-ray are such that it is mostly absorbed by metal objects and bones of a body but are transmitted by different body cells and tissues. The X-ray absorbing capability of tissue depends on its radiological density. X-ray imaging is a noninvasive, fast, and relatively inexpensive procedure. In this process, the professional sets the right parameter such as type and amount of X-ray subject to the patient's body part to be examined, size of the patient, and amount of image contrast needed [20]. The image that is generated is captured on a piece of the radiograph. In the case of GI tract examination, a special type of X-ray known as barium X-ray is used. Various abnormal conditions such as a tumor, ulcer, polyps, and hernias can be diagnosed using barium X-ray. Barium is a good X-ray absorber which when used with X-ray generated effective visibility of GI tract [21]. When barium is introduced into the GI region, it covers the wall of various organs such as the esophagus, stomach, small and large intestine from inside. This helps in clear visualization of different characteristics such as shape, size, openness, the contour of the GI tract. There are very minimal side effects of X-ray. Exposure to very high radiation may sometime cause vomiting, dizziness, and hair loss.

2.2.3 *Ultrasound*

Dr. Ian Donald and Tom Brown devised the first model of an ultrasound system which came into clinical use in the late 1950s and was regularly used in Glasgow Hospital [22]. This procedure uses the ultrasonic sound waves on the body parts, tissues, and sensing the echoes which return. The sound waves used in ultrasound are above the range of human audibility [22]. When the echoes bounce back at the transducer, it produces electrical signals which are directed to the scanner. The scanner's work is to calculate the distance from the transducer to the tissue boundary using the speed of sound and time of individual echo's bounce back. The distance that is calculated is helpful in generating a 2D image of the organs. Ultrasound is a noninvasive method of medical imaging. In the field of healthcare, it is an important technique to diagnose various medical conditions related to organs such as heart, blood vessels, kidney, liver, thyroid, abdominal organs, and a developing fetus as well. The most common use of ultrasound is to check the development and growth of the fetus. The major functionality in the case of the GI tract is a visualization of the intestinal wall, inflammation, abnormal wall structure, kidney stones [23], tumors, or fluid packed intestinal portions. This can be effectively done using transabdominal ultrasound. The very recent development in the medical usage of this method is done in the detection of diseases related to gall bladder and jaundice which is possible due to high-resolution gray-scale scanning.

2.2.4 *CT-scan*

Godfrey Hounsfield in 1971 [24] had built the first CT scanner with the help of which a detailed image of a living brain was captured. CT scanner directs several X-ray beams to the human system from different angles [24]. This process is called tomography. Detectors that are present inside the device record in what way the rays pass through the body and using complex mathematical calculations generated a 3D cross-sectional image of the corresponding organ on a monitor. CT scan is also a noninvasive, fast, painless procedure [25]. In the case of gastroenterology, a CT scan plays an important role in detecting the cause of pain in the abdomen or disease related to small bowel and colon. In some GI cases, the patient is required to take some barium liquid which helps to get a better visualization of the organs. A CT scan is useful in detecting various digestive system related problems such as infections, inflammation, cancer, stone, internal injuries, and even internal bleeding. This is also helpful in guiding various processes like biopsies, minimally invasive treatments, plan for surgeries, evaluation of after surgery effects, and also check on the response to chemotherapy [25]. In this imaging modality, a patient's body is largely exposed to radiation which may increase the chances of cancer. But its accuracy in diagnosis far overshadows the risk. Ultrasound may give a better result than a CT scan in case of detecting gallbladder stones.

2.2.5 MRI

The beginning of the MRI was in 1977 when the imaging technique was used to investigate a live human patient [26]. MRI uses radio waves, magnetic fields, and a computer system to scan the human body. In this process, the patient is positioned on a moveable bed that goes into a large spherical magnet. Inside the device, a strong magnetic field is created that helps to align the protons present in hydrogen atoms and finally they are exposed to radio waves. A faint signal is generated by this process which is captured by the receiver of the MRI scanner. An image is created as the signal processed by the computer. Like other modalities, MRI is noninvasive [27]. It is very effective for soft-tissue imaging which is very helpful for scanning the GI tract. Using MRI one can have a cross-sectional view of an organ which is quite effective for displaying the complete thickness of the gastric as well as the intestinal wall. Because there is no radiation involved in the scanning process there is no adverse effect on the human system. For this reason, an MRI scan is preferable over the US and CT. Patients who have to undergo CT scans multiple times or patients who are pregnant can opt for MRI for its no radiation property. Some of the medical conditions like choledocholithiasis [28] and stones can be better evaluated using MRI. The major disadvantage is that it is not cost-effective which cannot be afforded by all groups of patients. This examination is not available in all the medical centers and the time taken for the complete procedure is also longer as compared to other modalities.

2.2.6 PET

Unlike other imaging modalities, PET scan is used to produce a 3D image of the human body functions. The use of PET in medicine was first established by Micheal Ter-Pogossian an American scientist [29] along with a group of scientists in the 1970s. During the examination patient is given an injection or asked to swallow a benign radioactive [30] material which is known as "tracer." After positioning the patient in the PET machine, the radiotracer is tracked by the detectors in the device. The most commonly used radiotracer is the fluorodeoxyglucose (FDG) which is similar to simple glucose and it requires 30–60 min for FDG to distribute in the whole body. Excluding the intravenous injection, the PET scan is a noninvasive procedure. This is a very effective method to detect any abnormal conditions from a very early stage as well as it can evaluate the effect of the medication in patients [31]. Many types of cancer, tumors, gastrointestinal abnormalities, neurological conditions, heart diseases, and other various issues can be examined using PET. Nowadays, PET is being fused with CT or MRI for better outcomes. This helps the professionals to study and evaluate the medical condition in a more effective way. Particularly for diseases related to gastroenterology like esophageal cancer, small intestine tumors, colorectal cancer, and lymphomas in the gastric tract can be examined in an effective way using PET-CT [31]. The major benefit of opting PET is that it gives a detailed functioning of different organs and changes in the body at the cellular level. But the limitation of PET scan is that it is a costly and complex procedure. Not all medical institutes are able to provide a PET scan facility.

3. Conventional-ML in gastroenterology

MLs are AIs with the ability to learn without being explicitly programmed and show improved performance over time with higher exposure to data. Presently, ML is one of the highly acceptable or standard approaches under the umbrella of AI. The ML-based tools build and solve the mathematical algorithms automatically from given input data (training data), and the predictor makes decisions in ambiguous situations without human interference.

In the last several years, ML in image processing focused primarily on synthetic features, where the programmer coded several complex features extraction (FE) algorithms. In the course of training, an intelligent classifier learned to differentiate between the features of diverse classes and used these facts to predict the actual class of a tested image model. However, the paradigm of manmade features has transformed into DL approaches where the features are learned automatically (no need for extra FE tools) by itself.

This section reviewed several ML-based approaches-based GI-tract diagnosis system which is successfully implemented in the last decade (from 2010 to 2019). Table 5.2 shows a comparatively summarized the reviewed conventional ML-based GI-tract diagnosis system. It can be analyzed from the table that artificial neural networks (ANN) and support vector machine (SVM) are the two most popularly used ML approaches that have been implemented by several authors. Although several published articles have confirmed the high accuracy of ML-based computer-aided diagnosis systems, there are still a lot of confines from the general practitioner viewing platform.

4. DL based GI-tract diagnosis system

DL is a part of ML in which a group of computational techniques having several layers for processing tries to understand data with different stages of abstraction [68]. DL is a collection of techniques related to neural networks, variations of unsupervised, and supervised feature extraction methods as well as probabilistic techniques. There is a sudden rise of attention toward DL due to its better performances as compared to traditional methods [69] and the ability to process highly complex data from different sources. The goal of this section is to provide a formal introduction of different DL techniques-based applications in detecting GI-tract diseases. In the DL framework, several algorithms have been studied such as a deep neural network (DNN), Autoencoder (AE), stack autoencoder (SAE), Deep Belief Network (DBN), recurrent Boltzmann Machine (RBM) and convolution neural network (CNN). The detailed mathematical analysis of these algorithms has been stated by the authors

TABLE 5.2 Summary of ML application in GI (last decade, 2010–19).

ML model	Year	GI field	Applications on GI	Modality	Remark on dataset	Remark on outcome	Refs.	Article type (J/C)
ANN	2011	Upper GI	Model designed to predict the mortality rate in nonvariceal upper GI bleeding	Endoscopy	The database developed by collecting data from 23 different sites throughout Italy including the Progetto Nazionale Emorragie Digestive (PNED) study group	Specificity: 97.5%, Sensitivity: 83.8%, AUROC: 0.95, Accuracy: 96.8%	[32]	Journal
	2012	Upper GI	Model developed for diagnosing the depth of invasion in gastric cancer	White-light Endoscopy	902 endoscopic images were collected for 344 patients who had undergone endoscopic tumor resection or gastrectomy between 2001 and 2010	Accuracy: 77.2%, 49.1%, 51.0%, and 55.3% for T1 to T4 staging, respectively	[33]	Journal
	2015	Upper GI	Work done to predict prognosis in ulcerative colitis after cytoapheresis therapy	NA	Population study was carried out at the Keio University hospital for 90 ulcerative colitis patients having moderate to severe activity and had received CAP therapy	Specificity: 97.0%, Sensitivity: 96.0%,	[34]	Journal
	2015	Small bowel	System developed for predicting mucosal healing by azathioprine therapy in inflammatory bowel disease	Endoscopy	In the Turkiye Yuksek İhtisas Hospital inflammatory bowel disease clinic 2700 patient's data was collected from 1995 to 2014 with inflammatory bowel disease and out of those 129 who had undergone azathioprine treatment only were selected for the research work	Positive classification rate: 79.1% Area under the curve: 88.3%	[35]	Journal
	2015	Small bowel	Work done to predict the frequency of severity, onset, and relapse of inflammatory bowel disease	Radiology, endoscopy	Data of 901 patient's suffering from ulcerative colitis or Crohn's disease were collected from January 2003 to December 2011, at the Shanghai inflammatory bowel disease center	Accuracy error: MSE: 0.065 MAPE: 37.58% Severity error: MSE: 0.065 MAPE: 42.15%	[36]	Journal
	2016	Small bowel	System developed for the diagnosis of intestinal polyps	Wireless capsule endoscopy	A group of 54 patients was examined out of which 8 patients had small bowel polyps	Specificity: 91.38%, Sensitivity: 93.75%, Positive predicted value: 95.71%, Negative predicted value: 96.36%	[37]	Journal

SVM	2010	Colon and rectum	Model designed for classification of colorectal polyps based on vascularisation features	Narrow band imaging	Dataset was created for a prospective pilot study with 128 patients who were undergoing zoom NBI colonoscopy in which 209 detected polyps were visualized and thereafter removed for analysis	Accuracy: 91.9%, Sensitivity: 96.9%, Specificity: 71.4%, Positive classification rates: 91.9% (consensus decision) and 90.9% (safe decision)	[38]	Journal
	2010	Colon and rectum	System developed for classification of zoom-endoscopic images for identification of polyps	Colonoscopy	484 images were collected from the department of gastroenterology and Hepatology, the medical University of Vienna in 2005–09	Accuracy: 96.9, Sensitivity: 97.2, Specificity:96.0	[39]	Journal
	2011	Colon and rectum	Model developed for differentiating small colonic polyps of less than 10 mm	Narrow band imaging	555 of 1612 patients who underwent colonoscopy were considered in the study	Accuracy: 93.1%, Specificity: 90.3%, Sensitivity: 95.0%	[40]	Journal
	2013	Esophagus	Model designed for predicting postoperative distant metastasis in esophageal squamous cell carcinoma	NA	Data of 1071 patients were collected from Sun Yat-sen University in South China and Linzhou oesophageal cancer Hospital in North China who underwent surgical treatment between 1997 and 2004	Accuracy: 78.7%, Sensitivity: 56.6%, Specificity: 97.7%, Positive predictive value: 95.6%, Negative predictive value: 72.3%	[41]	Journal
	2013	Upper GI	Model developed for the classification of Barrett's esophagus using narrow-band imaging	Narrow band imaging using an endoscope	197 images were collected at Karolinska Universitessjukhuse, Sweden during regular clinical work where videos from 84 patients were recorded and analyzed for diagnosis purposes	Sensitivity: 91.8%, Specificity: 92.1%, Accuracy: 91.8%, Area under ROC: 0.96	[42]	Conference
	2013	Colon	System designed for detection of colonic polyps as prevention of colorectal cancer	Wireless capsule endoscopy	Database was a collection of 1500 images including 300 images of polyps and 1200 images of nonpolyps	Sensitivity: 91.0%, Specificity: 95.2%, Accuracy: 95.0%	[43]	Conference

(Continued)

TABLE 5.2 Summary of ML application in GI (last decade, 2010–19).—cont'd

ML model	Year	GI field	Applications on GI	Modality	Remark on dataset	Remark on outcome	Refs.	Article type (J/C)
	2013	Small bowel	Model to assess Crohn' s disease lesions	Wireless capsule endoscopy	The database included endoscopy images from 2 patient exams at the Johns Hopkins medical Institutions, where one contained the Crohn's lesions and another one included the normal small bowel. Total of 100 frames was extracted from the Crohn's lesions video and 1728 frames from the normal small bowel video	Sensitivity: 81.1%, Specificity: 93.6%, Accuracy: 80.2%	[44]	Conference
	2013	Small bowel	Model designed for risk prediction for inflammatory bowel disease	Wireless capsule endoscopy	Data were collected as a part of international inflammatory bowel disease Genetics Consortium's Immuno chip project which included 17,000 Crohn's disease cases, 13,000 ulcerative colitis cases as well as 22,000 controls from 15 European countries	Sensitivity: 95.2%, Specificity: 92.4%, Accuracy: 93.8% Areas under the curve for Crohn's disease: 0.86 and 0.83 for ulcerative colitis	[45]	Journal
	2016	Lower GI	Model designed for classifying histology of colorectal polyps	Narrow-band imaging using endoscopy	Dataset of 2247 cut out training images from 1262 colorectal lesions were collected prior to the study	Sensitivity: 93.0%, Specificity: 93.3%, Accuracy: 93.2%, PPV: 93%, NPV: 93.3%	[46]	Journal
	2016	Colon and rectum	Model of 2nd generation for discrimination of neoplastic changes in small polyps	Endocytoscopy	One endocytoscopic and one white-light image of 205 small colorectal polyps ($\leq$10 mm) from 123 patients was taken from web-based test	Accuracy: 89% for diminutive (<5 mm) as well as small (<10 mm) polyps	[47]	Journal
	2016	Lower GI	Model developed for the diagnosis of colorectal lesions using microvascular findings	Endocytoscopy with narrow-band imaging	An image database was formed having consecutive series of 1079 images using endocytoscopy with narrow-band imaging that consisted of 431 nonneoplasms and 648 neoplasms from 85 lesions	Accuracy: 90%	[48]	Journal

2016	Esophagus	Model designed for discrimination of early neoplastic lesions in Barrett's esophagus	White-light Endoscopy	Dataset was based on 100 images from 44 patients with Barrett's esophagus	Sensitivity: 83%, Specificity: 83% (per-image analysis)	[49]	Journal
2016	Colon and rectum	System developed for computer-aided classification of GI lesions	Colonoscopy	Publicly available image databases containing medical images or videos from the GI tract	Sensitivity: 72.7%, Specificity: 85.9%, Accuracy: 82.5%	[50]	Journal
2016	Small bowel	System designed for automatic tumor detection	Wireless capsule endoscope	An experimental dataset was taken which consisted of 400 normal images from 6 normal pathological cases and 400 tumor images from 6 pathological cases with different tumor appearances. These images were taken from PillCam Company	Sensitivity: 94.0%, Specificity: 93.0%, Accuracy: 93.5%	[51]	Journal
2016	Small bowel	Model designed for detection of the inflammatory lesion using hybrid adaptive filtering	Wireless capsule endoscope	A database consisting of 800 images were used which was acquired from 13 patients who undertook wireless capsule endoscope examinations	Sensitivity: 95.2%, Specificity: 92.4%, Accuracy: 93.8%	[52]	Journal
2017	Esophagus	System developed for identification of early Barrett's esophagus neoplasia on ex vivo volumetric laser endomicroscopy Images	Ex vivo volumetric laser Endomicroscopy	Data used consisted of 60 images in which 30 were nondysplastic Barrett's esophagus images and 30 neoplastic images	Sensitivity: 90%, Specificity: 93%	[53]	Journal
2017	Liver	System developed for the diagnosis of chronic liver disease	Ultrasound shear wave elastography Imaging with a Stiffness value clustering	The clinical dataset consisted of 126 patients which included 56 healthy controls and 70 with chronic liver disease	Sensitivity: 93.5%, Specificity: 81.2%, Accuracy: 87.3%, AUROC: 0.87	[54]	Journal
2017	Colon and rectum	Model designed for the diagnosis of invasive colorectal cancer	Endocytoscopy with narrow-band images and stained images	The image database consisted of a consecutive series of 5843 endocytoscopy images of 375 lesions	Sensitivity: 89.4%, Specificity: 98.9%, Accuracy: 94.1%, Positive predictive value: 98.8%, negative predictive value: 90.1%	[55]	Journal

(Continued)

TABLE 5.2 Summary of ML application in GI (last decade, 2010–19).—cont'd

ML model	Year	GI field	Applications on GI	Modality	Remark on dataset	Remark on outcome	Refs.	Article type (J/C)
	2017	Liver	Model developed for the diagnosis of liver cirrhosis	Ultrasound liver capsule images	Dataset consisted of 91 ultrasound images in which 44 images were from normal people and 47 images from people with cirrhosis	AUROC: 0.951	[56]	Journal
	2017	Colon and rectum	System designed for detection of GI polyps	Colonoscopy	Most of the data was taken from the Department of electronics, University of Alcala and endoscopic vision challenge	Sensitivity: 98.8%, Specificity: 98.5%, Accuracy: 98.7%	[57]	Journal
	2017	Colon and rectum	System developed for the diagnosis of colorectal lesion	Endocytoscopy	The system was validated with the help of 173 randomly chosen endocytoscopy images among which 49 were nonneoplasms and 124 were neoplasms	Sensitivity: 94.3%, Specificity: 71.4%, Accuracy: 87.8%	[58]	Journal
	2017	Lower GI	Model developed for classification of pediatric Inflammatory bowel disease	Endoscopy	Historical data of 287 children with pediatric Inflammatory bowel disease was considered	Accuracy: 83.3% Sensitivity: 83.0%	[59]	Journal
	2018	Lower GI	System designed to predict lymph node metastasis so as to minimize the requirement of surgery in T1 colorectal cancer	Endoscopy	Data on 690 consecutive patients with T1 colorectal cancer that were surgically resected from 2001 to 2016 were considered for the study	Sensitivity: 100%, Specificity: 66%, Accuracy: 69%,	[60]	Journal
	2018	Upper GI	Model developed for the diagnosis of early gastric cancer	Narrow-band imaging	Randomly 66 narrow-band images having early gastric cancer and 60 noncancer images were selected for training, whereas for testing 61 narrow-band images having early gastric cancer and 20 noncancer images were selected	Sensitivity: 96.7%, Specificity: 95%, Accuracy: 96.3%, Positive predictive value: 98.3%	[61]	Journal
	2018	Lower GI	System developed for the diagnosis of neoplastic diminutive polyp	Endocytoscopy with narrow-band images and stained images	Images of 791 consecutive patients undergoing colonoscopy were collected	Prediction rate: 98.1%	[62]	Journal

	2018	Lower GI	Model developed for predicting persistent histologic inflammation in ulcerative colitis patients	Endocytoscopy with narrow-band imaging	Endocytoscopy based images and biopsy samples of each patient were collected from six colorectal segments: Cecum, ascending colon, transverse colon, descending colon, sigmoid colon, and rectum. For validation samples, 525 validation sets of 525 autonomous segments were collected from 100 patients, and 12,900 endocytoscopy images from the remaining 87 patients were used for ML to construct the network	Sensitivity: 74%, Specificity: 97%, Accuracy: 91%	[63]	Journal
	2019	Upper GI	Model designed for the diagnosis of early gastric cancer	Chromoendoscopy	The model was trained using the 100 image samples per class and tested with 1900 image samples per each class	F1 score: 0.96	[64]	Journal
KNN	2011	Upper GI	Automated classification of gastrointestinal tissues	CT-scan	NA	Results on classification of tumor tissue: Sensitivity: 99.97%, Specificity: 99.85%, Accuracy: 99.88%	[65]	Journal
	2018	Upper GI	Classification of bleeding and normal frames in GI-tract	Wireless capsule endoscopy (WCE)	A total of 1000 WCE images were collected comprising 500 normal frames and 500 bleeding frames, from 30 dissimilar patients' videos. The training and testing data set were divided into 680 images and 320 images, including both bleeding and normal frames	Sensitivity: 98.51%, Specificity: 99.53%, Accuracy: 99.22%	[66]	Conference
	2020	Upper GI	Bleeding and Z-Line classification in GI-tract	Wireless capsule endoscopy (WCE)	The training and testing datasets comprising images from GI-tract are collected from KVASIR. It has 1000 images of bleeding and anatomical standards such as z-line, each group comprises 500 images of 40 patients	Sensitivity: 99.25%, Specificity: 96.76%, Accuracy: 96.71%	[67]	Conference

TABLE 5.3 Comparison of CNNs, DBNs/DBMs, and AE with respect to several properties.

Model properties	CNN	DBN or DBM	AE/SAE
Unsupervised learning	*	√	√
Training efficiency	*	*	√
Scale/rotation/translation invariance	√	*	*
Feature learning	√	*	*
Generalization	√	√	√

√, good performance; *, bad performance or complete lack thereof.

in Refs. [68–70]. In Table 5.3, a brief comparison of all these DL tools is presented based on different model properties, where it can be seen that CNN has generally outperformed in the current literature on standard computer vision dataset.

In the world of GE, with the increase in image modalities, the volume of image data is increasing. To handle the complex unstructured image data, CNN is found to be the most efficient DL technology. CNN is used to analyze the images, predict prognosis, and also in the diagnosis of various health conditions. A major advantage of the CNN based model is the automatic feature learning ability. However, the CNNs depend on the availability of labeled training dataset, while DBNs/DBMs and SAEs do not have this restriction and can accomplish the task in an unsupervised manner [71].

One of the important steps in the classification problem is the task of feature extraction. Feature extraction can be carried out by professionals, but it is a time-consuming process. In recent times a new technique has gained popularity which incorporates feature extraction within the training step. This technique is called CNN which has the ability to evaluate high dimensional data. It consists of a special type of neural network which not only includes the feature extractor but also determines the weights during the training of data. In CNN, the feature extraction layer is given the input data, and its output which is the extracted features are given as input to the classification layer. The classification layer works on the extracted features and gives the output. The feature extraction network comprises of stacks of convolution layer and pooling layer pairs. The convolution layers work as a digital filter that transforms the data using the convolution operation whereas the pooling layer works as a dimension reduction layer. Mostly CNN is used in image processing and so the operation of the convolution and pooling pair are in a two-dimension plane. As CNN incorporates the feature extraction process within the network it is a major advantage over other networks. Table 5.4 presents recent reviews on DL application in GI-tract. It can be analyzed from the table that CNN is the most popularly used DL techniques in GI-tract diseases diagnosis. Fig. 5.4 shows the general architecture of a CNN model consisting of its different layers.

4.1 Related works: application of CNN in gastroenterology

Takiyama et al. [72] designed a model based on CNN technique that could group ECD images into anatomical regions with AUC of 1.00 for esophagus, 1.00 for larynx, 0.99 for

TABLE 5.4 Summary of CNN application in GI.

Year	GI field	Applications on GI	Modality	Remark on outcome	Refs.	Article type (J/C)
2018	Upper GI	Model developed to group esophagogastroduodenoscopy images into anatomical regions	Esophagogastroduodenoscopy	AUC: 1.00 for esophagus, 1.00 for larynx, 0.99 for stomach and duodenum	[72]	Journal
2019	Esophagus	Model designed to detect esophageal cancer including squamous cell carcinoma and adenocarcinoma	Endoscopy using narrow-band imaging	Accuracy: 98%, Sensitivity: 98%, negative predictive value: 95%, Positive predictive value: 40%	[73]	Journal
2019	Esophagus	Model designed for detecting esophageal squamous cell carcinoma	Endocytoscopy	Accuracy: 90.9%, Sensitivity: 92.6%, Specificity: 89.3%	[74]	Journal
2017	Upper GI	Model designed to detect *H. pylori* through endoscopic images	Endoscopy	Accuracy: 87.7%, Specificity: 87.4%, Sensitivity: 87.4%, Time of diagnosis: 88.9%	[75]	Journal
2018	Upper GI	Model designed for detection of *H. pylori* infection	Endoscopy	Specificity: 86.7%, sensitivity: 86.7%	[76]	Journal
2018	Upper GI	Model designed for detection of gastrointestinal cancer	Endoscopy	Sensitivity: 92.2% rate of detection: 98.6% The performance was cancerous tissue diameter 6 mm	[77]	Journal
2018	Upper GI	Model developed for classifying the depth of gastrointestinal cancer	Endoscopy	Accuracy: 89.2%, Specificity: 95.6%	[78]	Journal
2018	Upper GI	Model designed for early gastric cancer	Endoscopy	Accuracy: 87.6%, Specificity: 94.8%, Sensitivity: 80.0%	[79]	Journal
2019	Upper GI	Model designed for detection of early gastric cancer without blind spots	Endoscopy	Accuracy: 92.5%, Sensitivity: 94.0%, Specificity: 91.0%, Positive predictive value: 91.3%, Negative predictive value: 93.8%	[80]	Journal

(Continued)

TABLE 5.4 Summary of CNN application in GI.—cont'd

Year	GI field	Applications on GI	Modality	Remark on outcome	Refs.	Article type (J/C)
2010	Lower GI	Model developed for computer assisted detection of tumorous lesions	Colonoscopy	Accuracy: 96.9% (two classes case) and 86.8% (six classes case)	[81]	Journal
2016	Colon	Model developed for detection of colonic polyps automatically	Colonoscopy	Sensitivity: 90.0%, Specificity: 63.3%	[82]	Journal
2017	Colon	Model developed for automatic colonic polyp detection	Colonoscopy	Accuracy: 89.45%, Specificity: 98.6%	[83]	Journal
2018	Colon	Model designed for identification and localization of colonic polyps	Colonoscopy	Accuracy: 96.4%, AUROC: 0.991	[84]	Journal
2017	Colon and rectum	Model developed for finding the variations of diminutive colorectal polyps	Narrow-band imaging	Accuracy: 94%, Specificity: 83%, Sensitivity: 98%	[85]	Journal
2018	Colon and rectum	Model developed for classification of diminutive colorectal polyps using computer-aided analysis	Narrow-band imaging	Negative predictive value: 91.5%, Positive predictive value: 89.6% The performance was for identifying hyperplastic or neoplastic polyps of less than 5 mm	[86]	Journal
2018	Colon and rectum	Model developed for classification of neoplastic colorectal polyps	Colonoscopy	Accuracy: 78.0%, Sensitivity: 92.3% Negative predictive value: 88.2%	[87]	Journal
2017	Colon and rectum	Model developed for colorectal polyp classification	White-light colonoscopy with narrow-band imaging and chromoendoscopy	Accuracy: 75.1%	[88]	Journal
2018	Colon and rectum	Model developed for generating a diagnostic support system for cT1b colorectal cancer	Colonoscopy	Accuracy: 81.2%, Sensitivity: 67.5%, Specificity: 89.0%, Area under ROC: 0.871	[89]	Journal

2019	Colon and rectum	Model designed for autonomous detection and localization of colorectal polyps	Colonoscopy	Accuracy: 96.4%, Sensitivity: 97.1%, Specificity: 93.3%	[90]	Journal
2019	Small bowel	Model designed for detection of gastrointestinal angiectasia using semantic segmentation images	Capsule endoscopy	Sensitivity: 100%, Specificity: 96%, Positive predictive value: 96%, Negative predictive value:100%.	[91]	Journal
2017	Small bowel	Model designed for classification of celiac disease	Capsule endoscopy	Sensitivity: 100%, Specificity: 100% performance mentioned is for the test dataset	[92]	Journal
2018	Lower GI	Model developed for detection of hookworm in the intestine	Capsule endoscopy	Sensitivity: 84.6%, Specificity: 88.6%	[93]	Journal
2016	Small bowel	Model designed for characterizing six different motility events in the small intestine	Wireless capsule endoscopy	Accuracy: 96%	[94]	Journal
2018	Colon	Model designed for detection of colonic polyps	White-light colonoscopy	Specificity: 95.2%, Sensitivity: 94.38%; Area under ROC: 0.984	[95]	Journal

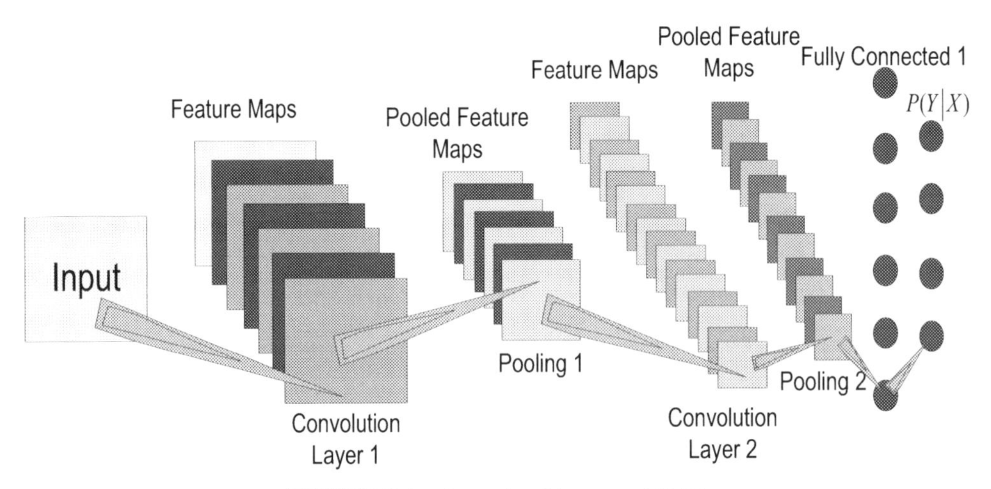

FIGURE 5.4 General architecture of CNN.

stomach and duodenum. The model is able to categorize the anatomical location of the stomach with AUC of 0.99 for three different regions of the stomach. Horie et al. [73] designed a CNN system to detect esophageal cancer including squamous cell carcinoma and adenocarcinoma. Endoscopy images using narrow-band imaging were used for the training and testing of the model which gave a significant performance. The model could recognize small cancer of size less than 10 mm and showed an accuracy, sensitivity, negative predictive value, and positive predictive value of 98%, 98%, 95%, and 40% respectively. A CNN model was designed for detecting esophageal squamous cell carcinoma by Kumagai et al. [74] which was proposed for substituting biopsy-based histology for esophageal squamous cell carcinoma using endocytoscopy. The system gave an accuracy of 90.9%, sensitivity of 92.6%, and specificity of 89.3%.

One of the most important factors of gastric and peptic ulcer is the *Helicobacter pylori* (*H. pylori*) infection. Being a research challenge for years two Japanese personnel developed a CNN based model to detect *H. pylori*. Shichijo et al. [75] gave a performance comparison of the model with 23 professional endoscopists to evaluate and detect *H. pylori* through endoscopic images. The model gave a better performance on the basis of accuracy, specificity, sensitivity, and time of diagnosis. The performance of CNN versus endoscopist are 87.7% versus 82.4%, 87.4% versus 83.2%, 87.4% versus 83.2%, 88.9% versus 79.0%, and 194s versus 230s for accuracy, specificity, sensitivity and time of diagnosis respectively. Itoh et al. [76] proposed a model that gave a promising output having a specificity of 86.7% and a sensitivity of 86.7%. The CNN model 596 endoscopy images after data augmentation of a previous set of 149 images for the detection of *H. pylori* infection.

Recently, several major works have been done in the field of gastrointestinal cancer. Hirasawa et al. [77] stated the good efficiency of a CNN model developed for the detection of gastrointestinal cancer through images obtained by endoscopy. The model is trained by 13,584 endoscopy images and tested with 2296 images. The performance measure is in terms of sensitivity (92.2%) and rate of detection (98.6%) for a diameter of 6 mm of the cancerous tissue. All invasive cancer was recognized by the model. Among the lesions, 69.4% were diagnosed as cancer but was benign and the misdiagnosis was due to gastritis having features like atrophy and redness. The depth of gastrointestinal cancer was classified using a CNN model proposed by Zhu et al. [78]. The model was fed with 790 endoscopic images for training and tested with another set of 203 images. The model generated a specificity of 95.6% and an accuracy of 89.2%. Sakai et al. [79] designed a CNN model for early gastric cancer using endoscopy. The accuracy, sensitivity, specificity given by the system is 87.6%, 94.8% and 80.0% respectively. A CNN model was developed for detection of early gastric cancer without blind spots by Wu and group [80]. In the study, endoscopic images were taken for developing the system which gave an accuracy of 92.5%, sensitivity of 94.0%, and specificity of 91.0%.

Detection of polyps is very common during colonoscopy, hence developing an AI model for a colonoscopy is a challenging task. By using 546 short video clips from a full-length video which was 73 in number, authors in Ref. [81] created a CNN system that could detect colonic polyps automatically. The total data was divided into two training data groups in which 105 were polyp positive videos and 306 were polyp negative videos. Likewise, test data was divided into 50 polyps positive and 85 polyp negative videos. The model gave a sensitivity of 90.0% and specificity of 63.3%. Bernal and group [84] designed a CNN model for automatic colonic polyp detection using colonoscopy which gave an accuracy of 89.45% and sensitivity of 98.6% [82]. In the way to detect colonic polyps, Urban et al. [83] developed a CNN model in which they took 8641 manually labeled images and 20 number of colonoscopy videos. The authors used these data in various combinations of training data and testing data. The system gave an accuracy of 96.4% and found polyps in real-time with AUROC of 0.991. In addition to it, the system did assist in the detection of another nine different polyps that were compared with professional endoscopists.

In 2017, by the help of only NBI videos Byrne et al. [84] designed a CNN system for the real-time variation of diminutive colorectal polyps. With an accuracy of 94%, the system differentiated hyperplastic polyps from adenomas, and in addition to that, it recognized the adenomas with a specificity of 83% and sensitivity of 98%. Another CNN model trained with the help of 2157 images was developed by Chen et al. [85]. The system identified hyperplastic or neoplastic polyps of less than 5 mm having an NPV of 91.5% and PPV of 89.6%. Likewise, in 2017, a CNN model differentiated adenomas from nonadenomas polyps. Renner et al. designed a model for using in the classification of neoplastic colorectal polyps using colonoscopy [86]. The performance of the system was given as accuracy of 78.0%, sensitivity of 92.3%, and a negative predictive value of 88.2%. The model was proposed by Komeda et al. [87] and trained using 1800 endoscopic images with NBI, chromoendoscopy, and WLI. The system performed with an accuracy of 75.1% with 10-fold cross-validation. Ito and group developed a model for the purpose of generating a diagnostic support system for cT1b colorectal cancer in which colonoscopy images were used [88]. The performance measure given by the system was having an accuracy of 81.2%, sensitivity of 67.5%, specificity of 89.0%. Blanes-Vidal had designed a model for autonomous detection and localization of colorectal polyps which gave a performance of accuracy: 96.4%, sensitivity: 97.1%, and specificity: 93.3% [89].

A group of Japanese researchers stated multiple articles on the application of AI with endocytoscopic images for the distinction of polyps. The articles state the observation of nuclei on-site and also reported diagnostic outputs that were comparable with the pathologic examinations. Authors in Ref. [90] developed a CNN model for the gastrointestinal angiectasia detection with the use of semantic segmentation images. They took 600 control images and another 600 number of typical angiectasia images so as to form 4166 capsule endoscopy videos of the small bowel. These groups of images were equally divided into training data and testing data. The model performed well with a specificity of 96%, NPV of 100%, the sensitivity of 100%, and PPV of 96%. Another CNN system designed by Zhou et al. [91] for the classification of celiac disease performed with specificity and sensitivity of 100%. The performance shown was on the test dataset. The classification was carried out from control with capsule endoscopy clips taken from five controls and six celiac patients.

One of the most challenging jobs is the detection of the presence of hookworms in the intestinal region due to its structure, color that is similar to the internal of the intestine. However, researchers have worked on developing various CNN models to detect the presence of intestinal hookworm. A CNN system developed by He et al. [92] helped to find hookworm in the intestine through images captured by capsule endoscopy. Some of the performance parameters like specificity and sensitivity are significant with a value of 88.6% and 84.6% respectively. A portion of the false detection is 11% of the non-hookworm image and 15% of hookworm images. Another CNN system was designed for small bowel motility characterization by Seguí et al. [93]. The small intestine was characterized by six different motility events with an accuracy rate of 96%. Wang et al. [94] proposed a CNN model for polyp detection using colonoscopy images. The author worked on colonoscopy images of 1290 patients and validation of the model was done using images collected from 1138 patients. The model gave a satisfying performance of 95.2% specificity, 94.38% sensitivity. The area under ROC was given as 0.984.

5. Critical analysis and discussions

Although several research articles published in recent years have shown the effectiveness of AI-based computer-aided diagnosis (CAD) system applied to GI, there are still numerous shortcomings from doctors' viewing platform. These limitations included several aspects for example datasets, experimental design, and result indicators. Generally, the images used for AI study are taken from the open-source/access database. Moreover, the researchers are more inclined about the collection of high-quality data to construct the training database and ignore the low-quality pictures, for example, obscure mucus, stain, or partial views of lesions. This causes overfitting and poor-generalization in models, which further leads to a reduction in the accuracy of the diagnostic system. Therefore, the analysis of real-world clinical images should be taken into considerations rather than optimally collected images. Moreover, a large number of clinical pictures can be used in an AI-based CAD system for better results. Artificial intelligence-based CAD system using DL has shown to be greater potential than the systems which rely on hand-crafted features only. Thus, the majority of AI studies on GI reported recently have made usage of DL tools. It can be analyzed from the above-mentioned review that CNN based tools are most popularly used DL methods in GI-tract diseases diagnosis. Even though DL and particularly deep CNN gives better performance in the field of medical imaging, there are few limitations to look into. CNN models require a large amount of computational power for its execution. Limited computational power will require more amount of time to train the model that will be subjected to the training data size. The CNN architecture needs labeled information for the implementation of supervised learning and manual labeling for the images which is tough work. These problems are gradually overcome as there is large computational power available, large data storage capacity, and also efficient deep CNN networks.

5.1 Miscellaneous DL based tool to detect the GI tract

Generally, the GI infectious diseases are initiated by food infection, and therefore, prediction of the disease can be extremely valuable for etiological factor controlling and clinical resource utilization. Song et al. [95] compared several ML-based tools to predict the morbidity as a result of food contaminations. Song et al. [96] presented deep-denoising-AE (DDAE) to model the effect of food contamination on GI infections, and consequently offer a valued tool for morbidity prediction. The experimental results show its effectiveness over ANN on real-time datasets. Similarly, authors in Ref. [95] established a DNN based on a DAE to predict GI-infection morbidity from the food-contamination dataset in four states of China during 2015–16. The obtained results show that the proposed model outperforms to other conventional ML-based approaches. GI infectious diseases are influenced by environmental pollutants, but forecasting their morbidity using contamination statistics is pretty challenging due to the complex relationship amongst the contaminants and the infections. Song et al. [97] studied a deep Boltzmann machine (DBM) based DNN model for reckoning the illness of GI infections using 129 kinds' data of pollutants confined in soil and water. Song et al. [98] presented a GI cancer morbidity prediction model which is useful for the following three functions: (1) predicting the morbidity of a diverse disease within the similar region; (2) predicting the morbidity of similar disease in a distinctive region; and (3) predicting the morbidity of an unlike disease in a different region.

5.2 Status of GI-tract diseases diagnosis versus DL versus ML

Fig. 5.5 shows the popularity curve regarding the research on GI-tract disease detection versus DL (especially CNN) and ML. ML approaches have widely applied for different applications under the area of GE in the last decade. It can also be seen that the research trend toward the application of DL in GI or GE have been increased in the last 4 years (83% of the total are achieved). The blue (gray in printed version) line shows the research trend describing the various application of the DL in GE. Similarly, the black line shows the application of CNN in GE. The curve indicates the popularity of CNN toward GI-tract disease detection and diagnoses in the last few years.

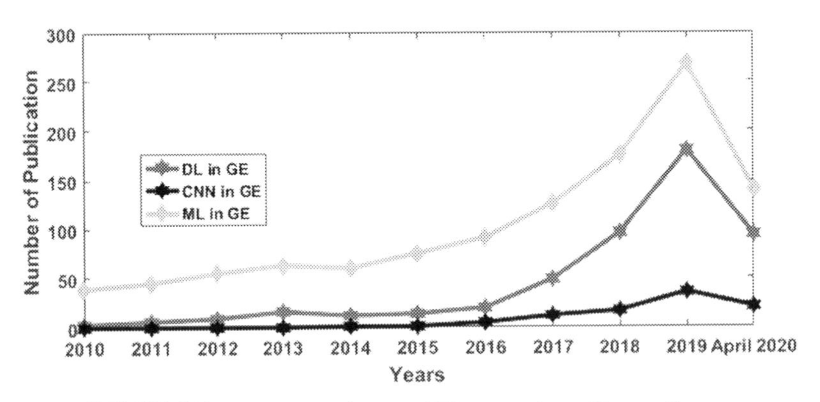

FIGURE 5.5 Popularity of ML and DL research on GI-tract diseases.

6. Conclusion

The application of AI has shown the incredible potential in diagnosing GI-tract diseases in recent years. However, the evolution of AI in the form of DL has acquired the space of ML-based AI applications in several fields of engineering and the medical domain. In this regard, the authors of this chapter have tried to provide a clear picture of DL application in a particular area of the medical domain such as "Gastroenterology." Moreover, several non-DL or conventional-ML based approaches for GI-tract applications are also analyzed briefly. The initial work of this chapter explains the several anomalies in GI-tract and Medical Image Modalities available for GE. Although several published articles have confirmed the high accuracy of the DL and ML-based computer-aided system, there are still a lot of confines from the general practitioner viewing platform. However, further studies and clinical assessments exhibiting the implementation of DL in real-life situations are required. As the application of DL and especially CNN is a recent research topic in the domain of GI imaging, it can be a great scope for researchers. Researchers can work on prospective data that can be more effective in developing universal models to detect and diagnose diseases related to the field of gastroenterology. We hope that this survey will helpful for the researchers/scientists working in the area of GE.

References

[1] R. Tandon, Progress of gastroenterology in India, Indian J. Gastroenterol. 26 (2007) S31.
[2] https://www.databridgemarketresearch.com/news/global-surgical-endoscopes-market. (Last accessed 1 March 2020).
[3] R. Miotto, F. Wang, S. Wang, X. Jiang, J.T. Dudley, Deep learning for healthcare: review, opportunities and challenges, Briefings Bioinf. 19 (6) (2018) 1236–1246.
[4] P. Mamoshina, A. Vieira, E. Putin, A. Zhavoronkov, Applications of deep learning in biomedicine, Mol. Pharm. 13 (5) (2016) 1445–1454.
[5] R. Nussinov, Advancements and challenges in computational biology, PLoS Comput. Biol. 11 (1) (2015).
[6] R. Bellazzi, B. Zupan, Predictive data mining in clinical medicine: current issues and guidelines, Int. J. Med. Inf. 77 (2) (2008) 81–97.
[7] Y. LeCun, Y. Bengio, G. Hinton, Deep learning, Nature 521 (7553) (2015) 436–444.

[8] O. Abdel-Hamid, A.R. Mohamed, H. Jiang, L. Deng, G. Penn, D. Yu, Convolutional neural networks for speech recognition, IEEE/ACM Trans. Audio Speech Lang. Process. 22 (10) (2014) 1533–1545.
[9] Enlitic Uses Deep Learning to Make Doctors Faster and More Accurate. http://www.enlitic.com/index.html. (Last accessed 1 March 2020).
[10] Google's DeepMind Forms Health Unit to Build Medical Software. https://www.bloomberg.com/news/articles/2016-02-24/google-s-deepmind-forms-health-unit-to-build-medical-software. (Last accessed 1 March 2020).
[11] Q.D. Chu, J.F. Gibbs, G.B. Zibari (Eds.), Surgical Oncology: A Practical and Comprehensive Approach, Springer, 2014.
[12] R. Miyahara, Y. Niwa, T. Matsuura, O. Maeda, T. Ando, N. Ohmiya, A. Itoh, Y. Hirooka, H. Goto, Prevalence and prognosis of gastric cancer detected by screening in a large Japanese population: data from a single institute over 30 years, J. Gastroenterol. Hepatol. 22 (9) (2007) 1435–1442.
[13] S. Angenent, E. Pichon, A. Tannenbaum, Mathematical methods in medical image processing, Bull. Am. Math. Soc. 43 (3) (2006) 365–396.
[14] H. Kasban, M.A.M. El-Bendary, D.H. Salama, A comparative study of medical imaging techniques, Int. J. Inf. Sci. Intell. Syst. 4 (2) (2015) 37–58.
[15] J.B. Frøkjær, A.M. Drewes, H. Gregersen, Imaging of the gastrointestinal tract-novel technologies, World J. Gastroenterol. 15 (2) (2009) 160.
[16] M. Camilleri, New imaging in neurogastroenterology: an overview, Neuro Gastroenterol. Motil. 18 (9) (2006) 805–812.
[17] https://www.olympus-global.com/technology/museum/endo/?page=technology_museum. (Last accessed 1 March 2020).
[18] https://www.medicalnewstoday.com/articles/153737#procedure. (Last accessed 1 March 2020).
[19] https://www.history.com/this-day-in-history/german-scientist-discovers-x-rays. (Last accessed 1 March 2020).
[20] https://www.medicalradiation.com/types-of-medical-imaging/imaging-using-x-rays/radiography-plain-x-rays/. (Last accessed 1 March 2020).
[21] https://www.hopkinsmedicine.org/health/conditions-and-diseases/barium-xrays-upper-and-lower-gi. (Last accessed 1 March 2020).
[22] https://www.livescience.com/32071-history-of-fetal-ultrasound.html. (Last accessed 1 March 2020).
[23] https://www.atlantagastro.com/services/abdominal-ultrasound/#1533764685836-7b038da1-675f6a80-5318. (Last accessed 1 March 2020).
[24] http://broughttolife.sciencemuseum.org.uk/broughttolife/techniques/ctcatscanner. (Last accessed 1 March 2020).
[25] https://www.radiologyinfo.org/en/info.cfm?pg=abdominct. (Last accessed 1 March 2020).
[26] https://www.aps.org/publications/apsnews/200607/history.cfm. (Last accessed 1 March 2020).
[27] B. Liu, M. Ramalho, M. AlObaidy, K.K. Busireddy, E. Altun, J. Kalubowila, R.C. Semelka, Gastrointestinal imaging-practical magnetic resonance imaging approach, World J. Radiol. 6 (8) (2014) 544.
[28] https://appliedradiology.com/articles/current-applications-of-mri-in-emergent-gastrointestinal-diseases. (Last accessed 1 March 2020).
[29] http://broughttolife.sciencemuseum.org.uk/broughttolife/techniques/pet. (Last accessed 1 March 2020).
[30] https://www.radiologyinfo.org/en/info.cfm?pg=pet. (Last accessed 1 March 2020).
[31] https://www.snmmi.org/AboutSNMMI/Content.aspx?ItemNumber=943. (Last accessed 1 March 2020).
[32] G. Rotondano, L. Cipolletta, E. Grossi, M. Koch, M. Intraligi, M. Buscema, R. Marmo, Italian Registry on Upper Gastrointestinal Bleeding (Progetto Nazionale Emorragie Digestive), Artificial neural networks accurately predict mortality in patients with nonvariceal upper GI bleeding, Gastrointest. Endosc. 73 (2) (2011) 218–226.
[33] K. Kubota, J. Kuroda, M. Yoshida, K. Ohta, M. Kitajima, Medical image analysis: computer-aided diagnosis of gastric cancer invasion on endoscopic images, Surg. Endosc. 26 (5) (2012) 1485–1489.
[34] T. Takayama, S. Okamoto, T. Hisamatsu, M. Naganuma, K. Matsuoka, S. Mizuno, R. Bessho, T. Hibi, T. Kanai, Computer-aided prediction of long-term prognosis of patients with ulcerative colitis after cytoapheresis therapy, PLoS One 10 (6) (2015).
[35] F. Hardalaç, M. Basaranoğlu, M. Yüksel, U. Kutbay, M. Kaplan, Y. Özderin Özin, Z.M. Kılıç, A.E. Demirbağ, O. Coskun, A. Aksoy, V. Gangarapu, The rate of mucosal healing by azathioprine therapy and prediction by artificial systems, Turk. J. Gastroenterol. 26 (2015) 315–321.

[36] J.C. Peng, Z.H. Ran, J. Shen, Seasonal variation in onset and relapse of IBD and a model to predict the frequency of onset, relapse, and severity of IBD based on artificial neural network, Int. J. Colorectal Dis. 30 (9) (2015) 1267–1273.
[37] A.F. Constantinescu, M. Ionescu, V.F. Iovănescu, M.E. Ciurea, A.G. Ionescu, C.T. Streba, M.G. Bunescu, I. Rogoveanu, C.C. Vere, A computer-aided diagnostic system for intestinal polyps identified by wireless capsule endoscopy, Rom. J. Morphol. Embryol. 57 (3) (2016) 979–984.
[38] J.J.W. Tischendorf, S. Gross, R. Winograd, H. Hecker, R. Auer, A. Behrens, C. Trautwein, T. Aach, T. Stehle, Computer-aided classification of colorectal polyps based on vascular patterns: a pilot study, Endoscopy 42 (03) (2010) 203–207.
[39] M. Hafner, L. Brunauer, H. Payer, R. Resch, A. Gangl, A. Uhl, F. Wrba, A. Vécsei, Computer-aided classification of zoom-endoscopical images using fourier filters, IEEE Trans. Inf. Technol. Biomed. 14 (4) (2010) 958–970.
[40] S. Gross, C. Trautwein, A. Behrens, R. Winograd, S. Palm, H.H. Lutz, R. Schirin-Sokhan, H. Hecker, T. Aach, J.J. Tischendorf, Computer-based classification of small colorectal polyps by using narrow-band imaging with optical magnification, Gastrointest. Endosc. 74 (6) (2011) 1354–1359.
[41] H.X. Yang, W. Feng, J.C. Wei, T.S. Zeng, Z.D. Li, L.J. Zhang, P. Lin, R.Z. Luo, J.H. He, J.H. Fu, Support vector machine-based nomogram predicts postoperative distant metastasis for patients with oesophageal squamous cell carcinoma, Br. J. Cancer 109 (5) (2013) 1109–1116.
[42] F. Riaz, M.D. Ribeiro, P. Pimentel-Nunes, M.T. Coimbra, Integral scale histogram local binary patterns for classification of narrow-band gastroenterology images, in: 2013 35th Annual International Conference of the IEEE Engineering in Medicine and Biology Society (EMBC), IEEE, July 2013, pp. 3714–3717.
[43] O. Romain, A. Histace, J. Silva, J. Ayoub, B. Granado, A. Pinna, X. Dray, P. Marteau, Towards a multimodal wireless video capsule for detection of colonic polyps as prevention of colorectal cancer, in: 13th IEEE International Conference on BioInformatics and BioEngineering, IEEE, November 2013, pp. 1–6.
[44] W.S.L. Jebarani, V.J. Daisy, Assessment of Crohn's disease lesions in wireless capsule endoscopy images using SVM based classification, in: 2013 International Conference on Signal Processing, Image Processing & Pattern Recognition, IEEE, February 2013, pp. 303–307.
[45] Z. Wei, W. Wang, J. Bradfield, J. Li, C. Cardinale, E. Frackelton, C. Kim, F. Mentch, K. Van Steen, P.M. Visscher, R.N. Baldassano, Large sample size, wide variant spectrum, and advanced machine-learning technique boost risk prediction for inflammatory bowel disease, Am. J. Hum. Genet. 92 (6) (2013) 1008–1012.
[46] Y. Kominami, S. Yoshida, S. Tanaka, Y. Sanomura, T. Hirakawa, B. Raytchev, T. Tamaki, T. Koide, K. Kaneda, K. Chayama, Computer-aided diagnosis of colorectal polyp histology by using a real-time image recognition system and narrow-band imaging magnifying colonoscopy, Gastrointest. Endosc. 83 (3) (2016) 643–649.
[47] Y. Mori, S.E. Kudo, P.W.Y. Chiu, R. Singh, M. Misawa, K. Wakamura, T. Kudo, T. Hayashi, A. Katagiri, H. Miyachi, F. Ishida, Impact of an automated system for endocytoscopic diagnosis of small colorectal lesions: an international web-based study, Endoscopy 48 (12) (2016) 1110–1118.
[48] M. Misawa, S.E. Kudo, Y. Mori, H. Nakamura, S. Kataoka, Y. Maeda, T. Kudo, T. Hayashi, K. Wakamura, H. Miyachi, A. Katagiri, Characterization of colorectal lesions using a computer-aided diagnostic system for narrow-band imaging endocytoscopy, Gastroenterology 150 (7) (2016) 1531–1532.
[49] F. van der Sommen, S. Zinger, W.L. Curvers, R. Bisschops, O. Pech, B.L. Weusten, J.J. Bergman, E.J. Schoon, Computer-aided detection of early neoplastic lesions in Barrett's esophagus, Endoscopy 48 (07) (2016) 617–624.
[50] P. Mesejo, D. Pizarro, A. Abergel, O. Rouquette, S. Beorchia, L. Poincloux, A. Bartoli, Computer-aided classification of gastrointestinal lesions in regular colonoscopy, IEEE Trans. Med. Imag. 35 (9) (2016) 2051–2063.
[51] V. Faghih Dinevari, G. Karimian Khosroshahi, M. Zolfy Lighvan, Singular value decomposition based features for automatic tumor detection in wireless capsule endoscopy images, Appl. Bionics Biomech. 2016 (2016).
[52] V.S. Charisis, L.J. Hadjileontiadis, Potential of hybrid adaptive filtering in inflammatory lesion detection from capsule endoscopy images, World J. Gastroenterol. 22 (39) (2016) 8641.
[53] A.F. Swager, F. van der Sommen, S.R. Klomp, S. Zinger, S.L. Meijer, E.J. Schoon, J.J. Bergman, H. Peter, W.L. Curvers, Computer-aided detection of early Barrett's neoplasia using volumetric laser endomicroscopy, Gastrointest. Endosc. 86 (5) (2017) 839–846.
[54] I. Gatos, S. Tsantis, S. Spiliopoulos, D. Karnabatidis, I. Theotokas, P. Zoumpoulis, T. Loupas, J.D. Hazle, G.C. Kagadis, A machine-learning algorithm toward color analysis for chronic liver disease classification, employing ultrasound shear wave elastography, Ultrasound Med. Biol. 43 (9) (2017) 1797–1810.

[55] K. Takeda, S.E. Kudo, Y. Mori, M. Misawa, T. Kudo, K. Wakamura, A. Katagiri, T. Baba, E. Hidaka, F. Ishida, H. Inoue, Accuracy of diagnosing invasive colorectal cancer using computer-aided endocytoscopy, Endoscopy 49 (08) (2017) 798–802.

[56] X. Liu, J.L. Song, S.H. Wang, J.W. Zhao, Y.Q. Chen, Learning to diagnose cirrhosis with liver capsule guided ultrasound image classification, Sensors 17 (1) (2017) 149.

[57] M. Billah, S. Waheed, M.M. Rahman, An automatic gastrointestinal polyp detection system in video endoscopy using fusion of color wavelet and convolutional neural network features, Int. J. Biomed. Imag. 2017 (2017).

[58] M. Misawa, S.E. Kudo, Y. Mori, K. Takeda, Y. Maeda, S. Kataoka, H. Nakamura, T. Kudo, K. Wakamura, T. Hayashi, A. Katagiri, Accuracy of computer-aided diagnosis based on narrow-band imaging endocytoscopy for diagnosing colorectal lesions: comparison with experts, Int. J. Comput. Assist. Radiol. Surg. 12 (5) (2017) 757–766.

[59] E. Mossotto, J.J. Ashton, T. Coelho, R.M. Beattie, B.D. MacArthur, S. Ennis, Classification of paediatric inflammatory bowel disease using machine learning, Sci. Rep. 7 (1) (2017) 1–10.

[60] K. Ichimasa, S.E. Kudo, Y. Mori, M. Misawa, S. Matsudaira, Y. Kouyama, T. Baba, E. Hidaka, K. Wakamura, T. Hayashi, T. Kudo, Artificial intelligence may help in predicting the need for additional surgery after endoscopic resection of T1 colorectal cancer, Endoscopy 50 (03) (2018) 230–240.

[61] T. Kanesaka, T.C. Lee, N. Uedo, K.P. Lin, H.Z. Chen, J.Y. Lee, H.P. Wang, H.T. Chang, Computer-aided diagnosis for identifying and delineating early gastric cancers in magnifying narrow-band imaging, Gastrointest. Endosc. 87 (5) (2018) 1339–1344.

[62] Y. Mori, S.E. Kudo, M. Misawa, Y. Saito, H. Ikematsu, K. Hotta, K. Ohtsuka, F. Urushibara, S. Kataoka, Y. Ogawa, Y. Maeda, Real-time use of artificial intelligence in identification of diminutive polyps during colonoscopy: a prospective study, Ann. Intern. Med. 169 (6) (2018) 357–366.

[63] Y. Maeda, S.E. Kudo, Y. Mori, M. Misawa, N. Ogata, S. Sasanuma, K. Wakamura, M. Oda, K. Mori, K. Ohtsuka, Fully automated diagnostic system with artificial intelligence using endocytoscopy to identify the presence of histologic inflammation associated with ulcerative colitis (with video), Gastrointest. Endosc. 89 (2) (2019) 408–415.

[64] R. Ogawa, J. Nishikawa, E. Hideura, A. Goto, Y. Koto, S. Ito, M. Unno, Y. Yamaoka, R. Kawasato, S. Hashimoto, T. Okamoto, Objective assessment of the utility of chromoendoscopy with a support vector machine, J. Gastrointest. Canc. 50 (3) (2019) 386–391.

[65] P.B. Garcia-Allende, I. Amygdalos, H. Dhanapala, R.D. Goldin, G.B. Hanna, D.S. Elson, Morphological analysis of optical coherence tomography images for automated classification of gastrointestinal tissues, Biomed. Optic Express 2 (10) (2011) 2821–2836.

[66] X. Xing, X. Jia, M.H. Meng, Bleeding detection in wireless capsule endoscopy image video using superpixel-color histogram and a subspace KNN classifier, in: 2018 40th Annual International Conference of the IEEE Engineering in Medicine and Biology Society (EMBC), IEEE, July 2018, pp. 1–4.

[67] R. Ponnusamy, S. Sathiamoorthy, Bleeding and Z-line classification by DWT based SIFT using KNN and SVM, in: International Conference on Computational Vision and Bio Inspired Computing, Springer, Cham, September 2019, pp. 679–688.

[68] J. Ahmad, H. Farman, Z. Jan, Deep learning methods and applications, in: Deep Learning: Convergence to Big Data Analytics, Springer, Singapore, 2019, pp. 31–42.

[69] J. Schmidhuber, Deep learning in neural networks: an overview, Neural Networks 61 (2015) 85–117.

[70] D. Ravì, C. Wong, F. Deligianni, M. Berthelot, J. Andreu-Perez, B. Lo, G.Z. Yang, Deep learning for health informatics, IEEE J. Biomed. Health Inform. 21 (1) (2016) 4–21.

[71] A. Voulodimos, N. Doulamis, A. Doulamis, E. Protopapadakis, Deep learning for computer vision: a brief review, Comput. Intell. Neurosci. 2018 (2018).

[72] H. Takiyama, T. Ozawa, S. Ishihara, M. Fujishiro, S. Shichijo, S. Nomura, M. Miura, T. Tada, Automatic anatomical classification of esophagogastroduodenoscopy images using deep convolutional neural networks, Sci. Rep. 8 (1) (2018) 1–8.

[73] Y. Horie, T. Yoshio, K. Aoyama, S. Yoshimizu, Y. Horiuchi, A. Ishiyama, T. Hirasawa, T. Tsuchida, T. Ozawa, S. Ishihara, Y. Kumagai, Diagnostic outcomes of esophageal cancer by artificial intelligence using convolutional neural networks, Gastrointest. Endosc. 89 (1) (2019) 25–32.

[74] Y. Kumagai, K. Takubo, K. Kawada, K. Aoyama, Y. Endo, T. Ozawa, T. Hirasawa, T. Yoshio, S. Ishihara, M. Fujishiro, J.I. Tamaru, Diagnosis using deep-learning artificial intelligence based on the endocytoscopic observation of the esophagus, Esophagus 16 (2) (2019) 180–187.
[75] S. Shichijo, S. Nomura, K. Aoyama, Y. Nishikawa, M. Miura, T. Shinagawa, H. Takiyama, T. Tanimoto, S. Ishihara, K. Matsuo, T. Tada, Application of convolutional neural networks in the diagnosis of *Helicobacter pylori* infection based on endoscopic images, EBioMedicine 25 (2017) 106–111.
[76] T. Itoh, H. Kawahira, H. Nakashima, N. Yata, Deep learning analyzes *Helicobacter pylori* infection by upper gastrointestinal endoscopy images, Endosc. Int. Open 6 (02) (2018) E139–E144.
[77] T. Hirasawa, K. Aoyama, T. Tanimoto, S. Ishihara, S. Shichijo, T. Ozawa, T. Ohnishi, M. Fujishiro, K. Matsuo, J. Fujisaki, T. Tada, Application of artificial intelligence using a convolutional neural network for detecting gastric cancer in endoscopic images, Gastric Cancer 21 (4) (2018) 653–660.
[78] Y. Zhu, Q.C. Wang, M.D. Xu, Z. Zhang, J. Cheng, Y.S. Zhong, Y.Q. Zhang, W.F. Chen, L.Q. Yao, P.H. Zhou, Q.L. Li, Application of convolutional neural network in the diagnosis of the invasion depth of gastric cancer based on conventional endoscopy, Gastrointest. Endosc. 89 (4) (2019) 806–815.
[79] Y. Sakai, S. Takemoto, K. Hori, M. Nishimura, H. Ikematsu, T. Yano, H. Yokota, Automatic detection of early gastric cancer in endoscopic images using a transferring convolutional neural network, in: 2018 40th Annual International Conference of the IEEE Engineering in Medicine and Biology Society (EMBC), IEEE, July 2018, pp. 4138–4141.
[80] L. Wu, W. Zhou, X. Wan, J. Zhang, L. Shen, S. Hu, Q. Ding, G. Mu, A. Yin, X. Huang, J. Liu, A deep neural network improves endoscopic detection of early gastric cancer without blind spots, Endoscopy 51 (06) (2019) 522–531.
[81] M. Misawa, S.E. Kudo, Y. Mori, T. Cho, S. Kataoka, A. Yamauchi, Y. Ogawa, Y. Maeda, K. Takeda, K. Ichimasa, H. Nakamura, Artificial intelligence-assisted polyp detection for colonoscopy: initial experience, Gastroenterology 154 (8) (2018) 2027–2029.
[82] J. Bernal, N. Tajkbaksh, F.J. Sánchez, B.J. Matuszewski, H. Chen, L. Yu, Q. Angermann, O. Romain, B. Rustad, I. Balasingham, K. Pogorelov, Comparative validation of polyp detection methods in video colonoscopy: results from the MICCAI 2015 endoscopic vision challenge, IEEE Trans. Med. Imag. 36 (6) (2017) 1231–1249.
[83] G. Urban, P. Tripathi, T. Alkayali, M. Mittal, F. Jalali, W. Karnes, P. Baldi, Deep learning localizes and identifies polyps in real time with 96% accuracy in screening colonoscopy, Gastroenterology 155 (4) (2018) 1069–1078.
[84] M.F. Byrne, N. Chapados, F. Soudan, C. Oertel, M.L. Pérez, R. Kelly, N. Iqbal, F. Chandelier, D.K. Rex, Real-time differentiation of adenomatous and hyperplastic diminutive colorectal polyps during analysis of unaltered videos of standard colonoscopy using a deep learning model, Gut 68 (1) (2019) 94–100.
[85] P.J. Chen, M.C. Lin, M.J. Lai, J.C. Lin, H.H.S. Lu, V.S. Tseng, Accurate classification of diminutive colorectal polyps using computer-aided analysis, Gastroenterology 154 (3) (2018) 568–575.
[86] J. Renner, H. Phlipsen, B. Haller, F. Navarro-Avila, Y. Saint-Hill-Febles, D. Mateus, T. Ponchon, A. Poszler, M. Abdelhafez, R.M. Schmid, S. Von Delius, Optical classification of neoplastic colorectal polyps—a computer-assisted approach (the COACH study), Scand. J. Gastroenterol. 53 (9) (2018) 1100–1106.
[87] Y. Komeda, H. Handa, T. Watanabe, T. Nomura, M. Kitahashi, T. Sakurai, A. Okamoto, T. Minami, M. Kono, T. Arizumi, M. Takenaka, Computer-aided diagnosis based on convolutional neural network system for colorectal polyp classification: preliminary experience, Oncology 93 (Suppl. 1) (2017) 30–34.
[88] N. Ito, H. Kawahira, H. Nakashima, M. Uesato, H. Miyauchi, H. Matsubara, Endoscopic diagnostic support system for ct1b colorectal cancer using deep learning, Oncology 96 (1) (2019) 44–50.
[89] V. Blanes-Vidal, G. Baatrup, E.S. Nadimi, Addressing priority challenges in the detection and assessment of colorectal polyps from capsule endoscopy and colonoscopy in colorectal cancer screening using machine learning, Acta Oncol. 58 (Suppl. 1) (2019) S29–S36.
[90] R. Leenhardt, P. Vasseur, C. Li, J.C. Saurin, G. Rahmi, F. Cholet, A. Becq, P. Marteau, A. Histace, X. Dray, S. Sacher-Huvelin, A neural network algorithm for detection of GI angiectasia during small-bowel capsule endoscopy, Gastrointest. Endosc. 89 (1) (2019) 189–194.
[91] T. Zhou, G. Han, B.N. Li, Z. Lin, E.J. Ciaccio, P.H. Green, J. Qin, Quantitative analysis of patients with celiac disease by video capsule endoscopy: a deep learning method, Comput. Biol. Med. 85 (2017) 1–6.
[92] J.Y. He, X. Wu, Y.G. Jiang, Q. Peng, R. Jain, Hookworm detection in wireless capsule endoscopy images with deep learning, IEEE Trans. Image Process. 27 (5) (2018) 2379–2392.

[93] S. Seguí, M. Drozdzal, G. Pascual, P. Radeva, C. Malagelada, F. Azpiroz, J. Vitrià, Generic feature learning for wireless capsule endoscopy analysis, Comput. Biol. Med. 79 (2016) 163–172.

[94] P. Wang, X. Xiao, J.R.G. Brown, T.M. Berzin, M. Tu, F. Xiong, X. Hu, P. Liu, Y. Song, D. Zhang, X. Yang, Development and validation of a deep-learning algorithm for the detection of polyps during colonoscopy, Nat. Biomed. Eng. 2 (10) (2018) 741–748.

[95] Q. Song, Y.J. Zheng, J. Yang, Effects of food contamination on gastrointestinal morbidity: comparison of different machine-learning methods, Int. J. Environ. Res. Publ. Health 16 (5) (2019) 838.

[96] Q. Song, Y.J. Zheng, Y. Xue, W.G. Sheng, M.R. Zhao, An evolutionary deep neural network for predicting morbidity of gastrointestinal infections by food contamination, Neurocomputing 226 (2017) 16–22.

[97] Q. Song, M.R. Zhao, X.H. Zhou, Y. Xue, Y.J. Zheng, Predicting gastrointestinal infection morbidity based on environmental pollutants: deep learning versus traditional models, Ecol. Indicat. 82 (2017) 76–81.

[98] Q. Song, Y.J. Zheng, W.G. Sheng, J. Yang, Tridirectional transfer learning for predicting gastric cancer morbidity, IEEE Trans. Neural Networks Learn. Syst. (2020), https://doi.org/10.1109/TNNLS.2020.2979486. (In press).

CHAPTER

6

Application of soft computing techniques to calculation of medicine dose during the treatment of patient: a fuzzy logic approach

Ramjeet Singh Yadav

Department of Computer Science and Engineering, Ashoka Institute of Technology and Management, Varanasi, Uttar Pradesh, India

1. Introduction

Modern India is making significant progress in the field of artificial intelligence. Soft computing is the current field of artificial intelligence. Today we are developing different types of intelligent systems through soft computing. The use of computers has increased enormously in modern India. Artificial intelligence has been used to diagnose diseases and cure patients, especially in the field of medical science. Real-life problems in the field of medical science have a lot of complexity and uncertainty. These types of issues can be solved very well by intelligent systems based on fuzzy logic, artificial neural networks, and genetic algorithms. Soft computing is the integration of fuzzy logic, artificial neural network, genetic algorithm, and probabilistic reasoning. Here I have developed a fuzzy logic—based intelligent system which will be helpful in calculating the drug dose given to the patient in the correct amount. The amount of drug dose for different types of patients depends on the age, weight, history of patients and blood sedimentation, etc. No automatic method of determining the dose of the drug according to these qualities of the patient is available till now. In the field of medicine, fuzzy logic-based approaches have been developed and used in many areas ranging from kidney disease to diagnosis of cancer and from asthma to the determination of the drug dose.

Handbook of Computational Intelligence in Biomedical Engineering and Healthcare
https://doi.org/10.1016/B978-0-12-822260-7.00003-0

For the last 15–20 years, an automated system based on the fuzzy method was developed using the input data of some dialysis patients to determine the dosage of the drug of those patients, and the results obtained by this method were also promising. This automatic system based on fuzzy logic gives the doctor very accurate information about the dosage of the medicine given to the patient considering all the factors of the patient and proves to help the doctor in deciding very efficiently. Through this method, various errors in the field of medical science are worked out, and possible complications are prevented. Various researchers have proved the reliability of this method more by the statistical method, and it has been recognized in the whole world [1,2]. Therefore, I am presenting the method of determining the dose of the patient's medicine in India also by this method. This method will help the doctor to determine the dosage of drug in various hospitals in India.

For different types of patients who are suffering from the same type of disease, the doctor gives them different doses of the same medicine for many reasons. In providing the dose of medicine to patients with long-term medical treatment, the doctor pays attention to its properties such as the age, weight, sex of the patient, and the history of the disease before it. Through this presented fuzzy logic method, I can solve such unstressed and complex problems in the field of medical science in a very efficient way. Thus on evaluating the results obtained from a system designed based on fuzzy logic, it was found that this system is very reliable in determining the dose of the drug. Especially where there is no concrete formulation to determine the drug dose, the possible improvement in the disease can safely achieved the above criteria using such systems [3]. The capacity of fuzzy logic–based methods to adjust the parameters of the pharmacokinetic and pharmacodynamic model-based controller for the conveyance of the muscle relaxant pancuronium was discovered [4]. The fuzzy logic–based method employments control the rate of medicine delivery and measures the muscle unwinding level [4,5].The fuzzy logic–based control system of nicardipine mixture for hypertension control amid anesthesia was displayed in Ref. [6]. In this system a robotized blood weight controller has introduced to conceive the palatable and secure strategies of regulating nicardipine [6]. A few studies are committed to utilize the artificial neural systems to predict drug disintegration profiles [7] and to create versatile forms of straight, nonlinear, and time-dependent models to test data sets collected from hemodialysis patients getting anticoagulant heparin amid treatment [8]. It has seen that the data set of drug parameters are not fresh, so I have proposed the method based on a shape of fuzzy logic. This will offer assistance to make a contemplated approach to decide a medicine dose that will allow administrative bodies to set fitting benchmarks for drugs (modern and maybe ancient) and helps medicine producers in creating a contemplated approach in item improvement.

Generally, in medical treatment, doctors decide the dosage of pharmaceuticals for a child agreeing to his weight and for a grown-up concurring to the outline. In any case, the measurements of the drug to be utilized by each should be decided concerning the patient's properties such as foundation, age, sex, weight, and changeless ailments. Due to the reasons like unpleasant working conditions and mental obstructions, etc., now and then the doctors cannot give sound choices which result in the utilization of off-base dosages. The drug is produced by the pharmaceutical manufacturing companies in a prescribed standard gram but the dosage of the drug is determined by the age, weight, sex and permanent diseases of the patient. For illustration, for a patient who requires single dose of 370 mg of the drug, the drug company produces 550 mg of this drug in the form of a packet. Since a few drugs are exceptionally costly and huge measurements are not reasonable for numerous patients, there's a need to diminish its measurements. With the help of the fuzzy logic method in numerous circumstances, it is conceivable to foresee the correct dosage of the drug. Within the nearness of capacities of information collected amid treatment inside a few periods, it is conceivable to construct prototypes shrewdly for both foreseeing patients' reactions and deciding a satisfactory drug measurement (dose), which permits for accomplishing craved medicate concentration or medicine or drug impact [5,8]. In this think about, a fuzzy logic-based framework is outlined to decide the dosage of salazopyrine medication utilized within the handling of unremitting intestinal infection—in connection to two criteria. However, it can be decided in connection to numerous criteria.

Many properties of a drug as an input data set do not form a crisp set. Therefore, to solve this type of problem, fuzzy logic needs to be used. At this time, fuzzy logic is proving very useful to solve the uncertain problem. The automated proposed method is an excellent way to determine the dose of the drug. There will be no need for a doctor in this proposed method. Sometimes the doctor does not know to give the right dose of medicine to his patient, and due to this, the patient's disease persists for a long time and does not cure quickly. Often, in medical treatment, the doctor determines the dose for a child according to his weight and the adult determines the dose of the drug according to the medication manual. While each patient is required to determine the drug by the environment, sex, weight, and age of the patient, the doctor should also prescribe the medicine according to the patient's permanent illness.

Sometimes the doctor does not give the right dose to the patient due to the monstrous working conditions of the patient and due to psychological constraints, due to which the wrong drug dose is used. Sometimes due to the number of drugs in grams as a standard by pharmaceutical companies, the patient needs another additional dose. Consider that a patient needs 371 Magdose, but the production of this medicine by the drug makers is 400 Mag. Some medicines made by companies are very expensive, and some medicines are very cheap. For this reason, the doctor gives this expensive medicine in small quantities to the poor patient. In such situations, predicting the correct dose of the drug using the fuzzy logic method would be very beneficial for the patient. The fuzzy logic method is an automated method by which determining the right amount of medicine to the low-income families of this modern age mother India will be a very useful step. In our country, i.e., India, the determination of medicine by this automated method, i.e., fuzzy logic would be a very bold step because in our country 40% to 60% of the population is very poor.

Based on the facts collected during therapy, it is possible to adapt the dose of the drug in the correct amount and to make an intelligent model by reasoning method to find out the correct position of the patients. This model permits achieving the drug effect [5,8]. In this study by the author, an expert system is created by a fuzzy logic method to determine the relation of the double standards of the dose of the drug used in the treatment of chronic intestines during infection although this model determines the relationship of several parameters.

The objective of this book chapter is to find out more refine application areas of fuzzy logic-based automated systems for the calculation of exact drug dose patients. The proposed intelligent system can be used to determine the dose of the drug in various diseases and good results can be obtained in the treatment of the patient.

2. Soft computing

Soft computing is the adjustment of "fuzzy logic, artificial neural networks, genetic algorithms, and probabilistic reasoning." Soft computing is a broad form of hard computing. Hard computing is a very old method based on zero and one logic. Due to the value of this computing only being zero and one, it is inequitable to find solutions to many real-life problems of human beings. Such types of real-life problems can be solved by soft computing methods. Therefore, here I can say that soft computing is an improved form of hard computing. Soft computing is an excellent method of solving real-life problems of humans. Today, by this method, I can get straightforward and correct answers to many types of complex problems in the modern era. Soft computing is an adjustment of the following four types of methods:

1. Fuzzy Logic.
2. Artificial Neural Network.
3. Genetic Algorithm.
4. Probabilistic Reasoning.

There are many types of problems in real life, which are very important to be solved by the logic method. Determining the "dose of medicine" given to the patient is also a problem based on real life. Hence, I have presented a method of calculation "the dose of medicine" given to the patient by the fuzzy logic method, which is an automated method. Through this automated intelligent fuzzy logic method, the patients can calculate the amount of drug dose.

3. Fuzzy logic

Fuzzy logic is an extended form of crisp-logic. The truth-value of classical logic is only 1 and 0. In other words, we can say that classical logic has value only 0 and 1. Classical logic is based on crisp set methods, and fuzzy logic is based on fuzzy sets. In comparison, the value of fuzzy logic is 0 or 1 or between 0 and 1. Therefore, fuzzy logic is also called multivalued logic. The fuzzy logic was invented by Prof. Lofti A. Zadeh in 1965. Probability theory has a very week relation with fuzzy logic. Fuzzy logic is a part of soft computing. Soft computing is an advanced topic of artificial intelligence. Now there are many applications of soft computing such as medical sciences, defenses, educational field, financial sector, and many other sectors.

3.1 Fuzzy set

The crisp set is a well-defined collection of objects with a crisp boundary. For example, let us consider a set $A = \{1,2,3,4,5\}$, if x belongs to set A (i.e., $x \in A$), which is denoted by 1 and if x does not belong to set A (i.e., $x \notin A$) which is denoted by 0. This formulation is denoted by characteristics function or membership function, which denoted by $\mu_A(x)$ and defined by the following formula [Eq. (6.1)]:

$$\mu_A(x) = \begin{cases} 1 & \text{if} \quad x \in A \\ 0 & \text{if} \quad x \notin A \end{cases} \tag{6.1}$$

The concept of man and ideas cannot be communicated by a crisp set as both are ambiguous. Therefore, to explain and solve these kinds of problems in real life of man, Professor Lofti A. Zadah was invented fuzzy logic in place of crisp sets in 1964. In the case of a fuzzy set, many ranges (degrees) of membership (in between 0 and 1) can be allowed. "Membership function of a fuzzy set is denoted by $\mu_A(x)$, where A is a fuzzy set and x is an element of fuzzy set A. Membership function of a fuzzy set maps every element of the universe of discourse X to the range zero and one [0, 1], i.e., $\mu_A(x) : X \rightarrow [0, 1]$. A fuzzy set A [Eq. (6.2)] is defined on X (universe of discourse) when x is an element of X as a collection of ordered pairs is as follows" [9]:

$$A = \{(x, \mu_A(x)) : x \in X\} \tag{6.2}$$

For example, $A = \{(1, 0.5), (2, 0.3), (3, 0.4), (4, 0.8), (5, 1)\}$.

The transformation from a crisp set or classical set into the fuzzy set with the help of this membership function is known as fuzzification [10].

3.1.1 Membership formulation

"Fuzzy set is fully defined by its membership function (MF). Generally, triangular, trapezoidal and Gaussian membership function has been used for converting the crisp set into the fuzzy set, which is as follows" [9]:

Triangular Membership Function: The triangular membership function [Eq. (6.3)] is defined as follows with three parameters (a, b, c) [9] (Fig. 6.1):

$$\text{Triangle}(x; a, b, c) = \max\left(\min\left(\frac{x-a}{b-a}, \frac{c-x}{c-b}\right), 0\right) \tag{6.3}$$

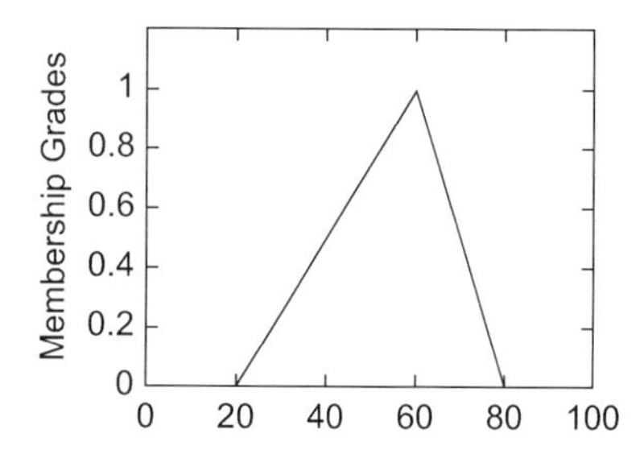

FIGURE 6.1 Triangular membership function.

Trapezoidal Membership Function: The trapezoidal membership function [Eq. (6.4)] is defined as follows with four parameters (a, b, c, d):

$$\text{Trapezoidal}(x; a, b, c, d) = \max\left(\min\left(\frac{x-a}{b-a}, 1, \frac{d-x}{d-c}\right), 0\right) \tag{6.4}$$

The trapezoidal "membership function" is visualized in Fig. 6.2.

The real-time implementation of fuzzy logic finds extensive use of "triangular and trapezoidal membership functions" due to straightforward computationally effective formulas.

Gaussian Membership Function: This function has two parameters (c, σ), which are given below [Eq. (6.5)]:

$$\text{Gaussian}(x; c, \sigma) = e^{-\frac{1}{2}\left(\frac{x-c}{\sigma}\right)^2} \tag{6.5}$$

Where $c =$ Membership Function center.
$\sigma =$ Width of the Membership Function.
Fig. 6.3 shows the shape of Gaussian membership.

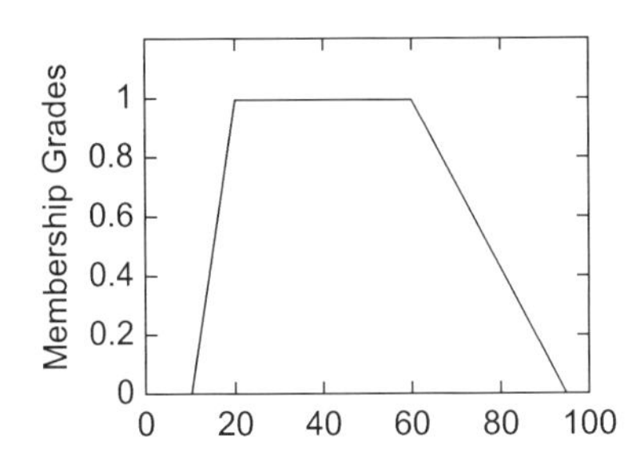

FIGURE 6.2 Trapezoidal membership function.

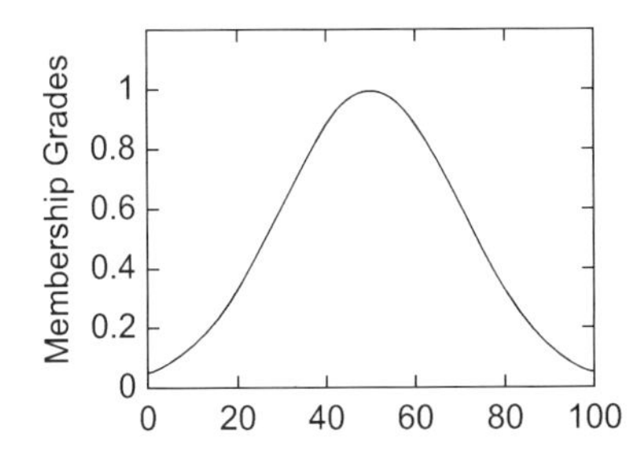

FIGURE 6.3 Gaussian membership function.

3.1.2 Fuzzy relation

The fuzzy relation is a fuzzy set that maps every element in $X \times Y$ to membership value between zero and one. Assume X and Y be two universes of discourse. The fuzzy relation [Eq. (6.6)] is defined by the following formula:

$$R = \{((x,y), \mu_R(x,y)) / (\in X \times Y)\} \tag{6.6}$$

3.1.3 Maximum-minimum composition

The maximum-minimum composition [Eq. (6.7)] of R_1 and R_2 is a fuzzy set defined by:

$$R_1 \circ R_2 = \{[(x,z), \text{maxmin}(\mu_{R_1}(x,y), \mu_{R_2}(y,z))] / x \in X, y \in Z, z \in Z\} \tag{6.7}$$

Where R_1 = Fuzzy relation on $X \times Y$
R_2 = Fuzzy relation on $Y \times Z$

3.1.4 Fuzzy IF-THEN rules

Linguistic variable (LV) is defined as a quintuple as follows [11]:
$LV = (x, T(x), X, G, M)$:
Where x = Name of the variable.
$T(x)$ = Set of its linguistic values.
X = Universe of discourse.
G = Syntactic rule, which generates the terms in $T(x)$: "It is in the form of Backus-Naur form to specify the set of possible natural language values for a linguistic variable" [11].
M = Semantic rule (associates with each linguistic value): "It is mapping from U to a fuzzy set about specific linguistic value, that is to say, for $u \in U, t \in T(X)$, assuming a value in [0,1] called degree of membership to u, which represent the confidence score that u satisfies the linguistic value t" [11].
$M(A)$ = Fuzzy set in X.
A fuzzy if-then rule is defined as follows:
If x is A then y is B
Where A and B = Linguistic variables which are defined by fuzzy sets on universal sets X and Y, respectively.

Often "x is A" and is known as the "antecedent or premise", while "y is B" which is known as the "consequence or conclusion".

In the fuzzy IF-THEN rule, the expression "if x is A then y is B" which is also denoted as $A \rightarrow B$. The "fuzzy IF-THEN rule" may be defined as a "fuzzy relation R on the product space. $X \times Y$".

There is two way to represent the fuzzy IF-THEN rule:

1. If $A \rightarrow B$ represented as A, which shows as follows (Eq. 6.8):

$$R = A \rightarrow B = A \times B = \int_{X \times Y} \mu_A(x) \tilde{*} \mu_B(y)/(x,y) \tag{6.8}$$

where $\tilde{*}$ = T-norm operator.
$A \rightarrow B$ = Fuzzy relation in R.

2. If $A \rightarrow B$ represented as A entails B which is shown as follows (Eq. 6.9):

$$R = A \rightarrow B = A \times B = \neg A \cup B \tag{6.9}$$

3.1.5 *Fuzzy reasoning*

"Fuzzy reasoning (approximate reasoning) is an inference process that derives conclusions from a set of fuzzy IF-THEN rules and facts (Eq. 6.10). Let us consider A, A' and B are the fuzzy sets of X, X, and Y, respectively. Suppose that the fuzzy implication $A \rightarrow B$ is spoken as a fuzzy relation R on $X \times Y$ then the fuzzy B induced by x is A and the fuzzy rule if x is A then y is B is defined by:"

$$\mu_{B'}(y) = \max_x \min[\mu_{A'}(x), \mu_R(x, y)] \tag{6.10}$$

When multiple rules are joined together by suitable conjunction in antecedent part of fuzzy IF-THEN rule (Eq. 6.11), then interpretation taken as the union of the fuzzy relation corresponding to the fuzzy rules. Fuzzy reasoning can be employed as a fuzzy inference method to obtain the resultant output fuzzy set C' (Fig. 6.4). Since the maximum-minimum composition operator o is distributive over the ∪ operator as given below:

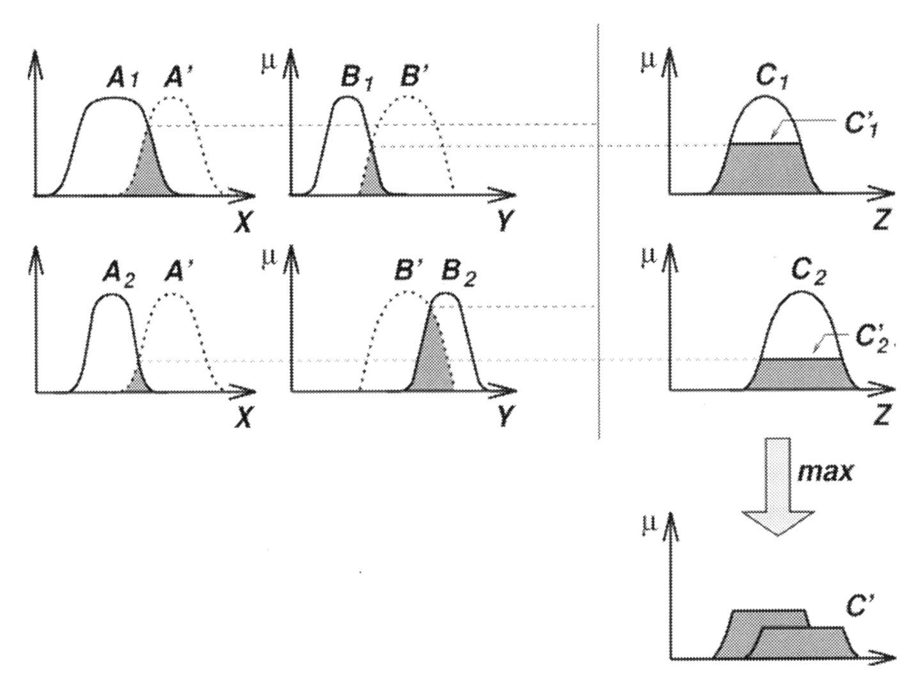

FIGURE 6.4 Multiple antecedents and multiple rules of fuzzy reasoning.

$$C' = (A' \times B') \circ (R_1 \cup R_2)$$
$$= [(A' \times B') \circ R_1] \cup [(A' \times B') \circ R_2] - C'_1 \cup C'_2 \tag{6.11}$$

where C'_1 and C'_2 are the inferred fuzzy set for rule 1 and rule 2, respectively. "Fuzzy inference systems" based on "fuzzy set theory" has two main components (1) Fuzzy IF-THEN rules (2) Fuzzy reasoning.

3.1.6 *Defuzzification*

Defuzzification is a method for the conversion of the fuzzy set (fuzzy output) to the crisp set or crisp output. Four methods of defuzzification are given below:

Maximum-Membership Method: This defuzzification technique is also called as the height method and given by Eq. (6.12):

$$\mu_C(z^*) \geq \mu_C(z), \quad \text{for all } z \in Z \tag{6.12}$$

Where z^* = Defuzzified value, as visualized in Fig. 6.5.

Center of Gravity Method: This method is defined by the following formula (Eq. 6.13):

$$z^* = \frac{\int \mu_C(z) \cdot z dz}{\int \mu_C(z) dz} \tag{6.13}$$

Where $\int$ = Algebraic integration, as shown in Fig. 6.6.

Weighted Average Method: This method is defined by the following formula (Eq. 6.14):

$$z^* = \frac{\sum \mu_C(\overline{z}) \cdot \overline{z}}{\sum \mu_C(\overline{z})} \tag{6.14}$$

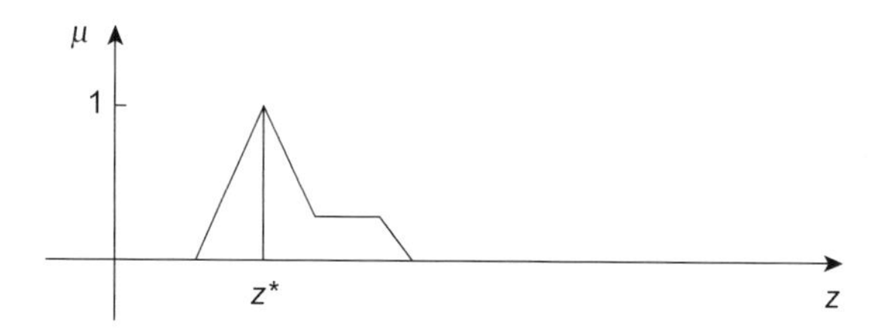

FIGURE 6.5 Defuzzification method of maximum membership.

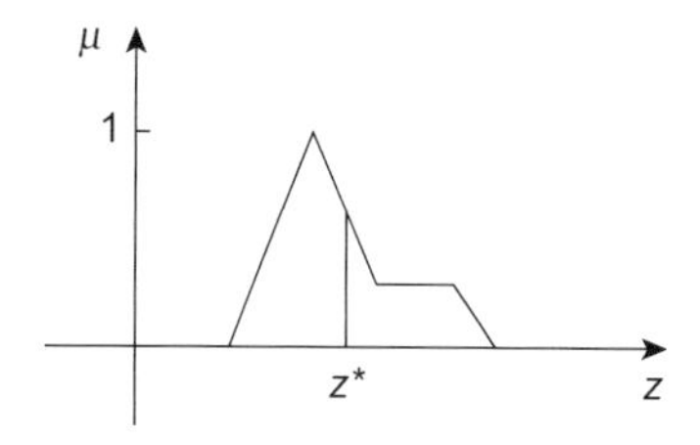

FIGURE 6.6 Centroid defuzzification method.

Where $\sum$ = Algebraic sum.

$\bar{z}$ = Centroid of every symmetric membership function (Fig. 6.7).

Center of Sums: Center of Sums is faster in comparison to other defuzzification techniques. It is given by the algebraic expression (Eq. 6.15):

$$z^* = \frac{\sum_{k=1}^{n} \mu_{C_k}(z) \int (\bar{z} dz)}{\sum_{k=1}^{n} \mu_{C_k}(z) \int dz} \tag{6.15}$$

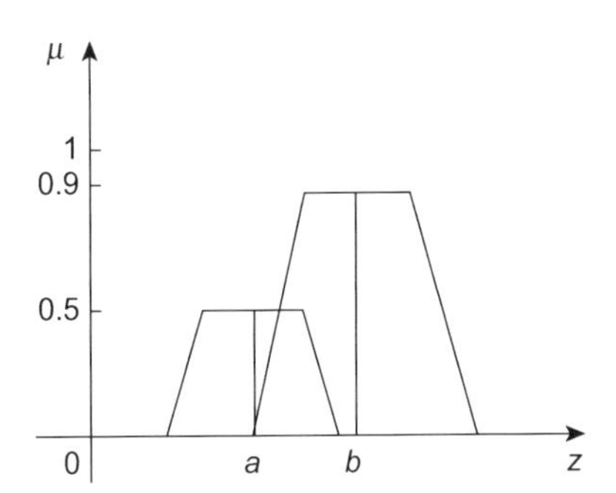

FIGURE 6.7 Average weighted method.

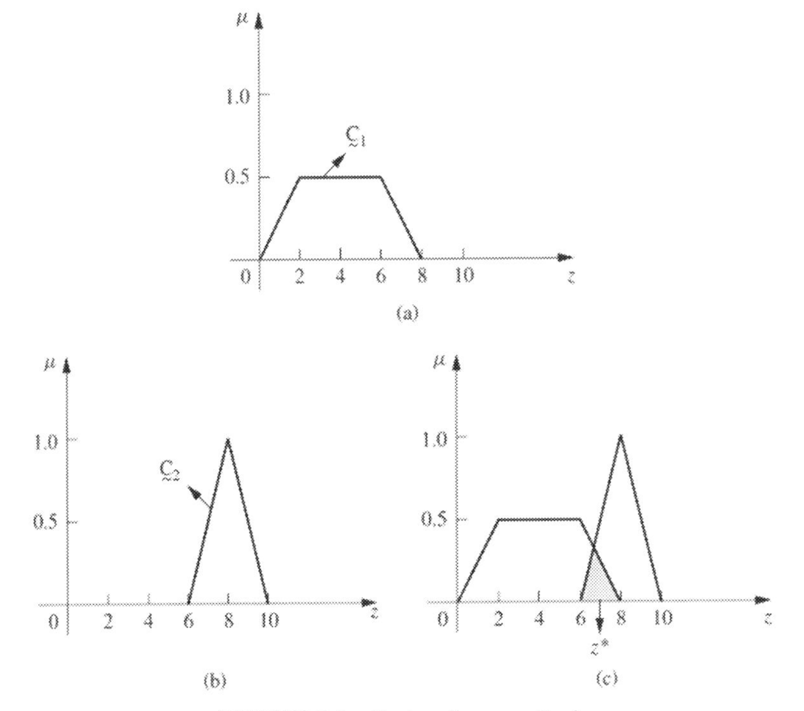

FIGURE 6.8 Center of sum method.

Where $\bar{z}$ = Distance to the centroid of every relevant MF (Fig. 6.8).

Mean of Max Membership: Mean of Max Membership is closely related to the Max Membership principle. Mean of Maximum membership function is defined by the following formula (Eq. 6.16):

$$z^* = \frac{a+b}{2} \tag{6.16}$$

Fig. 6.9 shows the definition of a and b.

Center of Largest Area: This method is defined by the following formula (Fig. 6.10) and Eq. (6.17):

$$z^* = \frac{\int \mu_{C_m}(z) \cdot z dz}{\int \mu_{C_m}(z) dz} \tag{6.17}$$

Where C_m = Convex subregion (largest area making up C_k).

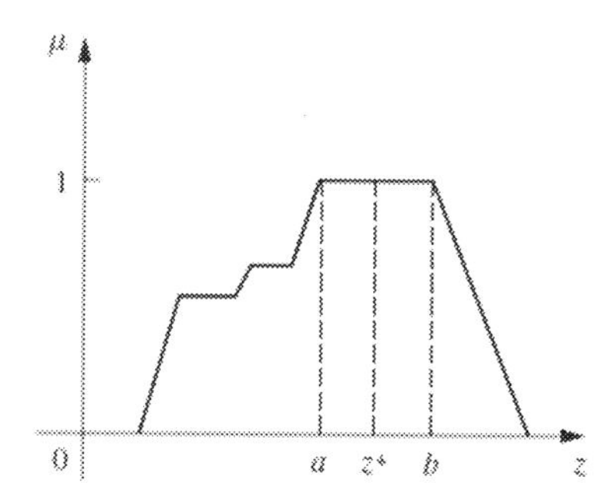

FIGURE 6.9 Mean max membership defuzzification method.

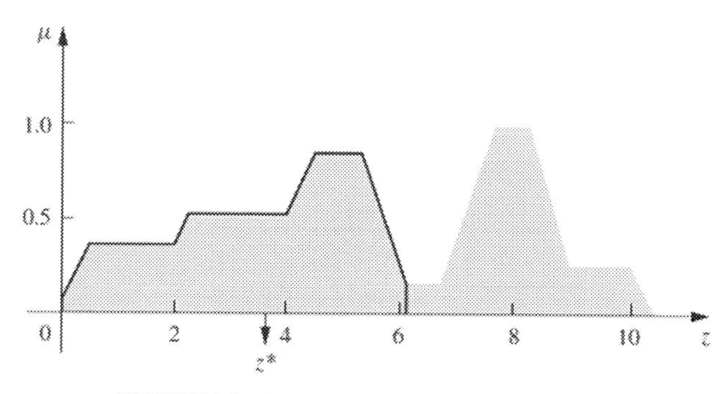

FIGURE 6.10 Center of largest area method.

3.1.7 Fuzzy inference system

The "Fuzzy Inference System (FIS)" is a most important "computing framework" which consists of: (1) Fuzzy set theory (2) Fuzzy reasoning (3) Fuzzy IF-THEN rules. The application areas of FIS are: (1) computer vision (2) expert systems (3) decision analysis (4) data classification (5) automatic control.

"Due to involvement of multidisciplinary approach fuzzy inference system is also variously known as:(1) Fuzzy rule-based system (2) Fuzzy expert system (3) Fuzzy model (4) Fuzzy logic controller (5) Fuzzy associative memory (6) Simply fuzzy system".

The fuzzy inference system consists of three main abstract components:

1. Rule base: This is a collection of fuzzy IF-THEN rules.
2. Database (dictionary): Database or dictionary defines the membership functions used in the fuzzy IF-THEN rules.
3. Reasoning mechanism: The reasoning mechanism performs the inference mechanism leading the fuzzy IF-THEN rules and produce realistic output.

A FIS is made up of five blocks (Fig. 6.11):

1. Rule base: Fuzzy IF-THEN rules made the rule base.
2. Database: "Fuzzy sets" used in the "fuzzy rule" is defined by a membership function of the membership database.
3. Decision-making unit: This unit performs the inference operation on the rule.
4. Fuzzification interface: This unit transforms crisp sets into fuzzy sets as per the degree of match with the given linguistic variable.
5. Defuzzification interface: This unit transforms fuzzy set into corresponding crisp output.

The knowledge base consists of a rule base and database. The FIS performs the following steps of fuzzy reasoning:

1. In this step fuzzification of the input, the variable is compared with membership function by applying the rule in the premise part to get the membership values of each linguistic level.

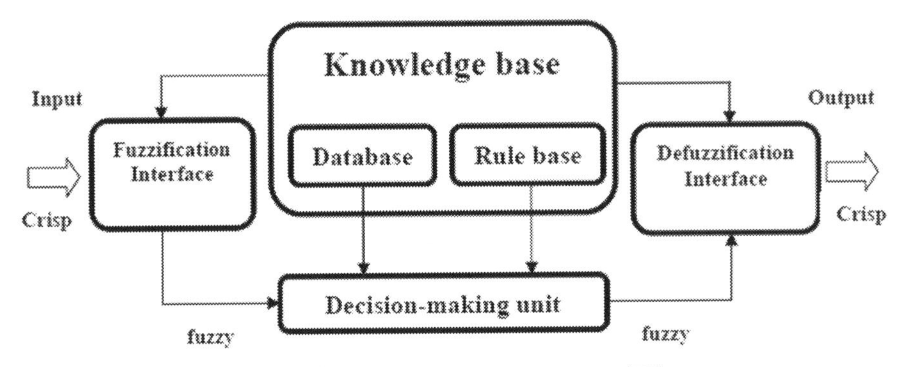

FIGURE 6.11 Fuzzy inference system (FIS).

2. A specific fuzzy T-norm operator is used to combine the membership values (multiplication or minimum on-premise part) to find the weight of (firing strength) of each rule.
3. Depending upon the weight, a qualified consequent is generated, which may be either fuzzy or crisp.
4. In the final step, which is called defuzzification, the qualified consequents are aggregated to build a crisp output.

4. Fuzzy logic based intelligent system

The development of the fuzzy Logic-based Intelligent System started in the 1970s, and to date, it has been converted into an intelligent system using innovative mathematical, artificial intelligence and logical methods. Due to its unique quality of intelligent systems and with the help of its logic, it can solve any real-life based problem and give the right decision. It was implemented for processing knowledge in a computer program to extract knowledge from human experts so that it could deal with quantitative and qualitative data. Compared to other traditional applications, which require a sequence of steps called an algorithm, intelligent systems are more knowledgeable as humans that allow fuzzy logic and can deal with incomplete data. The intelligent systems can be divided based on subject-specific fields and applications, for example, classification or identification control, simulation planning, and design, prediction, interpretation, instructions, repair and diagnosis, other areas like all disciplines of engineering branches.

All types of intelligent systems establish their respective codes, algorithms, relationships between users and programs and work according to their rules. The fuzzy based intelligent system presented in this book acts like an automated intelligent system. The main objective of the presented fuzzy based intelligent system will play a major role in determining the dose of the drug of a poor patient admitted to hospitals in India. With this help, the doctor can determine the correct amount of dose to be given to the patient.

4.1 Components of fuzzy logic intelligent system

Fig. 6.12 shows the major component of the fuzzy logic-based intelligent system. The main components of a fuzzy-based intelligent system are as follows:

1. **User Interface:** This interface establishes a good relationship between the user and the fuzzy-based intelligent system. A fuzzy-based intelligent system will feel good only when its user, as well as user interfaces, is very good. If there is no good user interface, then we will not be able to solve the problems of real-life correctly. Due to the user interface, the fuzzy-based intelligent system can understand the user's question and solve the problem correctly.
2. **Knowledge Base:** All the rules and facts of a problem are stored in the knowledge base part of the fuzzy-based intelligent system. The knowledge base is known to be the core component of the fuzzy-based intelligent system as it stores the knowledge of

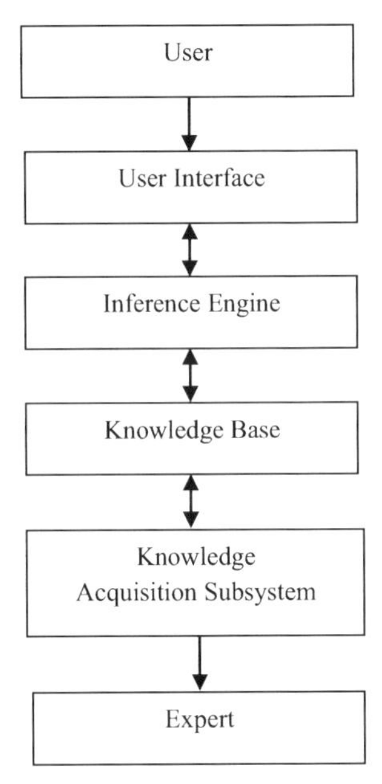

FIGURE 6.12 Major components of fuzzy intelligent system.

the expert to solve the problem. The knowledge gained by experts is also stored in this part of the intelligent system. The rules, facts, and descriptions of things are also stored in this repository of knowledge. The knowledge base is always stored in data when new fuzzy based intelligent systems are created. Due to this part of knowledge in a fuzzy-based intelligent system, it is necessary to understand, prepare and solve the problem. Representation of knowledge is done based on this knowledge; hence it is called the key to the knowledge base. Formal knowledge should be represented, which is obtained by experts. In this way, the representation of knowledge is related to the manipulation of information, knowledge, and structure of information.

3. **Inference Engine:** The inference engine is a part of the program. Through this, the store in the user interface applies the knowledge based on the facts and knowledge on the premises. Inference engines also provide better control strategies for problem-solving. Appropriate knowledge is found from the facts stored in the knowledge base by the inference engine. By this, solving the problem and applying the method of logical processing enhances new knowledge. The inference engine is called the brain of the fuzzy-based intelligent system. It uses the "IF-THEN" rule to obtain a solution to the problem and provide a better conclusion. This "IF-THEN" rule is the heart of the fuzzy-based intelligent system.
4. **Knowledge Acquisition Subsystem:** The knowledge gained by the data stored in the knowledge base is obtained by the experts. This work is done by a knowledge engineer. All the operations of the fuzzy-based intelligent system are very time consuming though it is efficient in doing the task. Time-consuming operation requires more time

and money to build a fuzzy-based intelligent system, and its work is also limited. The knowledge acquisition program is used by an individual. It has problems in adding, creating and building the knowledge base. All sources of knowledge include the user's own experience, databases, books, research reports, and human experts.

5. **Experimental Results:** The fuzzy based intelligent system is created using fuzzy set theory and artificial intelligence methods. This intelligent system is called a fuzzy-based intelligent system. This fuzzy based intelligent system works like an automated machine. Due to this reason, this fuzzy-based is called an intelligent system. This fuzzy based intelligent system is also used to implement a linguistic control strategy dependent on human knowledge. Fuzzy rules are laid down when creating a fuzzy-based intelligent system. This type of fuzzy rule is called "IF-THEN." The output received by the fuzzy-based intelligent system is also in the form of a fuzzy output. It is necessary to convert this fuzzy output to a crisp output. To convert fuzzy output to crisp output, mathematical function is used. The process of converting fuzzy output to crisp output is called defuzzification [12,13].

After deciding to build a fuzzy-based intelligent system, the first step is to produce "if-then" rules. Producing "if-then" rules is a time-consuming and very difficult task. We need domain experts to produce these types of rules. Determination of fuzzy rules without domain experts is very difficult and the solution to the problem achieved by it will be wrong. Due to this, the fuzzy-based intelligent system will not be able to answer the problem correctly [12–15].In the presented book chapter, the data of 20 patients and knowledge used in this fuzzy based intelligent system, provided by Dr. S.P. Sharma, Sar Sundar Lal Hospital, Banaras Hindu University, Varanasi and modern drug guides. For this experiment, I have used an observational data of 20 patients.

The input and output parameters are given below:

Input Parameter: There are two input parameters of a fuzzy-based intelligent system, namely: (1) Prostate-Specific Antigen (PSA) and (2) Sedimentation (SD). These two input parameters are further categorizing in form linguistic variable namely Low, Medium and High in both cases.

Output Parameter: There is only one output parameter called salazopyrine (SL). This parameter further divided into linguistic form, namely: Low, medium, high and very high.

Symptoms of chronic intestinal infection to patients considering the movement of "Prostate-Specific Antigen (PSA), sedimentation (SD)" and the dose of salazopyrine (SL) are linguistically shown in Table 6.1. The sample size of the data set is 20.

The fuzzy set of input variable of "Prostate-Specific Antigen" is shown in Table 6.2.

The value of Prostate Specific Antigen varies from 0 to 50. The descriptions of various type membership function of input variable of Prostate Specific Antigen are given by Eqs. (6.18)–(6.20):

$$\mu_{\text{Low}}(x) = \begin{cases} 0 & \text{if } x < 0 \\ \dfrac{5-x}{5} & \text{if } 0 \le x \le 5 \\ 0 & \text{if } x > 5 \end{cases} \tag{6.18}$$

TABLE 6.1 Data sets of "Prostate-Specific Antigen (PSA), sedimentation (SD)" (mm/h) and salazopyrine (SL) doses obtained by the expert doctor.

Patient	Input variables		Output variable (SL mg)
	Prostate-Specific Antigen (PSA) (ng/mL)	Sedimentation (SD) (mm/h)	Drug dose suggested by the Doctor/Day
1	0.50	10	500 (1 × 1)
2	0.80	16	500 (1 × 1)
3	1.00	20	500 (1 × 1)
4	1.00	24	500 (1 × 1)
5	2.00	25	500 (1 × 1)
6	2.00	81	1000 (2 × 1)
7	2.00	70	1000 (2 × 1)
8	3.00	40	1000 (2 × 1)
9	3.00	50	1000 (2 × 1)
10	4.00	42	1000 (2 × 1)
11	5.00	50	1000 (2 × 1)
12	6.00	67	1000 (2 × 1)
13	7.00	81	1500 (3 × 1)
14	7.50	75	1500 (3 × 1)
15	8.00	25	1500 (3 × 1)
16	8.50	25	1500 (3 × 1)
17	9.00	75	1500 (3 × 1)
18	12.0	80	1500 (3 × 1)
19	13.0	90	1500 (3 × 1)
20	15.0	90	1500 (3 × 1)

TABLE 6.2 Fuzzy set of input variable of "Prostate-Specific Antigen".

S.No.	Name of Membership Function	Input Variable of Prostate Specific Antigen in the form of Linguistic Variable	Interval
1	Triangular membership function	Low	(0, 0, 5)
2	Triangular membership function	Medium	(0, 5, 10)
3	Trapezoidal membership function	High	(5, 10, 50, 50)

$$\mu_{\text{Medium}}(x) = \begin{cases} 0 & \text{if } x < 0 \\ \dfrac{x}{5} & \text{if } 0 \leq x \leq 5 \\ \dfrac{10-x}{5} & \text{if } 5 \leq x \leq 10 \\ 0 & \text{if } x > 10 \end{cases} \tag{6.19}$$

$$\mu_{\text{High}}(x) = \begin{cases} 0 & \text{if } x < 0 \\ \dfrac{x-5}{5} & \text{if } 5 \leq x \leq 10 \\ 1 & \text{if } 10 \leq x > 50 \end{cases} \tag{6.20}$$

Various type of input variable Prostate Specific Antigen of "membership functions" is shown in Fig. 6.13.

Here, I convert the crisp data set of Prostate Specific Antigen into fuzzy data set using Eqs. (6.18)–(6.20). The fuzzy sets of Prostate Specific Antigen are given bellow:

$$\text{Low} = \{(0.5, 0.9), (0.8, 0.84), (1, 0.8), (2, 0.6), (3, 0.4), (5, 0)\}$$

$$\text{Medium} = \{(0.5, 0.1), (0.8, 0.16), (1, 0.2), (2, 0.4), (3, 0.6), (4, 0.8), (5, 1)\}$$

$$\text{High} = \{(0.5, 0), (0.8, 0), (1, 0), (2, 0), (3, 0), (4, 0), (5, 0), (6, 0.2), (7, 0.4), (7.5, 0.5), (8, 0.6), (8.5, 0.7), (9, 0.8), (12, 1), (13, 1), (15, 1)\}$$

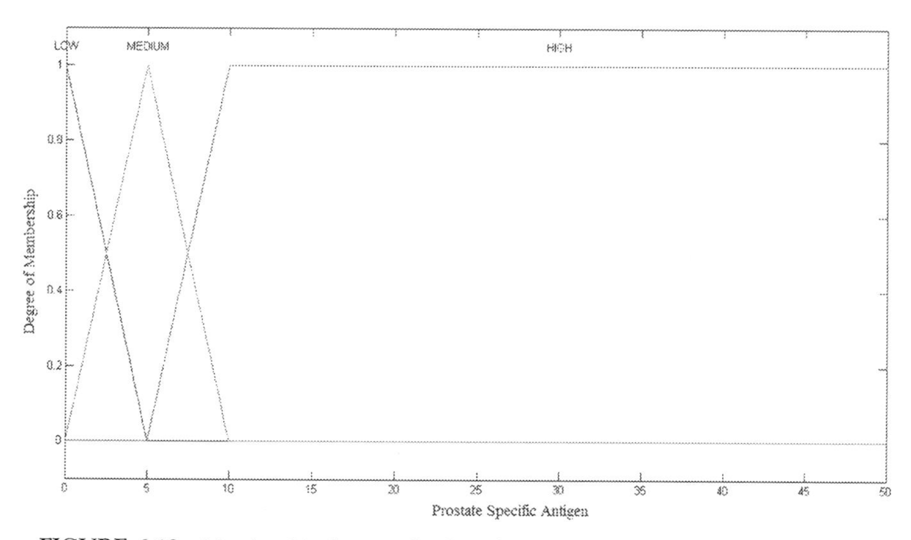

FIGURE 6.13 Membership function for three fuzzy values of Prostate Specific Antigen.

The fuzzy set of input variable of sedimentation is shown in Table 6.3.

The value of sedimentation varies from 0 to 91. The description of various type membership function of the second input variable of sedimentation is given by Eqs. (6.21)–(6.24):

$$\mu_{\text{Low}}(x) = \begin{cases} 0 & \text{if } x < 0 \\ \dfrac{50 - x}{50} & \text{if } 0 \leq x \leq 50 \\ 0 & \text{if } x > 50 \end{cases} \tag{6.21}$$

$$\mu_{\text{Medium}}(x) = \begin{cases} 0 & \text{if } x < 0 \\ \dfrac{x - 50}{50} & \text{if } 0 \leq x \leq 50 \\ \dfrac{81 - x}{31} & \text{if } 50 \leq x \leq 81 \\ 0 & \text{if } x > 81 \end{cases} \tag{6.22}$$

$$\mu_{\text{High}}(x) = \begin{cases} 0 & \text{if } x < 50 \\ \dfrac{x - 50}{31} & \text{if } 0 \leq x \leq 81 \\ \dfrac{90 - x}{9} & \text{if } 81 \leq x \leq 90 \\ 0 & \text{if } x > 90 \end{cases} \tag{6.23}$$

$$\mu_{\text{High}}(x) = \begin{cases} 0 & \text{if } x < 50 \\ \dfrac{50 - x}{31} & \text{if } 50 \leq x \leq 81 \\ 1 & \text{if } 81 \leq x > 90 \end{cases} \tag{6.27}$$

TABLE 6.3 Fuzzy set of input variable of sedimentation.

S.No.	Name of Membership Function	input variable of Sedimentation in the form of Linguistic Variable	Interval
1	Triangular membership function	Low	(0, 0, 50)
2	Triangular membership function	Medium	(0, 50, 81)
3	Trapezoidal membership function	High	(50, 81, 90, 90)

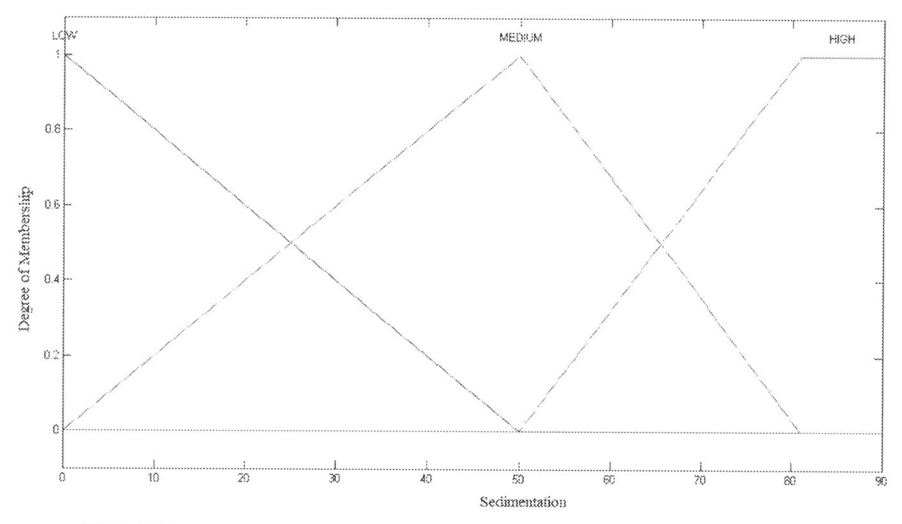

FIGURE 6.14 Membership function for three fuzzy values of sedimentation.

Various type of input variable sedimentation of membership functions are shown in Fig. 6.14:

Here, I convert the crisp data set of sedimentation into fuzzy data set using Eqs. (6.21)–(6.23). The fuzzy sets of sedimentation are given bellow:

$$\text{Low} = \{(10, 0.8), (16, 0.68), (20, 0.6), (24, 0.52), (25, 0.5), (40, 0.2), (41, 0.18), (50, 0)\}$$

$$\text{Medium} = \{(10, 0.8), (16, 0.68), (20, 0.6), (24, 0.52), (25, 0.5), (40, 0.2), (41, 0.18), (50, 1), (67, 0.45), (75, 0.19), (80, 0.03)\}$$

$$\text{High} = \{(50, 0), (67, 0.55), (75, 0.81), (80, 0.97), (90, 1)\}$$

The fuzzy set of output variable of doses of salazopyrine are shown in Table 6.4.

TABLE 6.4 Fuzzy set of output variable of doses of salazopyrine.

S.No.	Name of Membership Function	Output variable of drug doses of Salazopyrine (mg/day)in the form of Linguistic Variable	Interval
1	Trapezoidal membership function	Low	(0, 0, 200, 600)
2	Triangular membership function	Medium	(200, 600, 1000])
3	Triangular membership function	High	(600, 1000, 1400)
4	Trapezoidal membership function	Very high	(1000, 1400, 1400, 100,000)

The value of doses of Salazopyrine (mg/day) varies from zero to 91. The description of various type membership function of output variable of doses of Salazopyrine (mg/day) is given by Eqs. (6.25)–(6.28):

$$\mu_{\text{Low}}(y) = \begin{cases} 0 & \text{if } y < 0 \\ 1 & \text{if } 0 \le y \le 200 \\ \dfrac{600 - y}{30} & \text{if } 200 \le x \le 600 \\ 0 & \text{if } y > 600 \end{cases} \tag{6.25}$$

$$\mu_{\text{Medium}}(y) = \begin{cases} 0 & \text{if } y < 200 \\ \dfrac{x - 200}{400} & \text{if } 200 \le y \le 600 \\ \dfrac{1000 - y}{30} & \text{if } 600 \le y \le 1000 \\ 0 & \text{if } y > 1000 \end{cases} \tag{6.26}$$

$$\mu_{\text{High}}(y) = \begin{cases} 0 & \text{if } y < 600 \\ \dfrac{y - 600}{400} & \text{if } 600 \le y \le 1000 \\ \dfrac{1400 - x}{400} & \text{if } 1000 \le y \le 1400 \\ 0 & \text{if } y > 1400 \end{cases} \tag{6.27}$$

$$\mu_{\text{VeryHigh}}(y) = \begin{cases} 0 & \text{if } y < 1000 \\ \dfrac{y - 1000}{400} & \text{if } 1000 \le y \le 1400 \\ 1 & \text{if } 1400 \le y > 100000 \end{cases} \tag{6.28}$$

Various type of Output of doses of salazopyrine (mg/day) of membership functions are shown in Fig. 6.15.

Here, I convert the crisp data set of doses of salazopyrine into fuzzy data set using Eqs. (6.25)–(6.28). The fuzzy sets of doses of salazopyrine are given bellow:

$$\text{Low} = \{(1,0),(200,1),(300,0.81),(400,0.5),(500,0.3),(600,0)\}$$

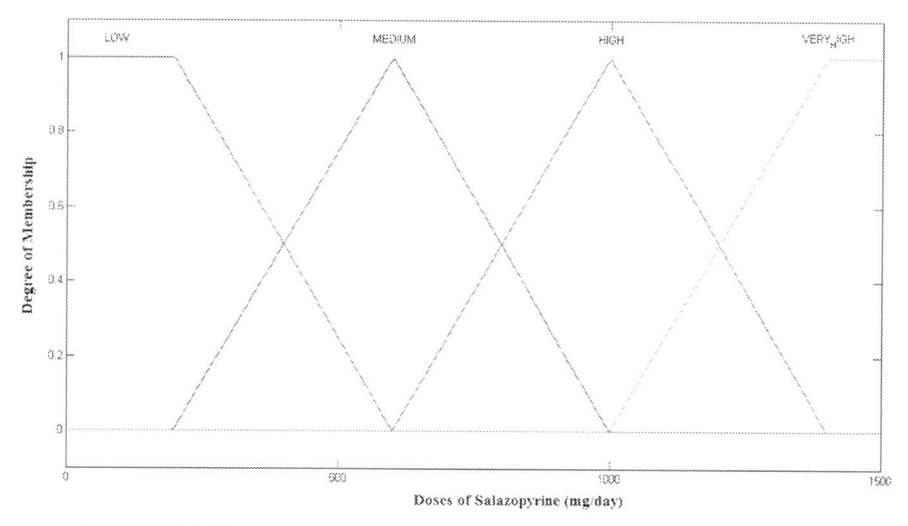

FIGURE 6.15 Membership function for the doses of salazopyrine (mg/day).

$$\text{Medium} = \{(200,0),(300,0.25),(400.0.5),(600,1),(800,0.5),(900,0.5),(1000,0)\}$$

$$\text{High} = \{(600,0),(700,0.25),(800,0.5),(1000,1),(1200,0.5),(1300,0.25),(1400,0)\}$$

$$\text{VeryHigh} = \{(1000,0),(1100,0.25),(1200,0.5),(1300,0.75),(1400,1),(1500,1)\}$$

The experimental result of fuzzy based intelligent system is shown in Table 6.5.

The architecture of the fuzzy-based intelligent system is shown in Fig. 6.16.

In this chapter, I have designed nine fuzzy rules for the calculation of the correct medicine dose of the patient. The nine fuzzy rules are as follows:

1. If Prostate Specific Antigen is LOW and sedimentation is LOW then doses of salazopyrine is LOW.
2. If Prostate Specific Antigen is LOW and sedimentation is MEDIUM then doses of salazopyrine is LOW.
3. If Prostate Specific Antigen is LOW and sedimentation is HIGH) then doses of salazopyrine is MEDIUM.
4. If Prostate Specific Antigen is MEDIUM and sedimentation is LOW then doses of salazopyrine is LOW.
5. If Prostate Specific Antigen is MEDIUM and sedimentation is MEDIUM then doses of salazopyrine is MEDIUM.
6. If Prostate Specific Antigen is MEDIUM and sedimentation is HIGH then doses of salazopyrine is HIGH.
7. If Prostate Specific Antigen is HIGH and sedimentation is LOW then doses of salazopyrine is MEDIUM.

TABLE 6.5 Experimental result obtained by expert doctor and proposed fuzzy based intelligent system.

	Input variables		Output variable (mg)
Patient	Prostate-Specific Antigen (PSA) (ng/mL)	Sedimentation (SD) (mm/h)	Drug dose suggested by the Proposed Method/Day
1	0.50	10	276.00
2	0.80	16	334.00
3	1.00	20	369.00
4	1.00	24	352.00
5	2.00	25	407.00
6	2.00	81	765.00
7	2.00	70	685.00
8	3.00	40	479.00
9	3.00	50	595.00
10	4.00	41	548.00
11	5.00	50	705.00
12	6.00	67	887.00
13	7.00	80	1080.0
14	7.50	75	1040.0
15	8.00	25	742.00
16	8.50	25	783.00
17	9.00	75	1203.0
18	12.0	80	1303.0
19	13.0	90	1330.0
20	15.0	90	1330.0

8. If Prostate Specific Antigen is HIGH and sedimentation is MEDIUM then doses of salazopyrine is HIGH.
9. If Prostate Specific Antigen is HIGH and sedimentation is HIGH then Doses of Salazopyrineis VERY HIGH.

The doses of salazopyrine obtained according to the amount of prostate specific antigen and sedimentation is shown in Fig. 6.17.

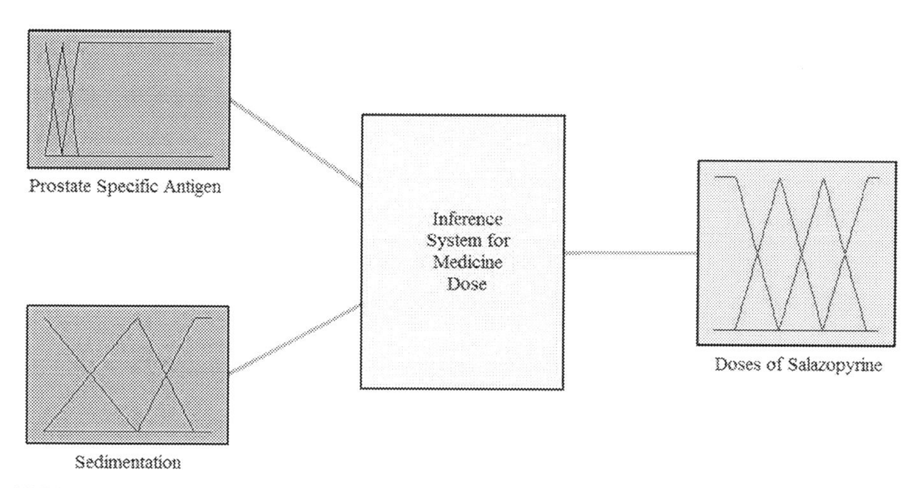

FIGURE 6.16 Architecture of fuzzy based intelligent system for calculation of medicine dose.

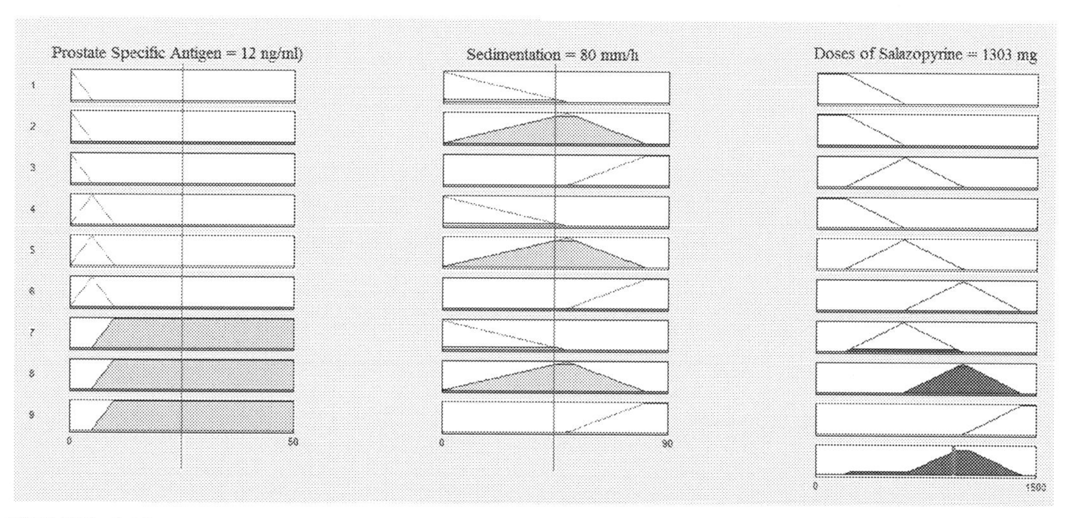

FIGURE 6.17 The doses of salazopyrine obtained according to the amount of prostate specific antigen and sedimentation.

Adjust the doses of the drug according to the prostate specific antigen and sedimentation is shown in Fig. 6.18.

Several types of variables based on temperature of the patient's disease, amount of drug in the blood, gender, weight, height, and age of patient are considered for the treatment. Therefore, all these variables play an important role in determining the drug dose of the patient.

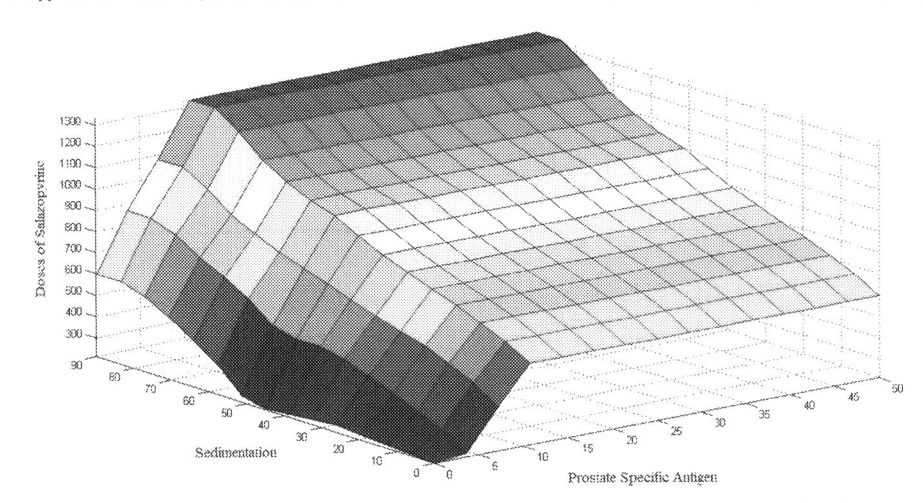

FIGURE 6.18 Adjust the doses of drug according to the prostate specific antigen and sedimentation.

5. Comparison of drug doses suggested by expert doctor and proposed fuzzy based intelligent system

Table 6.6 shows the comparative study of doses of drugs suggested by the physician and suggested by the proposed fuzzy-based intelligent system.

According to the expert doctor, finding sick patients with mild chronic intestines is a difficult task. I have divided the patients here into two parts as per the convenience.

Both parts of patients with chronic intestines are described as follows:

1. The patient with the first type of chronic intestine has been given a daily dose of the Salazopyrine drug by the doctor.
2. The second type of chronic bowel patient is given a daily dose of the Salazopyrine drug by a fuzzy expert logic based intelligent system.

From this study, it is found that it took about 1 year to 2 years to treat the patients in the first group. But in the second type of group of patients who were treated by a fuzzy logic-based intelligent system, all patients recovered in about three to 4 months. Thus, I can say here that determining the dose of the drug in patients with a fuzzy logic-based intelligent system can be a bold step. In the future, we can have this intelligent system installed in various hospitals in India to determine the dosage of medicine.

It is clear from Figs. 6.19 and 6.20 that the proposed fuzzy logic based intelligent system gives a very accurate answer to determine the dose of the drug of the patient, rather than the dose given by the doctor. It is also clear from this study that in many patients, the dose of the drug prescribed by the proposed intelligent system is much lower than that prescribed by the doctor. Therefore, I want to state here that the proposed method prevents the

TABLE 6.6 Experimental result obtained by expert doctor and proposed fuzzy based intelligent system.

	Input variables		Output variable (SL mg)	
Patient	Prostate-Specific Antigen (PSA) (ng/mL)	Sedimentation (SD) (mm/h)	Drug dose suggested by the Doctor/Day	Drug dose suggested by the Proposed Method/Day
1	0.50	10	500 (1 × 1)	276.00
2*	0.80	16	500 (1 × 1)	334.00
3	1.00	20	500 (1 × 1)	369.00
4	1.00	24	500 (1 × 1)	352.00
5*	2.00	25	500 (1 × 1)	407.00
6*	2.00	81	1000 (2 × 1)	765.00
7	2.00	70	1000 (2 × 1)	685.00
8*	3.00	40	1000 (2 × 1)	479.00
9*	3.00	50	1000 (2 × 1)	595.00
10	4.00	41	1000 (2 × 1)	548.00
11*	5.00	50	1000 (2 × 1)	705.00
12	6.00	67	1000 (2 × 1)	887.00
13*	7.00	80	1500 (3 × 1)	1080.0
14	7.50	75	1500 (3 × 1)	1040.0
15*	8.00	25	1500 (3 × 1)	742.00
16	8.50	25	1500 (3 × 1)	783.00
17*	9.00	75	1500 (3 × 1)	1203.0
18	12.0	80	1500 (3 × 1)	1303.0
19*	13.0	90	1500 (3 × 1)	1330.0
20	15.0	90	1500 (3 × 1)	1330.0

**From the above Table 6.6, out of the 20 patients, 10 chronic intestinal infections have been randomly selected to test the results obtained from the fuzzy logic-based intelligent system.*

patient from taking the wrong medication dose. In this way, the patient takes very little time to recover, and the patients also take less money to do medicine. Prescribing the dose of the drug by this method leads to fewer side effects on the patient. Therefore, it is said that in hospitals of India, using this type of intelligent based software will prove very beneficial for determining drug dose of the patient. Often in areas of complex and uncertain problems, intelligent systems based on fuzzy logic methods are used to properly treat diseases, analyze and evaluate laboratory and clinical data, and determine the correct dose of medicine to the patient.

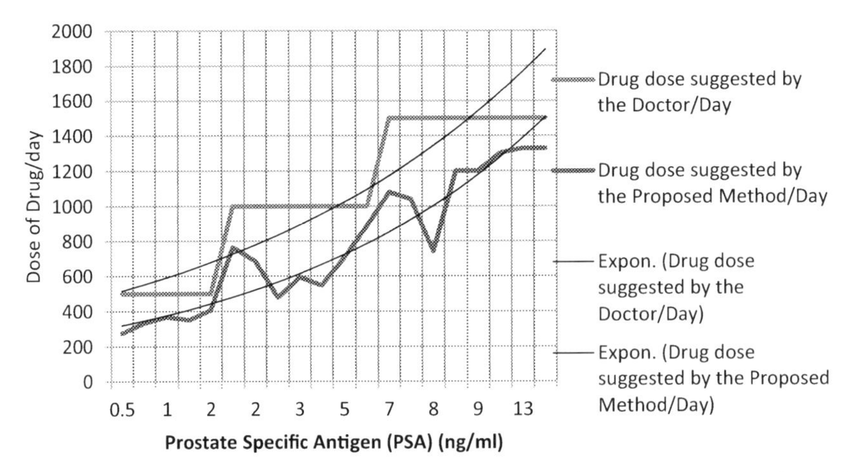

FIGURE 6.19 The doses of drug suggested by the expert doctor and proposed fuzzy logic-based intelligent system concerning Prostate-Specific Antigen (PSA) (ng/mL).

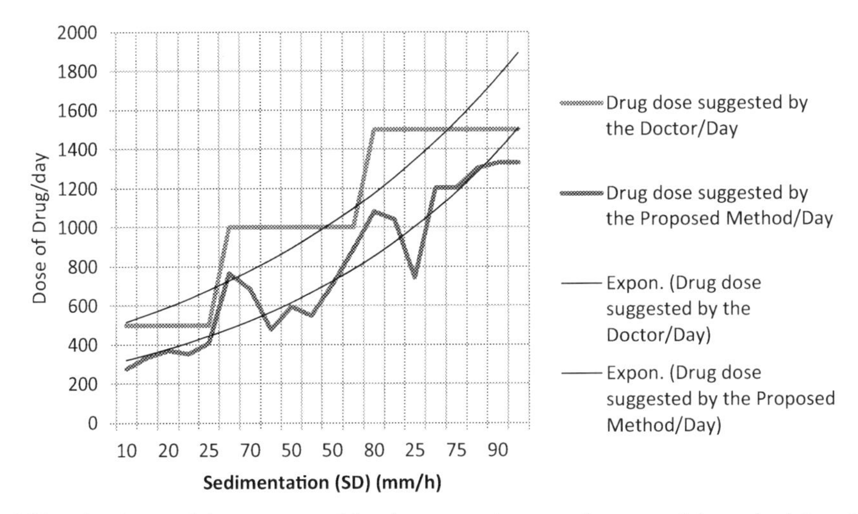

FIGURE 6.20 The doses of drug suggested by the expert doctor and proposed fuzzy logic-based intelligent system concerning sedimentation (SD) (mm/h).

The adjustment of the drug's dose is shown in Fig. 6.21 for the Prostate-Specific Antigen and Sedimentation. Fig. 6.10 also shows that the proposed fuzzy logic based intelligent system gives better results in comparison to doses of drugs prescribed by the physician.

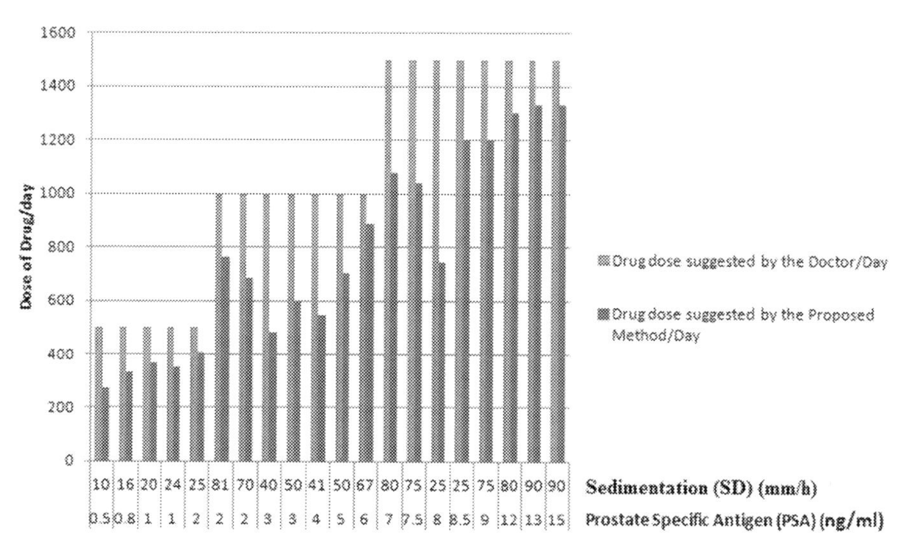

FIGURE 6.21 Adjust the doses of drug according to the prostate specific antigen and sedimentation.

6. Conclusion

The proposed intelligent system based on the fuzzy logic will prove very helpful in determining the dose of medicine given to patients with chronic intestinal infections. It is currently a prototype for prescribing a drug dose. Later on, intelligent systems based on the fuzzy logic will installed as software in various hospitals in India. Based on the results obtained from the proposed intelligent system, it can be said here that this method will in determining the prove to be very helpful dose of the drug in the future. Prescribing is an uncertain and complex the dose of medicine problem. Therefore, it can say here that with the help of this intelligent system, the solution to this complex and uncertain problem will be very good and the result will be very good. Through this intelligent system, in other diseases, the dose of medicine can be determined very well. From this study, it was known how the dose of medicine to be prescribed by medicine makers is determined. It has often seen that pharmaceutical companies prefer to make high capacity medicines to earn more money. This study also suggests that the production of different doses of medicines for diseases is much more efficient. Besides, the determination of the drug dose of patients admitted to intensive care is very important. Hence, the proposed intelligent system can be used to determine the dose of the drug in various diseases and good results can be obtained in the treatment of the patient.

In the future, if the number of input parameters and linguistic variables are increased, then an intelligent system based on a fuzzy logic method can be created that gives better results compared to this proposed method. A very good result will be obtained in determining the dose of the drug through this intelligent system.

In this proposed intelligent system, the experts have determined the fuzzy membership function. For this reason, sometimes the result obtained by this intelligent system is not effective for the patient. This is a major drawback of this intelligent system. In the future, to overcome this problem, fuzzy C-Means is used to create an intelligent system that will generate automatic membership function for the determination of drug doses of any patent. This automatic generation based intelligence will give better results as compared to the proposed intelligent system for determining drug doses of the patient in Indian hospitals.

References

[1] A.E. Gawedal, M.E. Brier, J.M. Zurada, Soft Computing Methods for Drug Dosing in Renal Anemia, Department of Medicine. University of Louisville, 2007, p. 177.

[2] A.E. Gawedal, A.A. Jacobs, M.E. Brierl, Fuzzy rule-based approach to automatic drug dosing in renal failure, in: IEEE International Conference on Fuzzy Systems, 2003, pp. 1206–1209.

[3] R. Bellazzi, C. Siviero, Predictive fuzzy controllers for drug delivery, Intell. Syst. Eng. (1994) 262–267, 5–9 September, Hamburg-Harburg, Germany.

[4] S.E. Kern, J.O. Johnson, D.R. Westenskow, Fuzzy logic for model adaptation of a pharmacokinetic-based closed loop delivery system for pancuronium, Artif. Intell. Med. 11 (1) (1997) 9–31.

[5] K. Kilic, B.A. Sproule, I.B. Türksen, C.A. Naranjo, Pharmacokinetic application of fuzzy structure identification and reasoning, Inf. Sci. 162 (2) (2004) 121–137.

[6] S. Oshita, K. Nakakimura, T. Sakabe, Hypertension control during anesthesia, IEEE Eng. Med. Biol. (1994) 667–670. November–December.

[7] K.K. Per, C.P. Lim, S.S. Quek, K.H. Khoh, Use of artificial neural networks to predict drug dissolution profiles and evaluation of network performance using similarity factor, Pharmaceut. Res. 17 (11) (2000) 1384–1388.

[8] P. Lada, M.E. Brier, J.M. Zurada, Therapeutic drug dosing prediction using adaptive models and artificial neural networks, in: IEEE International Joint Conference on Neural Networks, Washington, DC, USA, pp. 3699–3673, 1999.

[9] J.-S.R. Jang, C.-T. Sun, E. Mizutani, Neuro-Fuzzy and Soft Computing: A Computational Approach to Learning and Machine Intelligence, Prentic Hall, United State of America, 1997, pp. 165–166.

[10] R. Biswas, An application of fuzzy sets in students' evaluation, Fuzzy Sets Syst. 74 (2) (1995) 187–194.

[11] M. Nachtegael, D. van der Weken, E.E. Kerre, W. Philips, Soft Computing in Image Processing: Recent Advances, Sprnger, 2007, pp. 105–107.

[12] N. Allahverdi, Expert Systems. An Artificial Intelligence Application, Istanbul, Atlas, 2002, p. 248.

[13] N. Allahverdi, Fuzzy logic and systems, Retrieved March 21, 2007 from: http://farabi.selcuk.edu.tr/egitim/blanik/index.html, 2007.

[14] T.J. Ross, Fuzzy Logic with Engineering Applications, McGraw-Hill, New York, 1995, p. 151.

[15] L.-X. Wang, A Course in Fuzzy Systems and Control, Prentice-Hall, New York, 1997, pp. 424–425.

CHAPTER

7

Multiobjective optimization technique for gene selection and sample categorization

Sunanda Das[1], *Asit Kumar Das*[2]

[1]Department of Computer Science and Engineering, SVCET, Chittoor, Andhra Pradesh, India;
[2]Department of Computer Science and Technology, IIEST, Howrah, West Bengal, India

1. Introduction

Different data-mining techniques face a challenging problem, known as the curse of dimensionality [1], due to the high dimension of the dataset. To overcome such a problem, feature selection and dimension reduction techniques act as an important prior step in data mining and pattern recognition research area to handle the data efficiently. Feature selection algorithm selects only a subset of features that provides the same probability distribution as obtained considering all features of the dataset. Thus the goal of the proposed feature selection method is to find out the least number of genes necessary to provide the maximum accuracy while classifying the samples using those genes only. Feature selection algorithms are mainly of two types: wrappers and filters methods. The filter based feature selection method is independent to the learning algorithm and it selects the important feature subset using only the dataset. RELIEF [2] and its enhanced versions [3] are all the filter based feature selection techniques. Nonparametric methods [4] are used specially for feature selection based on mutual information between continuous valued features. The wrapper feature selection method [5,6] uses different supervised learning algorithms [7–9] to find out the important feature subset. Though the wrapper feature selection methods are expensive compared to the filtering techniques, they are more useful for providing better accuracy of the learning process.

Selection of important genes from a high dimensional microarray dataset is a kind of feature selection problem, which selects the genes relevant to the sample categorization or disease identification. In literature [10], many gene selection methods are available with their own pros and cons. One paper [10] uses the concepts of rough set theory (RST) [11,12] and

Handbook of Computational Intelligence in Biomedical Engineering and Healthcare
https://doi.org/10.1016/B978-0-12-822260-7.00008-X

generates a minimal spanning tree to select important genes. The RST is used to handle the incomplete, vague or imprecise data. Another gene subset selection method in paper [13] is proposed integrating RST and genetic algorithm (GA). In Ref. [14], a feature selection algorithm is proposed to select statistical rank based informative gene subset and RST is used on this informative gene subset to select only the genes with high sample classification accuracy. Unlike other multiobjective GA [16], another paper [15] proposed a Multiobjective GA based gene selection technique with relatively less time which does not require any global calculation. In this method [15], nonlinear uniform hybrid cellular automata [17] are used for the creation of population and steady state is chosen as selection method. In Ref. [18], a stepwise forward selection algorithm shows the graph generation with correlation coefficient as the weight of the edge between two terminal nodes of the edge. Correlation calculation and feature selection are the two phases of this method. In the first phase, correlation among the features is calculated and set it as the weight of the edge in the graph. In the second phase, the most informative feature (i.e., node) out of current set of features is selected and removes from the graph. The process is repeated until the graph becomes a null graph, resulting a set of selected features as the reduct. The method may contain some extraneous features, as the combination of two or more selected features may imply the other selected feature. In Ref. [19], multiobjective genetic algorithm is used for simultaneous feature selection and cluster formation with known cluster numbers. Ref. [20] uses the particle swarm optimization [21] algorithm for the same purpose. In Ref. [22], a comprehensive survey on feature selection approaches for clustering. Several mixture-based models are mentioned in Refs. [23,24]. In Ref. [25], both feature selection and cluster formation mixture based model is proposed where the features are independent. In Ref. [26], the Bernoulli model [27–29] is introduced for developing an efficient mixture based model for simultaneous clustering and feature selection.

In this proposed work, Improved Strength Pareto Evolutionary Algorithm (SPEA2) [30] is used to select the minimum number of genes sufficient for sample clustering. The flowchart of the proposed work is illustrated in Fig. 7.1. Here, the samples are partitioned based on their class label. Next, a population of chromosomes is randomly generated using "0" and "1" so that the size of each chromosome is equal to the number of genes in the samples. This population is used as the initial solution of the proposed SPEA2 [30] algorithm. To compute the fitness of the chromosomes, a subdataset is prepared selecting the genes corresponding to which the binary bit "1" is present in the chromosome. Then, a Cosine Similarity Matrix is generated for the subdataset based on cosine similarity values between every pair of samples for each chromosome. Then using the Enhanced Cluster Affinity Search Technique (E-CAST) [31] clustering algorithm, samples are clustered finding the clique graph [32,33]. In the clique graph, each complete graph represents a cluster with every node (i.e., sample) of similar type. Cosine similarity calculations of the samples are the measurement for calculating similarity. There is *connectivity threshold* in each cluster and *affinity* of each member/sample to a particular cluster according to E-CAST algorithm. The connectivity of a node/sample is dependent on affinity of that node. Within the cluster, if the affinity of a node is greater than or equal to the connectivity threshold of that cluster, then that node/sample becomes high connectivity node to a cluster otherwise it is a low connectivity node to that cluster. The formation of cluster will go on until no changes of samples will occur in the cluster.

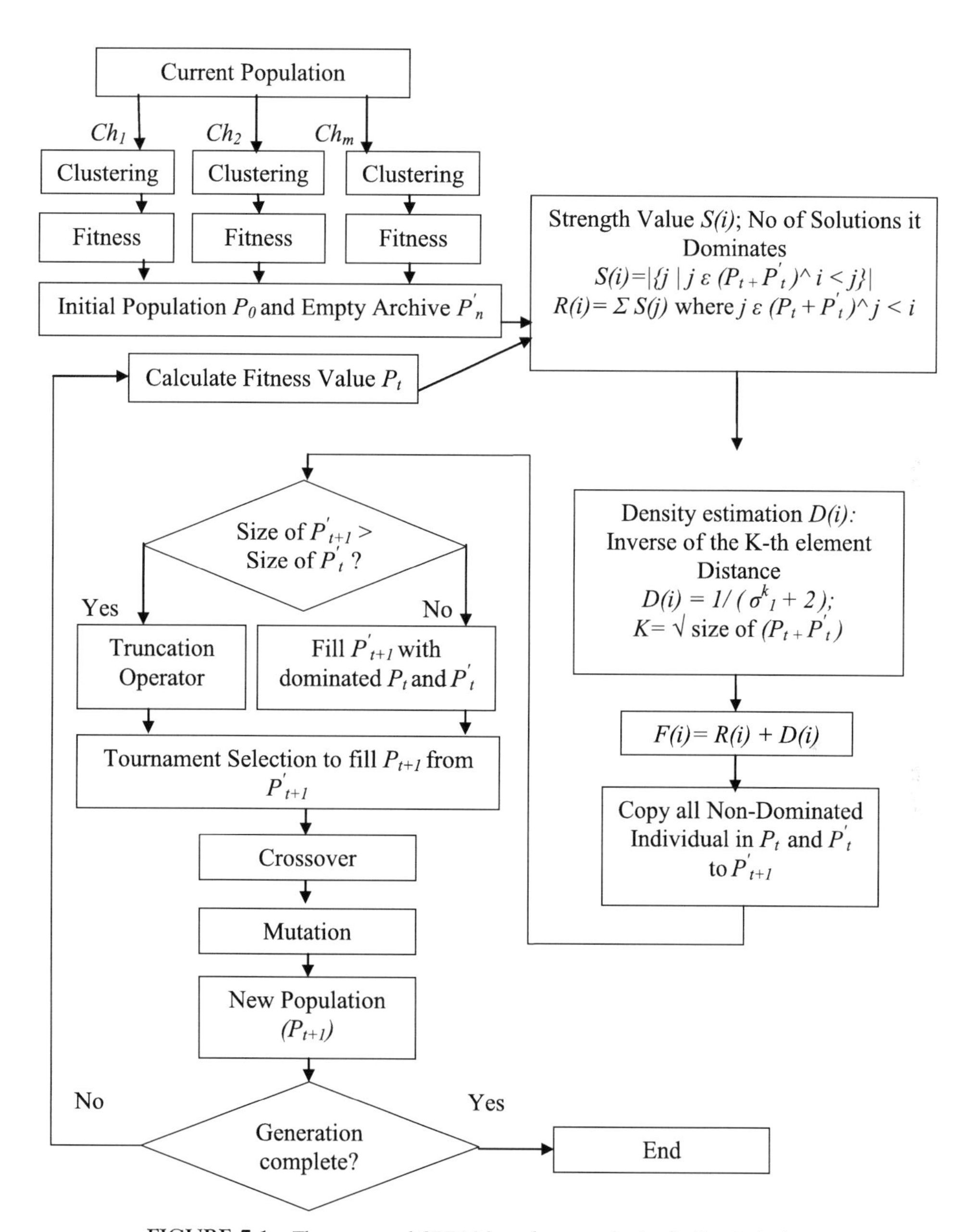

FIGURE 7.1 The proposed SPEA2 based gene subset selection technique.

The proposed SPEA2 based gene selection method uses three compatible objective functions to evaluate the fitness of chromosomes using: (1) gene numbers of each data subset, (2) F-Measure, which is an external cluster validation index between two sets of clusters, and (3) intracorrelation among the genes of the associated data subset. Among these objective functions, the first one has to minimize, the second one has to maximize and third one has to minimize. The basic operations of the GA will continue until it becomes converge to the optimal solution. After the convergence, the best chromosome is chosen from nondominated front of external population. In the best chromosome, the genes having binary bits "1" are taken as mostly significant genes. And the clusters generated for this best chromosome are the optimal clusters of samples.

The remaining tasks of the proposed method are mentioned as: The details of the proposed method for optimal gene selection are described in Section 2. The experimental results obtained from the proposed method are mentioned in Section 3. Then finally, a summary is concluded in Section 4.

2. Gene subset selection

Here, a SPEA2 [16] based novel feature selection algorithm is drawn to find optimal features. At first, a set of initial population is generated randomly. Each chromosome is a binary string. A nondominated pareto front is selected to identify important genes in chromosomes having minimum number of genes. The minimization of gene numbers for sample categorization, maximization of the external cluster index, and minimization of intracorrelation among predicting attributes are the three objective functions of SPEA2.

2.1 Initial population

The chromosomes of initial population are considered as initial solutions. Then in every iteration chromosomes are modified and thus the entire range of possible solutions in search space is obtained. The population size is determined under consideration from different viewpoints by many researchers, though the main focus is directed toward a trade-off between efficiency and effectiveness. The larger population size signifies sufficient space for exploring the search space effectively, whereas, small the size would not allow the efficiency of the proposed method. A population is randomly generated where the chromosomes as same length as sample size are taken as binary. The binary value "1" means that gene is present otherwise absent. So, each chromosome represents a data subset considering genes with corresponding bit in chromosome as "1" by applying vertical projection on the original dataset. This data subset is used for estimating the fitness of the associated chromosome. The chromosome having best fitness value is the optimal solution and the genes in present in that chromosome are the important genes.

2.2 Fitness function

The quality of the chromosomes in the population is estimated mainly by the fitness function. It helps for finding the optimal solution of the problem and for achieving the fast convergence of the GA. The fitness functions used here are: (1) the minimization function, i.e., the number of genes corresponding to the associated subdataset of the chromosome, (2) the maximization function, i.e., external index of two set of clusters, and (3) the minimization function, i.e., intracorrelation among the reduced features of data.

The computation of external cluster validation index is mentioned here:

(1) Each data subset is categorized firstly into several groups based on the sample types; say, $CL_1 = \{Cl_{11}, Cl_{12}, \ldots, Cl_{1s}\}$ cluster set is generated, where s denotes sample types.
(2) Another cluster set CL_2 is generated among sample pairs using Cosine Similarity Matrix as in Eq. (7.1).

$$Cos\_similarity(A,\ B) = \frac{\sum_{i=1}^{n} A_i B_i}{\sqrt{\sum_{i=1}^{n} A_i 2}\sqrt{\sum_{i=1}^{n} B_i 2}} \tag{7.1}$$

Where, A_i and B_i are i- th gene in A and B samples respectively. Here, the higher similarity value indicates more similarity between the samples and the lower similarity value indicates dissimilarity between the samples. Euclidean distance and Pearson correlation are also belongs to similarity measures like cosine similarity measures. Here, the E-CAST clustering algorithm is used to cluster the samples using the concept of the clique graph and divisive clustering algorithm. An affinity threshold T is taken as input which helps the size and quality of cluster. Affinity threshold T having hard coded predetermined value, (as in CAST [34]) can also be considered,but the dynamic computation is more preferable. In this work, dynamic affinity threshold is generated using the similarity values of unclustered objects. Based on similarity matrix S ($S(i,j)$ ε [1, 0]), we compute the affinity of a node/sample x to a particular cluster C as$a(x) = \sum_{k \varepsilon C} S(x, k)$. In each cluster there is a connectivity threshold $\chi = T\ |C|$, where $|C|$ indicates element numbers within the cluster C. Higher or equal affinity value of a node in compared with connectivity threshold in same cluster implies high connectivity node in that cluster. Then clusters are generated based on three operations mentioned here:

(a) Node Addition: Anode which is of highest connectivity, has to be added to the cluster.
(b) Node Deletion: Then a node having low connectivity is needed to remove from a cluster.
(c) Cluster Clearing: Any node belongs to low connectivity is also needed to remove and added to another cluster where it has highest connectivity or affinity. Though, every time this cluster clearing is not required.

Thus, a set of clusters $CL_2 = \{Cl_{21}, Cl_{22}, \ldots, Cl_{2t}\}$ are generated after *E-CAST*, where t is the cluster numbers.

(3) Next, F-Measure, the external cluster validation index, is calculated among two cluster sets, CL_1 and CL_2.The value of F-Measure is taken in range of [0, 1]. F-Measure having more high values implies that both the clustering algorithms have generated similar clusters.

The third fitness function is computed using Eqs. (7.2) and (7.3) to minimize the intracorrelation among the data objects.

$$f3(x) = \frac{1}{MC(x)} \tag{7.2}$$

$$MC(x) = \frac{1}{C(|X|,2)} \sum_{i=1}^{|X|} \sum_{j=i+1}^{|X|} |corr(xi, xj)| \tag{7.3}$$

Where, $f3(x)$ is the nonredundancy, $MC(x)$ is the intracorrelation in X, $C(|X|,2)$ is the number of 2-combinations of the data subset X associated to the chromosome, and $corr\ (x_i,x_j)$ is the correlation between two features x_i and x_j of the associated data subset.

2.3 External population

In SPEA2 [16], the external population P' is generated along with the current population P generation to get the best chromosome (s) for next generation. This helps to maintain the quality of GA instead of degrading. But, the size of the internal population is always higher than the external population. SPEA2 is advanced form of SPEA. In SPEA2, truncation operation is used and external population number is constant here.

Firstly, all the nondominated solutions are considered for the external population. Then, if the number of solution taken into the external population exceeds the size of the external population then truncation operation is used. Otherwise, best dominated solutions comes in external population from the previous and the external population. To find all nondominated members, we have to find out all members from the combined population and external population where fitness value $F(i) < 1$ i.e., $P't + 1 = \{i|i\varepsilon Pt + P't\hat{F}(i) < 1\}$

If $|\mathbf{P}'_{t+1}|$ value equals the External Population Size ($\mathbf{N}'$), then no need to do anything.

But, if $|\mathbf{P}'_{t+1}|$ value is less than the External Population Size ($\mathbf{N}'$), then the best dominated solutions will come in $\mathbf{N}'$ - $|\mathbf{P}'_{t+1}|$ from the previous population and the external population, i.e., copy the first $\mathbf{N}'$-$|\mathbf{P}'_{t+1}|$ members with $F(i) \geq 1$. But, If $|\mathbf{P}'_{t+1}| >$ External Population Size ($\mathbf{N}'$), external population size is reduced by the truncation operator mentioned in subsection E.

2.4 Selection

A part of the current population is participated (https://en.wikipedia.org/wiki/Selection_%28genetic_algorithm%29) in each generation for a new population generation. There are lots of selection methods like rank selection, roulette wheel selection, tournament selection, etc. The objective of such selection methods are to select the best chromosomes mating pool generation. Here, a binary tournament selection method is considered based on the chromosome's strength as likely as the SPEA2 algorithm. Strength of a member i, i.e., $S(i)$ signifies the number of solutions in both current population and external population to which it dominates, as mentioned in Eq. (7.4).

$$S(i) = \{j|j\varepsilon Pt + P't \wedge i \neq j\} \tag{7.4}$$

"$i \neq j$" implies that the member i dominates member j.

Then we calculate the raw fitness value $R(i)$ for each solution using Eq. (7.5).

$$R(i) = \sum_{j \in Pt+Pt' \wedge j \neq i} S(i) \tag{7.5}$$

"$j \neq i$" implies that the member j dominates member i, i.e., raw fitness is determined by its dominators in both current and external population. If $R(i)$ value is less, then it means that it is dominated by very few of other solutions.

Finally, the density estimation technique helps to handle the situation when most of the solutions are nondominated to each other. Here, the distance of every i-th solution to other solutions which are present in both initial population and external population is calculated and placed in increasing order of their distance values. Then, the distance σ_i^K at K^{th} entry of the ordered list is considered where, $K = \sqrt{N + N'}$, where $\mathbf{N}$ and $\mathbf{N'}$ are the current and external population sizes respectively. The density of each solution i is generated by Eq. (7.6), where, addition of "2" is for keeping the value of $D(i)$ within the range [0, 1].

$$D(i) = \frac{1}{\sigma_i^k + 2} \tag{7.6}$$

The fitness function is defined by Eq. (7.7).

$$F(i) = R(i) + D(i) \tag{7.7}$$

In SPEA2, binary tournament selection with replacement is performed only from the external population to create the mating pool.

2.5 Truncation operator

Here, in every iteration, the individual solution i is considered for removal for which $i \leq_d j$ for all j ε $P'_{t+1}$with

$$i \leq_d j : \Leftrightarrow \forall 0 < k < \left|P'_{t+1}\right| : \sigma_i^k = \sigma_j^k \vee$$

$$\exists 0 < k < \left|P'_{t+1}\right| : \left[\left(\forall 0 < l < k : \sigma_i^l = \sigma_j^l \right) \wedge \sigma_i^k < \sigma_j^k \right]$$

Where, σ_k^i indicates the distance of i-th solution to its k-th nearest neighbor in P'_{t+1}. In another way, the individual with the minimum distance is considered at each stage. For several individuals having minimum distance, the second smallest distance is chosen and so on which is shown in Figs. 7.2A and B.

2.6 Crossover

The fittest chromosomes only from the current population P are taken in the selection process. The offsprings generated from crossover operations helps for searching the better solutions toward local optima in the whole search space. Crossover is one of the basic operations of GA. Crossover operation is mainly occurred in every generation for convergence.

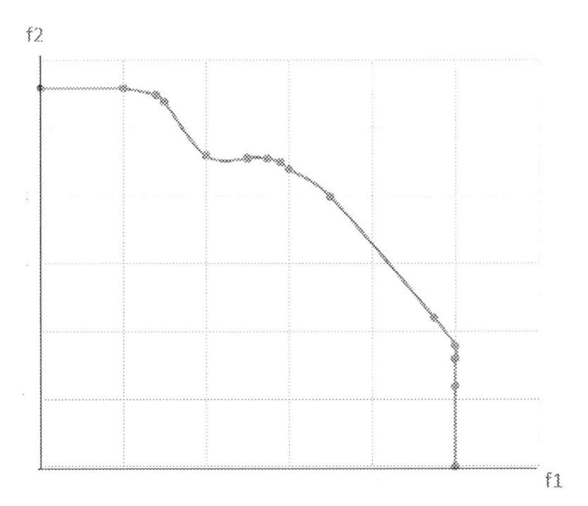

FIGURE 7.2-A SPEA2 with list of nondominated solutions.

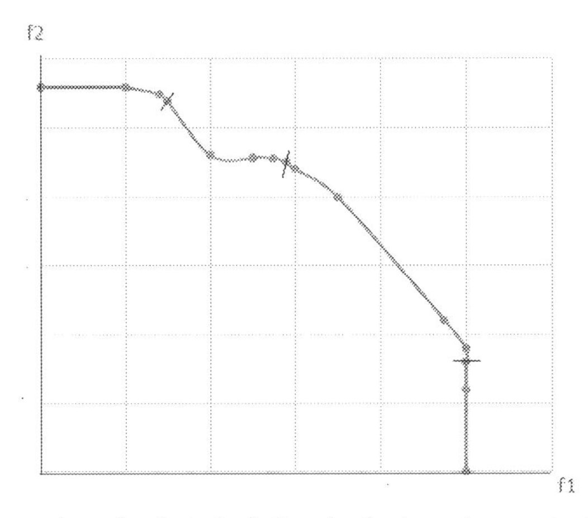

FIGURE 7.2-B Removed nondominated solutions by the truncate operator (assuming that N′ = 5).

Crossover and mutation processes are the main reason for newly chromosome generation. One point crossover, uniform crossover, multi point crossover belong to several crossover techniques. In SPEA2 we have used two-point crossover on the members present in the mating pool to generate the offspring.

2.7 Mutation

Mutation is another basic operation after crossover operation. It is a divergence operation. It may be required sometime to move few solutions from local optimum space to find better optima space. As mutation operator does not happen frequently, so it performs on very less number of chromosomes in the population. In SPEA2 we have applied mutation operator on the members present only in the mating pool.

2.8 Elitism

Elitism causes the GA to converge on local optima instead of the global optimum, so pure elitism is a race to the nearest local optimum. Here, the better chromosomes are put into the next population directly. So, it reduces the time by skipping the finding of previously discarded partial solutions. In this method, elitism generates best nondominated solutions of chromosomes in external population set.

The proposed GSPEA2 algorithm is mentioned here,

Algorithm: GSPEA2
/*Let, $D = (S, A)$ = Dataset, where S is Sample and A is Gene set */
Input: P = Initial Population of M x N, N' = External Population
G_{Max} = Maximum generation number
$P_{Crossover}$ = Probability of crossover and $P_{Mutation}$ = Probability of mutation
Output: Optimal_gene_subset and Optimal_clusters
BEGIN
Initialize the binary population P randomly of size M x N, where N denotes chromosome length.
Do {
For every chromosome $ch \in P$ **do**
{
Let, n_{ch} = Total 1's present in ch;
F_{ch} = Gene subset having ch-value as 1 in ch;
D_{ch} = Data subset from vertical projection of D for F_{ch};
Compute f_1 = *Number of 1's in* D_{ch};//first objective function
Generate a set of clusters $CL_1 = \{Cl_{11}, Cl_{12}, \ldots, Cl_{1|S|}\}$ from D_{ch} based on the given sample types;
Apply E-CAST clustering algorithm to generate clusters $CL_2 = \{Cl_{21}, Cl_{22}, \ldots, Cl_{2t}\}$ using cosine and cluster similarity indices;
Compute f_2 = Validation index for external clusters among CL_1 and CL_2.
Compute f_3 = Correlation between the features of D_{ch};
}
Take GenerationCount as 0;
Using the fitness functions f_1, f_2, and f_3, search for non-dominated solutions;
In the external population P'_{t+1}, place all non-dominated solutions from current population P_t and external population P'_t;
If (Non-dominated solutions in $P'_{t+1} < N'$) {
Fill N'-$|P'_{t+1}|$ members from $|P'_{t+1}|$ by best dominated solutions from P_t and P'_t
}
Else {
Reduced the size of the external population by truncation operator;
Do {
Apply binary tournament selection with replacement on external population P'_{t+1} to fill/create new mating pool/population P_{new} of selected chromosomes;
} **Until** (|P| = = |Pnew|);
Do {

Try two-point crossover with probability $P_{crossover}$ on any two random chromosomes;
Generate two offspring chromosomes;
For the parent and offspring chromosomes, calculate fitness values;
Choose two chromosomes having the best fitness values;
} **Until** (| P_{New} | /2 time crossover operation);
For every *ch* in P_{New} **do** {
Try mutation having probability $P_{Mutation}$;
Calculate fitness value on offspring chromosome after mutation;
Choose the best chromosome among parent and muted chromosome;
}
Increase the *GenerationCount* by 1;
} **Until** (*GenerationCount* == G_{Max});
Take the non-dominated individuals in the external population P'_{t+1};
OptimalGeneSubset = Genes present in ch_{Best};
OptimalClusters = Clusters obtained using E-CAST method for ch_{Best};
END

3. Results and discussions

The proposed GSPEA2 method uses the concept of improved Strength Pareto Evolutionary Algorithm (SPEA2), which is modified version of Strength Pareto Evolutionary Algorithm (SPEA). There are some differences of SPEA2 from SPEA [34], these are:

- ➢ The fitness assignment is improved here, by which numbers of individuals are accounted on which individuals it dominates and it is dominated by.
- ➢ An improved nearest neighbor density estimation technique is considered which improves the searching process.
- ➢ Another feature of SPEA2 is the new archive truncation method which assures the preservation of boundary solutions.

Table 7.1 shows the environmental set up for running the proposed GSPEA2 method.

Feature selection method actually searches the best feature subset among all possible combination of features in the data. It uses different attribute evaluator and search methods for selection of important features. There are variety of attribute evaluator and search methods; any combination of these can be used to determine the score and rank of attributes in the dataset. Here, various attribute evaluation techniques are considered for the comparison

TABLE 7.1 Parameter of Proposed GSPEA2 method.

Parameter	Value
Population size (**M**)	200
Probability of crossover ($P_{Crossover}$)	0.80
Probability of mutation ($P_{Mutation}$)	0.10
Maximum No. of successive generation (G_{Max})	500

purpose like Symmetrical Uncertain (SYU), Relief (RE), OneR (OR), Information Gain (IG), Gain Ratio (GR), and Correlation (CO) based techniques for attribute evaluation, which allow arbitrary combinations of search and evaluation methods. The proposed gene subset selection method (GSPEA2) shows how efficiently it can generates the best reduced gene subset for correct prediction of cancer. A comparison of the proposed method with other methods is listed in Table 7.2.

Summary of these datasets have been presented in Table 7.3.

WEKA tool [35] is used here to calculate the classification accuracy of all the feature selection techniques along with the proposed method. The 10-fold cross-validation results obtained by different methods for Case I to IV are listed in Tables 7.4–7.7, respectively. It should be noted that a classifier is removed when all subset selection methods provide same results. From the tables, it is observed that, on average, better accuracy is obtained after reducing the datasets by the proposed GSPEA2 method, which indicates that the selected genes are important. Also from Table 7.2, we notice that the proposed algorithm removes more than 90% genes as redundant. Therefore, we can conclude that, the algorithm selects very few genes as important genes and gives better accuracy by most of the classifiers compare to considering whole genes in the datasets and gene subset obtained by various feature selection algorithms. This demonstrates that the proposed gene selection algorithm is effective. At the same time, to measure the effectiveness of the algorithm for sample clustering, the F-Measure values for all datasets are computed and listed in Table 7.8. This result shows that *GSPEA2* method gives better F-Measure values for all datasets. Thus, it proves that *GSPEA2* method is more powerful method with respect to statistical analysis for simultaneous gene subset selection and sample categorization.

4. Conclusion and future work

A new gene subset selection and sample categorization method based on improved strength pareto evolutionary algorithm is proposed (GSPEA2). The method is experimented on various datasets and results are compared with various feature selection and classification algorithms. The results show that the proposed method selects very less number of genes and categorizes the samples more accurately, which demonstrates the effectiveness of the method. The proposed method can be extended for dynamic clustering in case of streaming dataset. When a new chunk of data comes, if the whole data, i.e., both the previous and new data are considered, the GA will take much more time for gene selection and clustering of samples. We can select the representatives of all previously generated clusters effectively using some statistical methods and consider all newly generated data to run the GSPEA2 algorithm and organize the outcomes to obtain the important genes and clusters of samples for the whole dataset, which is the future scope of the chapter.

TABLE 7.2 Summary of microarray dataset and the subset selection results by the GSPEA2 method.

	Subset selection method						
Dataset	**GSPEA2**	**SYU**	**RE**	**OR**	**IG**	**GR**	**CO**
Colon	**165**[a]	204	186	191	209	213	188
Lukemia	**244**	267	287	284	258	246	264
Lung cancer	**256**	289	312	316	301	307	342
Pros-train	212	247	238	322	210	**208**	222

[a]*Bold set is the minimum number of subset selected for each dataset.*

TABLE 7.3 Datasets summary.

Dataset	No. Genes	No. Samples	No. Classes
Colon tumor	2,000	62	2
Leukemia	7,129	38	2
Lung cancer	12,533	32	2
Pros-train	12,600	102	2

TABLE 7.4 Cross-validation results for colon tumor.

	Subset selection method						
Classifier	**GSPEA2 (%)**	**SYU (%)**	**RE (%)**	**OR (%)**	**IG (%)**	**GR (%)**	**CO (%)**
Nbmn	**88.64**[a]	**88.64**	85.64	85.48	**88.64**	**88.64**	**88.64**
Nbmnu	**87.09**	**87.09**	**87.09**	85.48	**87.09**	**87.09**	**87.09**
Logistic	**84.12**	72.58	83.87	75.80	72.58	72.58	72.58
Smo	86.87	83.87	82.26	83.87	83.87	83.87	**87.09**
Lwl	77.19	**77.41**	74.19	**77.41**	**77.41**	**77.41**	72.58
Kstar	65.61	64.52	64.52	64.52	**77.42**	64.52	64.52
Ibk	88.34	80.64	82.25	**88.70**	80.64	80.64	72.58
Mulcls	**85.87**	72.58	83.87	75.80	72.58	72.58	72.58
Vote	**64.52**	64.52	64.52	64.52	64.52	64.52	64.52
Decision stumb	**79.26**	77.41	70.96	79.03	77.41	77.41	70.96
Reep tree	**81.31**	77.41	79.03	75.80	77.41	77.41	79.03
Lmt	**87.24**	80.64	75.80	79.03	80.64	80.64	83.87
Jrip	**80.65**	**80.65**	83.87	79.03	79.03	64.51	70.96
OneR	**74.19**	**74.19**	69.36	**74.19**	**74.19**	**74.19**	69.35

[a]*Bold set(s) represent the best results(s) for each classifier.*

TABLE 7.5 Cross-validation results for leukemia.

Classifier	Subset selection method						
	GSPEA2 (%)	SYU (%)	RE (%)	OR (%)	IG (%)	GR (%)	CO (%)
Logistic	100	94.73	97.36	97.36	94.73	100	100
Smo	100	100	97.36	97.36	97.36	97.36	100
Lwl	92.57	89.47	92.10	89.47	92.10	94.73	94.73
Ibk	97.36	97.36	94.73	100	100	97.36	100
Mulcls	100	94.73	97.36	97.36	94.73	100	100
Decision stumb	98.36	97.37	94.74	94.74	97.37	97.37	97.37
Reep tree	96.84	94.73	89.47	92.10	94.73	94.73	92.10
Lmt	90.84	94.73	92.10	92.10	94.73	94.73	94.73
Jrip	91.57	97.36	94.73	94.73	97.36	97.36	97.36
Oner	84.21	97.37	94.75	94.75	97.36	97.36	97.36

TABLE 7.6 Cross-validation results for lung cancer (32 samples, 256 Genes).

Classifier	Subset selection method						
	GSPEA2 (%)	SYU (%)	RE (%)	OR (%)	IG (%)	GR (%)	CO (%)
Lwl	**97.87**	93.75	96.87	87.50	87.50	84.37	90.62
Decision stumb	**96.98**	93.75	93.75	96.87	96.87	96.87	93.75
Reep tree	**97.07**	93.75	93.75	96.87	96.87	96.87	93.75
Lmt	**100**	96.87	96.87	**100**	96.87	96.87	96.87
Jrip	97.87	96.87	96.87	96.87	93.75	96.87	**100**
Oner	97.75	93.75	93.75	100	96.87	**100**	93.75

The bold indicates the best classification accuracy. Suppose in Table 7.6, GSPEA2 gives best classification accuracy (97.87%) for the classifier LWL than other methods.

TABLE 7.7 Cross-validation results for pros-train.

	Subset selection method						
Classifier	GSPEA2 (%)	SYU (%)	RE (%)	OR (%)	IG (%)	GR (%)	CO (%)
Nbmn	82.35	82.35	**83.33**	82.35	80.23	82.35	82.35
Nbmnu	**86.15**	**86.15**	83.33	**86.15**	**86.15**	81.42	**86.15**
Logistic	**91.21**	88.23	87.25	83.33	83.33	91.17	89.21
Smo	96.11	94.11	96.07	94.11	92.15	**97.05**	94.11
Lwl	**87.28**	86.27	86.27	86.27	87.25	84.31	85.29
Ibk	91.79	94.11	93.13	90.19	93.13	**96.07**	93.13
Mulcls	**91.21**	88.23	87.25	83.33	83.33	91.17	89.21
Decision stumb	**87.25**	86.27	86.27	86.27	86.27	86.27	86.27
Reep tree	**88.23**	85.29	83.33	85.29	85.29	86.27	82.35
Lmt	**96.17**	92.15	94.11	91.17	91.17	91.17	96.07
Jrip	**89.25**	88.23	85.29	89.21	84.31	86.27	89.21
Oner	**83.35**	83.33	82.35	83.33	82.35	83.33	82.35

The bold indicates the best classification accuracy.

TABLE 7.8 Statistical analysis to measure cluster quality using the GSPEA2 method.

Dataset	F-measure	Dataset	F-measure
Colon tumor	96.67	Lukemia	95.89
Lung cancer	97.54	Pros-train	97.73

References

[1] T. Hastie, R. Tibshirani, J. Friedman, The Elements of Statistical Learning, Springer, 2001.

[2] K. Kira, L. Rendell, The feature selection problem: traditional methods and a new algorithm, in: Proc. 10th Nat'l Conf. Artificial Intelligence (AAAI-92), 1992, pp. 129–134.

[3] I. Kononenko, Estimating attributes: analysis and extensions of RELIEF, in: Proc. Seventh European Conf. Machine Learning, 1994, pp. 171–182.

[4] N. Kwak, C.H. Choi, Input feature selection by mutual information based on Parzen Window, IEEE Trans. Pattern Anal. Mach. Intell. 24 (12) (2002) 1667–1671.

[5] R. Kohavi, G. John, Wrappers for feature subset selection, Artif. Intell. 97 (1–2) (1997) 273–324.

[6] G. John, R. Kohavi, K. Pfleger, Irrelevant features and the subset selection problem, in: Proceedings Fifth International Conference on Machine Learning, New Brunswick, NJ (Morgan Kaufmann, Los Altos, CA, 1994, pp. 121–129.

[7] R. Duda, P. Hart, Pattern Classification and Scene Analysis Wiley, 1973 (New York).

[8] I.J. Good, The Estimation of Probabilities: an Essay on Modern Bayesian Methods, MIT Press, Cambridge, MA, 1965.

[9] P. Langley, W. Iba, K. Thompson, An analysis of Bayesian classifiers, in: Proceedings AAAI-94, AAAI Press and MIT Press, Seattle, WA, 1992, pp. 223–228.
[10] S.K. Pati, A.K. Das, Constructing minimal spanning tree based on rough set theory for gene selection, Int. J. Artif. Intell. & Appl. (IJAIA) 4 (1) (January 2013).
[11] S.K. Pal, A. Skowron, Rough Fuzzy Hybridization, Springer, 1999.
[12] Z. Pawlak, Rough Sets – Theoretical Aspects of Reasoning about Data, Kluwer, Boston, London, Dordrecht, 1991.
[13] A.K. Das, et al., Appling restrained genetic algorithm for attribute reduction using attribute dependency and discernibility matrix, in: ICIP 2012, CCIS, vol. 292, 2012, pp. 292–308.
[14] A.K. Das, S.K. Pati, Gene Subset selection for cancer classification using statistical and rough set approach, in: SEMCCO 2012, LNCS, vol. 7677, 2012, pp. 294–302.
[15] S.K. Pati, A.K. Das, A. Ghosh, Gene selection using multi-objective genetic algorithm integrating cellular automata and rough set theory, in: SEMCCO 2013, Part II, LNCS, vol. 8298, 2013, pp. 144–155.
[16] E. Zitzler, L. Thiele, Multiobjective evolutionary algorithms: a comparative case study and the strength Pareto approach, IEEE Trans. Evol. Comput. 3 (4) (2007).
[17] C. Moore, Predicting nonlinear cellular automata quickly by decomposing them into linear ones, Physica D 111 (1–4) (1998) 27–41.
[18] A. Basak, A.K. Das, A graph based feature selection algorithm utilizing attribute intercorrelation, in: Information Technology, Electronics and Mobile Communication Conference (IEMCON), 2016 IEEE 7th Annual, 2016, pp. 1–9.
[19] D. Dutta, P. Dutta, J. Sil, Simultaneous continuous feature selection and k clustering by multi objective genetic algorithm, in: 3rd IEEE International Advance Computing Conference (IACC), 2013, pp. 937–942.
[20] K.P. Swetha, V. Susheela, Simultaneous feature selection and clustering using particle swarm optimization, in: ICONIP 2012, Part I, LNCS, vol. 7663, 2012, pp. 509–515.
[21] R. Poli, J. Kennedy, T. Blackwel, Particle swarm optimization, Swarm Intell. 1 (2007) 33–57.
[22] E. Hancer, B. Xue, M. Zhang, A survey on feature selection approaches for clustering, Artif. Intell. Rev. 53 (2020) 4519–4545.
[23] P.R. Clarkson, A.J. Robinson, 'Language model adaptation using mixtures and an exponentially decaying cache, in: 1997 IEEE International Conference on Acoustics, Speech, and Signal Processing, Munich, vol. 2, 1997, pp. 799–802.
[24] C. Fraley, A.E. Raftery, Model-based clustering, discriminant analysis, and density estimation, J. Am. Stat. Assoc. (2011) 611–631.
[25] M.H.C. Law, M.A.T. Figueiredo, A.K. Jain, Simultaneous feature selection and clustering using mixture models, IEEE Trans. Pattern Anal. Mach. Intell. 26 (9) (2004) 1154–1166.
[26] C. Laclau, M. Nadif, Fast simultaneous clustering and feature selection for binary data, in: IDA 2014, LNCS, vol. 8819, 2014, pp. 192–202.
[27] B. Bollobas, Combinatorics: Set Systems, Hypergraphs, Families of Vectors, and Combinatorial Probability, Cambridge University Press, New York, 1986.
[28] J. Bourgain, On localization for lattice Schrödinger operators involving Bernoulli variables, Lect. Notes Math. 1850 (2004) 77–100.
[29] R. Carmona, A. Klein, F. Martinelli, Anderson localization for Bernoulli and other singular potentials, Commun. Math. Phys. 108 (1987) 41–66.
[30] E. Zitzler, M. Laumanns, L. Thiele, SPEA2: Improving the Performance of the Strength Pareto Evolutionary Algorithm, Technical Report 103, Computer Engineering and Communication Networks Lab (TIK), Swiss Federal Institute of Technology (ETH), Zurich, 2001.
[31] A. Bellaachia, D. Portnoy, Y. Chen, A.G. Elkahloun, E-CAST: a data mining algorithm for gene expression data, in: Proceedings of the BIOKDD02: Workshop on Data Mining in Bioinformatics (With SIGKDD02 Conference), Canada, Edmonton, Alberta, 2002.
[32] E. Balas, C.S. Yu, Finding a maximum clique in an arbitrary graph, SIAM J. Comput. 15 (4) (1986) 1054–1068.
[33] S. Butenko, W.E. Wilhelm, Clique-detection models in computational biochemistry and genomics, Eur. J. Oper. Res. 173 (1) (2006) 1–17.
[34] A. Ben-Dor, R. Shamir, Z. Yakhini, Clustering gene expression patterns, J. Comput. Biol. 6 (3–4) (1999) 281–297.
[35] n.d. http://www.cs.waikato.ac.nz/ml/weka/downloading.html.

CHAPTER

8

Medical decision support system using data mining semicircular-based angle-oriented facial recognition using neutrosophic logic

R.N.V. Jagan Mohan

Department of Computer Science and Engineering, Sagi Rama Krishnam Raju Engineering College, Bhimavaram, Andhra Pradesh, India

1. Introduction

An important aspect covered in this chapter is to apply the optimization technique for face recognition to minimize the rejections in the recognitions of size. The inspiration of this chapter covers the major areas, namely, optimization. Statistics are being used for optimization in software engineering, which helps in reducing the size of the data invention and saves time for the execution of the process [1]. In this process, facial detection is considered as one of the most prominent approaches to image processing as suggested by X. Zhu [2]. Y. Wen et al. [3] has proposed the drawbacks in facial recognition with in-depth pose verification. To solve this disadvantage, the angle-oriented facial recognition procedure is adopted and considered particularly in the semicircular approach in which the face can be moved or rotated 90 degrees of either direction. Directional Images used for identifying the features of image is suggested by P. Viola [4] and then correlating the picture of human being with the picture in the database [5]. The facial detection database system has individual facial features along with their geometric relationships is optional (B. Yang et al. 2015) [6]. This has attracted considerable attention for decades, where variation in brightness is an important factor in alerting the appearance of the face [7]. The facial features for the input images were compared to the database images. If the images are matched using any of the techniques, then the person is said to be recognized [8]; otherwise try with another input image. The feature extraction competence of the Discrete Cosine Transform (DCT) have been well-thought-out and initiated by some

Handbook of Computational Intelligence in Biomedical Engineering and Healthcare
https://doi.org/10.1016/B978-0-12-822260-7.00002-9

generalization techniques in which it enhances the robust image detection [9]. However, the research area is highly unsolved, largely due to variations in pose, illumination, expression, and several problems that occur as soon as using the pose orientation [10]. To solve this difficulty, angle-oriented images are considered which improves the identification process recommended by N. Ahmed [11]. Using this model, the feature pictures of the image angles classified even if they are having variance in the angles. When there is no input image of an angle of 90°, rotate the image to 90°, and then utilize the generalization techniques such as geometric and illumination methods. It is easy to identify the image face by using the rotation axis. If the input image spins rom horizontal axis to vertical axis, then the face spins anticlockwise. Similarly, if the input image spins from the vertical axis to the horizontal axis, then the face rotates clockwise. Therefore, if the input image angle is oriented, then the posture is altered or the angle is changed and then compared using the rotation axis. Apart from the angle orientation, new techniques known as neutrosophic logic is considered for facial identification, which is described in this chapter.

A study that assesses the uncertain image angles data processing advancement in neutrosophic logic related to a common usage of reason where various angles of image has to separate values for truth, falsehood, and indeterminacy for facial recognition is able to speedily classify [12], a law enforcement audiovisual (a trial that established the value of such knowledge). L. Wolf et al. [13] has developed a facial recognition technology to positively handle some cases in which facial images extracted from an audiovisual is captured below favorable conditions. Once the face is posed to the camera and if the illumination is good then this technology is correct up to 99%. Face recognition can speedily ascribe a name to a face by penetrating an outsizes database of face images and finding the neighboring match [14]. The advanced technique proposed by Y. Wen et al. [3] is used in programmed and actual appearance of facial recognition and it has used in many medical decision support systems as well as in the applications of sophisticated human-computer interaction, biometric recognition, and communications. For instance, a machine able to recognize human passions may speedily recognize the pose. The technique combines indication and appearance data in a way that evades interference from illumination variations and image transformations such as rotation. The method is hypothetically and strictly easy to implement.

Optimization is a procedure of obtaining the good possible solution to concern the below attention. It may be achieved through focusing on selecting suitable image angles parameter, i.e., minimum value or most utility from person yearning. In this proposed application, one of the analyzed optimization hassles is to compare numerous angles of images with various databases. In a state-of-the-art, study the images, different techniques are used for identifying the perfect measurements with the help of various techniques for processing the data [15]. Many indistinguishable facts made by the training image within the usefulness of area of application. To reach the goal, this concentrates on improving minimal iterations for pose-oriented reputation implemented using semicircular model. Using this approach, it takes an angle-oriented recognition system, which can input numerous angles of images from the sensed facts for comparing with the database picture in an integrated fashion. The variety of image information communications and receptions by means of duplicating the statistics at diverse stages of angle-oriented clusters are reduced. However, at the same time, the timeliness has to be recognized because of integrity constraint nature of detection system. Therefore, reducing the redundancy of the sensed data is focused by means of applying numerous optimization strategies to optimize the sensed records without dropping the generality such that the extent of the transmitted data from the application field to the end user may be decreased.

2. Semicircular model based angle oriented images

The density of the functions is obtained in the form of infinite series, which are difficult for computation. In any kind of distribution, the construction of well-defined angular can be done by implementing angle-oriented method on linear distribution. In fact, a random variable of a model for angular data when restricted to an arc of length, where L is an integer on the unit circle, then it is called L-arc model. If the above technique is applied on a linear model, then it results in a circular model. For $L = 1$, circular random variable generates circular model and $L = 2$, the circular random variable is confined to semicircular arc i.e., $\vartheta \in \left(\frac{-\pi}{2}, \frac{\pi}{2} \right)$. As discussed by Byoung and Hyoung [10], the semicircular model is used in the aircraft lost problem, where its point of departure and its initial headings are known. Some linear models such as Weibull, Exponential, and Gamma distributions are used for defining positive real numbers [16]. By using angle orientation, the image on such models can be classified; new distributions mapped onto $[0, \pi]$ are derived.

In this work, the model for describing the rotational object has been proposed. The object spins around the axis using fixed-axis model, which does not change in time. The classification of the angular position of the object is performed based on its rotation axis and an angle that gives the degree of the rotation of the object with respect to a reference point. The existence of ambiguous models occurs, if the rotation axis alters in time. The fixed-angle model that describes the sector for the time-varying orientations of the object are represented in a two-dimensional circle on the surface of the unit sphere having four dimensions. Estimators are provided with the common rotation axis and the angles are used for classifying the time varying orientation of the object [17]. It is assumed that testing of the fixed axis model can be designed with the score in the statistics. The errors at neighboring data points are produced by the methods used for handling the auto-correlation.

The proposed model, L-axial distribution is derived from semicircular normal distribution. L-axial distribution is appropriate to any arc of arbitrary length $2\pi/L$ as shown in Eq. (8.1). Therefore, it is preferable to expand the semicircular normal distribution. To create the L-axial normal distribution [18] in CN distribution, consider $L = 4$ and the transformation $\theta^* = \theta/L$.

The PDF $\theta*$ is given by

$$g(\theta) = \frac{1}{2\pi I_0(x)} e^{x \sin[90^\circ + L(\theta - \mu)]} \tag{8.1}$$

where $\theta > 0, \frac{-2\pi}{L} < \theta < \frac{2\pi}{L}$

For $L = 4$

$$g(\theta) = \frac{1}{2\pi I_0(x)} e^{x \sin[90^\circ + 4(\theta - \mu)]} \tag{8.2}$$

where $\theta > 0, \frac{-\pi}{2} < \theta < \frac{\pi}{2}$

The above Eq. (8.2) is used for angle orientation, which is very important for image recognition system. Even if the images are angle oriented, the feature images of the faces can be classified using Eq. (8.2). If the angle of the input image is not 90°, spin the image to 90°, and then employ normalization techniques such as geometric and illumination techniques. Any input given for the face recognition should be brought to the upright and frontal pose by using the novel based angle orientation approach before comparing to the database image by S. Sengupta [19]. In advance, the input which is given for face recognition having angular orientation should be rotated to appropriate angle in clockwise or anticlockwise to bring the face to equivalent position (upright and frontal pose) as in the actual database and then compared with the database image. Rotational axis is used to recognize an image which makes easier for identifying the face. If the input image spins from horizontal axis to vertical axis, then the face spins anticlockwise. Similarly, if the input image rotates from vertical axis to horizontal axis, the face rotates clockwise. Therefore, if input image is angle oriented, then the pose angle is altered using rotational axis and compared with the pose of the database image.

In Fig. 8.1, red (dark gray in printed version) line shows anticlockwise direction (substituting $\theta = 0-90°$) and the other shows clockwise direction. The mean value of $\sin\theta$, θ varies from 0° to 90° are between 0 and 1. The two forms of angular rotations such as anticlockwise and clockwise with various angles are depicted in Figs. 8.2 and 8.3.

2.1 Anticlockwise rotation

The images captured by the devices at various instants may consist of the faces that are to be recognized in various angles, which are different from the poses of the image present in the database. It will be efficient if our system is able to recognize the face even with these images having different angle orientations. To achieve this, the input image is rotated in anticlockwise direction from horizontal axis to vertical axis such that the face spins anticlockwise

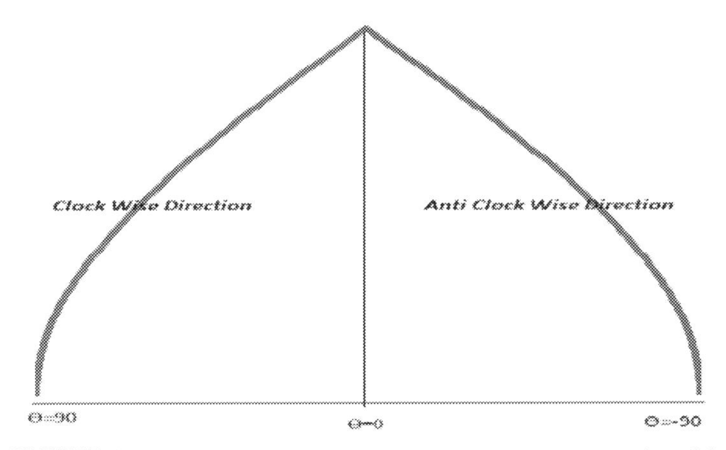

FIGURE 8.1 Semicircular model of angle orientation using L-axial model.

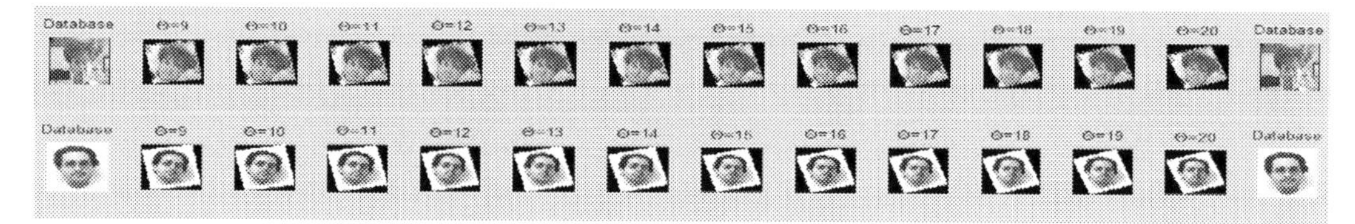

FIGURE 8.2 Angle rotation in anticlockwise direction.

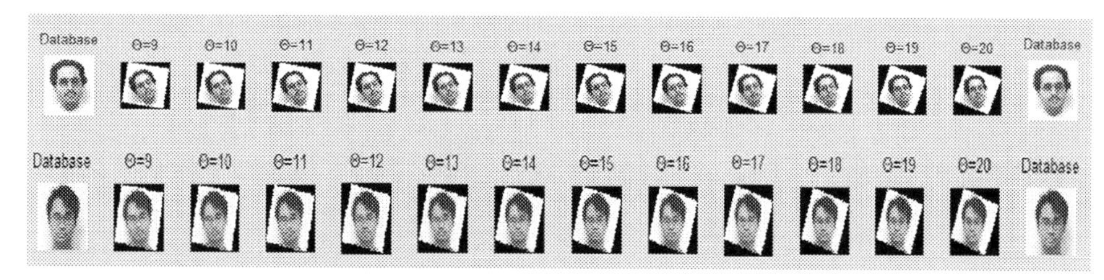

FIGURE 8.3 Angle rotation in clockwise direction.

and the face looks same as the database pose. With this, the object or the face is easily identified. This phenomenon is illustrated with the sample data set of the images considered as shown in Fig. 8.2. Here, the database image has the different angle when compared to the captured images. Therefore, the angle orientation approach has been applied by rotating the captured images in the anticlockwise direction in order to match with the database image. With a simple technique of anticlockwise rotation of the image, it is easy to recognize the face [20]. This technique is therefore cost effective in terms of computational calculations, processing time, resource consumption, etc.

2.2 Clockwise rotation

The input images captured from devices at various instants may consist of the faces that are to be rotated in clockwise direction in order to match with database pose. To achieve this, the input image is rotated in clockwise direction from horizontal axis to vertical axis such that the face rotates clockwise and the face resembles the same as the database pose [20], expanded the idea of increasing the reliability of angle oriented recognition system. With this, the object or the face is easily recognized. This phenomenon is illustrated with the sample dataset of the images considered as shown in Fig. 8.3. Here, the database image has the different angle when compared to the captured images. Therefore, the angle orientation approach has been implemented by rotating the captured images in the clockwise direction to correlate the input image with the database image. With a simple technique of clockwise rotation of the image, it is easy to identify the face. This technique is therefore cost effective in terms of computational calculations, processing time, resource consumption, etc. Therefore, if input image is angle-oriented (i.e., the pose is changed and angle is also altered with the rotational axis), then it is to be compared with the database image.

3. Angle-oriented fuzzy rough sets

Initially, chose an input image and correlate it with the image of the database. If the size of input image is not similar to the size of the database image, then resize the input image to match with the size of database image. The pose of the database image used in both input and database images are compared. If the angle of the input image is not at an angle of 90°, then images cannot be compared. To identify such images, eye coordinates technique has been utilized by Ziad M. Hafed and Martine D. Levine [21]. In this method, the feature images of the faces can be identified even if the images are angle oriented. If the angle of the input image is other than 90°, then normalization technique such as geometric and illumination approach have been applied. The identification of an image using rotational axis is easy to achieve or recognize the face. If the input image spins from horizontal axis to vertical axis, then the face rotates anticlockwise and it looks same as the database pose. Similarly, when the input images spins from vertical axis to horizontal axis, then the face rotates clockwise and the face looks same as the database pose. Therefore, if input image is angle-oriented, i.e., the pose is altered or angle is altered using rotational axis either in clockwise or anticlockwise, then compare it with the pose of the database image [20].

According to Pawlak's famous framework for the construction of lower and upper approximations of any orientation based incomplete information, models have been extended in two ways:

1. Any set of input ("I") images can be normalized to a fuzzy set in X, allowing the input images to be oriented (i.e., meet its characteristics) with different degrees.
2. In place of model rudiments in distinguish ability, one may measure their resemblance (input images are alike to a certain degree) that symbolized by a fuzzy relation R. Consequently, the input images are well thought-out into classes using "soft" boundaries based on their resemblance to one another [22–27].

Accordingly, the complete input image (tuple) can be divided into two parts namely anticlockwise and/or clockwise methods. For each method, there can be two clusters $0° - 45°$ and $45° - 90°$ which has been mathematically defined as follows.

Accordingly, the whole of input image (tuple) can be divided into two parts namely anticlockwise and/or clockwise methods [20] and for each method can be two clusters $0° - 45°$ and $45° - 90°$. Mathematically it is defined as follows as in Eq. (8.3).

$$\sum_{\theta=0°}^{45°} \sin\theta + \sum_{\theta=45°}^{90°} \sin\theta \tag{8.3}$$

The entire tuple, i.e., clockwise or anticlockwise that consist of the lower and upper approximations is known as rough set. Therefore, a rough set consist of two crisp sets, one is used for depicting the lower boundary of target X and the other is used for describing the upper boundary of X as shown in Eqs. (8.4)–(8.6).

$$\alpha_p(x) = \sum_{\theta=0°}^{45°} \sin\theta + \sum_{\theta=45°}^{90°} \sin\theta \tag{8.4}$$

When

$$\theta = \sum_{\theta=0^\circ}^{45^\circ} \sin\theta = 0 \ \Theta \ \text{Lower approximation is empty} \tag{8.5}$$

When

$$\theta = \sum_{\theta=45^\circ}^{90^\circ} \sin\theta = 1 \ \Theta \ \text{Upper approximation is one} \tag{8.6}$$

4. Ternary relationship with angle-oriented face recognition

The way of attitude oriented face popularity is intended toward understanding the database images and are carried out at three distinct strategies.

4.1 Face normalization

Face normalization is the basic degree for all phase popularity structures. Facial recognition technology was invented to start with the face area [28]. The face normalization method displayed in Fig. 8.4 records the photo of the length N × N and is compared with the size of the database photograph. If the size of input photograph and the database photograph are not same, then it must resize the entire image to get the same size as the database photograph. If the pose of the preferred image needs to be rotated to get the database photo, then the face must be rotated (0°−90°) until the database matches the photograph. Depending on the selected posture of the image, the rotation of the image may be bilateral, i.e., either clockwise or anticlockwise. The above concept of face recognition was discussed by Turk et al. [8].

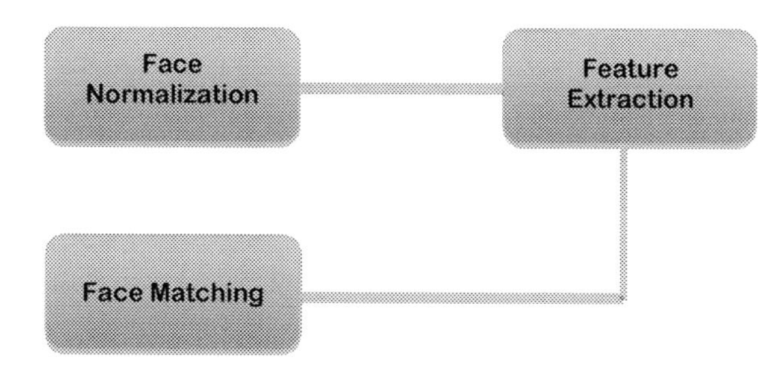

FIGURE 8.4 Ternary association with angle-oriented face recognition.

4.2 Feature extraction

In feature extraction, the feature is defined as a function of one or more dimensions; each of them specifies some quantitative property of an object. It is calculated with some important properties of the object. All features can be classified as low-level and high-end features. The extraction of low-level features can be obtained from the original images, but the extraction of high-level features depends on low-level features [29]. Once the normalized face is achieved, then it can be correlated with other faces, under the same nominal size, orientation, position, and brightness conditions [30]. This comparison is done using the properties acquired in the transformation method. One such popular transformation is the DCT [31], a characteristic extraction step in various studies on facial recognition. The division of input images into N × N blocks is performed to determine the local areas of processing. The N × N block is a two-dimensional image utilized to convert the data into a frequency domain. After that, block-level coefficient is generated by using statistical operators that calculates distinct functions of the spatial frequency in blocks [32]. This is a feature extraction method for angle-oriented face recognition.

4.3 Face matching

The last part of Ternary relationship shown in Fig. 8.4 is face matching. In this procedure, if you want to understand a specific entire photo, then diverse remote and nearest neighborhood classifiers [33] are used to evaluate the input photograph feature vector with the database feature vectors. After obtaining the distances for the N × N matrix, one may want to find the averages of each column of the matrix. With the overall average is 0 or −ve, there may be a suit among the input photo and database picture.

5. K-means fuzzy rough angle-oriented clusters

Consider a set of angle-oriented fuzzy rough images whose feature vectors $\theta = (0°, 1°, \ldots 90°)$. Each value in angle annotations is in n dimensions of the actual image vectors, which are used to find out the mean m_i of the angles based on image feature vectors in cluster K. The aim is to separate the n findings into K $(<=n)$ sets $S = \{S_1, S_2 \ldots S_K\}$ to reduce the variance within the angle-oriented cluster sum of squares. If the angle-oriented clusters are well alienated, then use a minimum distance Tanimoto distance method for the mean and variance calculation [34] and classification of the angle-oriented images. For example, if θ is in cluster S then $\|\theta - m_i\|$ is used to approximate minimum distances for angle-oriented clusters. The cluster images are stated using Tanimoto distance method proposed by Tanimoto et al. [35].

6. Neutrosophic logic

Trustworthiness assessment of series can be performed using distributed and parallelized system. To implement that study for the successful operation of a system having n images for example, it is essential that all n components must perform their intended function successfully in the case of series system, while in the case of parallel and distributed system only one needs to work effectively. However, the present system needs to have at least k successful operations ($1 <= k <= n$) out of n similar images. Such system is a well-known as k-out-of-n system or partial redundant systems. Moreover, in parallel system all n components work concurrently. For the successful operation of a parallel system only one needs to work profitably. This is known as lively redundancy. Software reliability can be improved if the software failures are well predicted beforehand. In support of software engineering wished-for, the classical unit interval used is [0, 1]. It is intended for $t, i, f \in [0, 1]$, that means in single valued neutrosophic logic, the sum of the images is $0 <= t + i + f <= 3$, if all the three images are independent of each other. It is $0 <= \max \{t + i, t + f, i + f\} <= 2$, if two components are dependent on each other and the third one is independent from them. It is $0 <= \max \{t + i + f\} <= 1$, if all the three images are dependent on each other. If three or two of the image components i.e., T, I, F are independent, then dissimilar one assigns space for incomplete information (sum < 1) or paraconsistent and contradictory information (sum > 1), or complete information (sum $= 1$). If all three components T, I, F are dependent, then similarity one assigns space for incomplete information (sum < 1) or complete information (sum $= 1$).

Usually, the sum of two components p and q that differ in the unitary interval [0, 1] is given by $0 <= p + q <= 2 - d(p, q)$, where the dependence distance among p and q is given by $d(p, q)$, while the independence distance among p and q is represented by $1 - d(p, q)$.

Normally, neutrosophic logic uses matching of angle-oriented images for a face recognition system, which depends on three principle components as specified below.

1. If the direction of the input angle-oriented image is same as the direction of the database Image then treat (T) as Match T, which can be further split into subcomponents T_1, T_2, and T_3.
2. If the Input Image is not having the similar direction as that of the database image, even though they are similar treat it as match indeterminacy of the images (I), which can be further split into subcomponents I_1, I_2, I_3 ... I_r.
3. If the Input Angle Oriented Image direction is not same as database image and they are dissimilar, then treat it as Not Matched (F), which can be split into subcomponents F_1, F_2, F_3 ... F_s Where $i + r + s = n >= 1$.

Note: It is a communal structure for amalgamation of many existing logics such as fuzzy logic, paraconsistent logic, intuitionistic logic, etc. The logic of neutrosophic is to exemplify all the logical statement into a three-dimensional neutrosophic space, where each dimension of the neutrosophic space symbolizes the truth (T), the falsehood (F), and the indeterminacy (I) of the statement under consideration. The T, I, and F are standard or nonstandard real subsets of]"0,1+[that does not necessarily have any relation between them. The analysis of events in the neutrosophic Statistics is represented with neutrosophic probability. The function that designs the neutrosophic probability of a random variable "x" is known as neutrosophic distribution which is given by $NP(x) = (T(x), I(x), F(x))$, where $T(x)$ symbolize the probability that value x appears, $F(x)$ describes the probability that value x does not occur, and $I(x)$ shows the indeterminate/unknown probability of value x.

7. Hyperplane

Hyperplane is used as a familiar tactic for the classification of both linear and nonlinear data. The data of angle-oriented images is an attribute, which consist of in-degree, out-degree, level, incidence etc. of the angle-oriented clusters. Let "W" be a set/cluster with the angle-oriented images cluster namely clockwise or anticlockwise. In each cluster, either clockwise or anticlockwise collection of θ angles like [0°−45°], [46°−90°] are denoted as $\{(P_1, I_1), (P_2, I_2) \ldots (P_n, I_n)\}$, where P_i is a tuple of angle-oriented clusters images and "i" with related class label I_i. Each I_i can obtain one of two values either positive class (+1/Yes) or negative class (−1/No) as shown in Fig. 8.5. For angle-oriented images, set with two attributes/dimensions can be used to draw infinite number of separators and can be extended to n−dimensions/attributes. This optimal separator is known as hyperplane. Hyperplane with a larger margin is more accurate than with a smaller margin.

Initially, the input images that have been taken are captured from input devices (user request image). Then classification technique is applied on these input images i.e., image can be rotated into clockwise or anticlockwise and measures the feature extraction of each image which are stored in the local database "W" [36]. The database "I" already consists of feature-extracted images. Now, when a user passes a query image as an input then the hyperplane calculates the similarity among the images of the query and images of the database based on Tanimoto distance. The hyperplane now moves the image that is having minimum distance by rotating either clockwise or anticlockwise to the database.

It is already known that separate convex sets that does not intersect are used to designate the use of hyperplane. Assume that "W" and "I" are two convex sets that do not intersect, i.e., $I \cap W = \varnothing$. The input image set represented as I is the collection of angle-oriented clusters of input images; "W" has the collection of those database image set.

Then $\exists\, a \neq 0$ and b such that $a^T x \leq b$ for all $x \in C$ and $a^T x \geq b$ for all $x \in W$. In this function, $a^T x - b$ is nonpositive on "I" and nonnegative on W. A separating hyperplane exists for the sets "I" and "W" if and only if $\{x \mid a^T x = b\}$ [37].

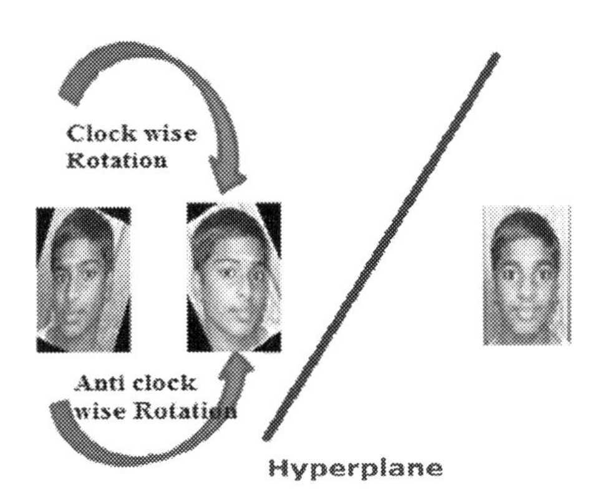

FIGURE 8.5 Hyperplane of angle-oriented image recognition.

8. Evolutionary optimization method

Evolutionary Optimization is an effortless optimization method. This algorithm needs $(2^N + 1)$ points of which 2^N are angle-corner dots of an N-dimensional $(\text{hypercube})^2$ centered on the other point. All $(2^N + 1)$ function values are compared and the best point is identified. In the next iteration, the roughly produced hypercube is considered as the best point. At any iteration, if an improved point is not found then the size of the hypercube is reduced and with this process hypercube may become negligible.

The algorithm consists of the following steps:

Input: Angle oriented images, i.e., $\theta = 0°–90°$

Output: Equivalent Pose Image

1. Select an initial point $i = 0$ and reduction size parameters Δ_i for all the values of "i" angle-oriented image variables.
2. Choose a termination parameter $\in$. Set $\mathring{i}$.
3. If $\|\Delta\| < \in$, Terminate.

   Else, create $2^{\mathrm{Sin}\theta}$ angle oriented image points by subtracting 1 from each θ variable $2^{\mathrm{Sin}\theta}$.

4. Compute the angle-oriented images at all $2^{\sin\theta}$ to find out the point having the minimum function value.
5. Select the best suitable image i.e., which has minimum θ value.
6. If $\left[2^{\sin 0°} - 1 < i < 2^{\sin 90°} - 1\right]$, reduce size parameter, $2^{\sin\theta}$, go to Step 2, Set $\mathring{i}$, go to step 2.

   Note: $\mathrm{i} = 2^{\mathrm{Sin}\theta - 1}$, where $\theta = 0°–90°$, $\mp 2\sin 0° - 1 < i < 2\sin 90° - 1, (\mp 0 < i < 1)$

From the aforementioned algorithm, if $\mathring{i}$ is the present best point then at the end of simulation, $\mathring{\mathrm{i}}$ becomes the obtained optimum point. It is clear from the algorithm that at most 2^N functions are evaluated in each level. Therefore, the necessary number of function evaluations rises exponentially with N.

9. Rotation and reduction procedure (R^2 procedure)

The intent of R^2 procedure is to show the data transformation for execution of the rotation and reduction outline.

The **rotational-map and reduction procedure** is described below:

Initially, the entire angle-oriented input images is split into two major clusters as shown above (Section 2) in angle-oriented fuzzy rough image cluster classifications. This will disseminate the process among all the nodes of the rotational-based maps. Then, the various angles of images between 0° and 90° that is clockwise or anticlockwise in each of the mapper is demonstrated and given angle-oriented image (1/Yes) to each of the images. The basis behind giving an angle-oriented image equal to 1/Yes is that every image occurs only once.

At this moment, a cluster (series of angles-value) will be formed with a key, where the key represents the individual angles and its corresponding value is one. The procedure of the mapping is similar for all the (nodes) images. The partition process that occurs after the rotational based mapper phase sent all the tuples having same keys to the reduction process using sorting and shuffling.

Therefore, in reduction process which occurs after the sorting and shuffling phase, each reducer will use a different angle in addition to a list of images corresponding to key-value. Every reducer in the reduction process are used for counting the angles that exist in that list of values. As depicted in Fig. 8.5, reducer gets a list of images in which [7] is utilized to represent the angle value. Then, the final output is obtained by counting the number of ones in the original list. Lastly, all the output of angle-oriented image cluster pairs is collected and written to the output file.

10. Experimental result

The standard databases are used for the analysis of angle-oriented recognition purpose. It uses three variety of the popular and standard datasets such as MIT, FERET, and Yale which are generally used as sample datasets for analyzing any research findings in the image processing and especially for face recognition applications. Further, to have more insight into the functionality and the performance analysis of the developed angle-oriented recognition system the local database has been synthesized, apart from the above, covering all possible cases of the face angles that may appear before the recognition system in practical scenario. The synthesized database named as SRKR Engineering College Student Database consists of sample images that satisfies the requirements to test most of the possible cases for which the system has to be subjected for analysis. Predict the possible angle orientations that usually appear in a general face recognition system and then synthesize the database accordingly. The synthesized image database used in the experiment consists of 6000 anterior (frontal) images in which 243 face images are randomly selected for the experiment. Only few images from database are shown in below figures. These images deviate in facial expression and illumination. The face images ($N \times N$) in the database were resized to 8×8 for angle oriented experiment. The experiments were divided into two parts and have been performed for 10 persons preferred from the dataset.

In the first part, one sample per person is used in the training process and the numbers of test samples are generated per person with angles ranging between $0°-90°$.The second parts has 10 samples for every individual and $0°-90°$ orientations for each person are used in testing process. Discrete Cosine Transform technique is used for recognizing the faces. Apply the angle orientation technique, to test each input for all image angles ranging between $0°-90°$ (anticlock or clockwise directions) in the above-mentioned database images and then apply the DCT based feature extraction method proposed by Rao et al. [38] to generate the image features of angle orientation system.

10.1 Clockwise rotation

The portrayal of recognition system has major drawbacks such as to test the multiple scales, orientation etc. It is more difficult to incorporate solutions to all these problems in any single autonomous face recognition system. The efforts in addressing these problems have given considerable contributions to solve these problems with the proposed angle-oriented recognition system. The scope of this part of experiment in the work is to achieve the clockwise rotation in angle orientation technique. The tested sample images and the

FIGURE 8.6 The first row is the clockwise input images in the test database, while the second row shows database images of MIT, FERET, Yale, and college students.

FIGURE 8.7 Mean values of the database images.

respective angle-oriented outputs for various images from various kinds of standard and synthesized databases as mentioned above are shown in the below Figs. 8.6 and 8.7. Here, the images with multiple orientations from three different databases are considered and generated the images with various angles between 0°−90°, which form an input data set for the clustering and classification part of the recognition system.

Various angles i.e., between 0°−90° in the clockwise with distinct orientations are shown in the above Figs. 8.8 and 8.9. The angle-oriented pose recognition system checks the

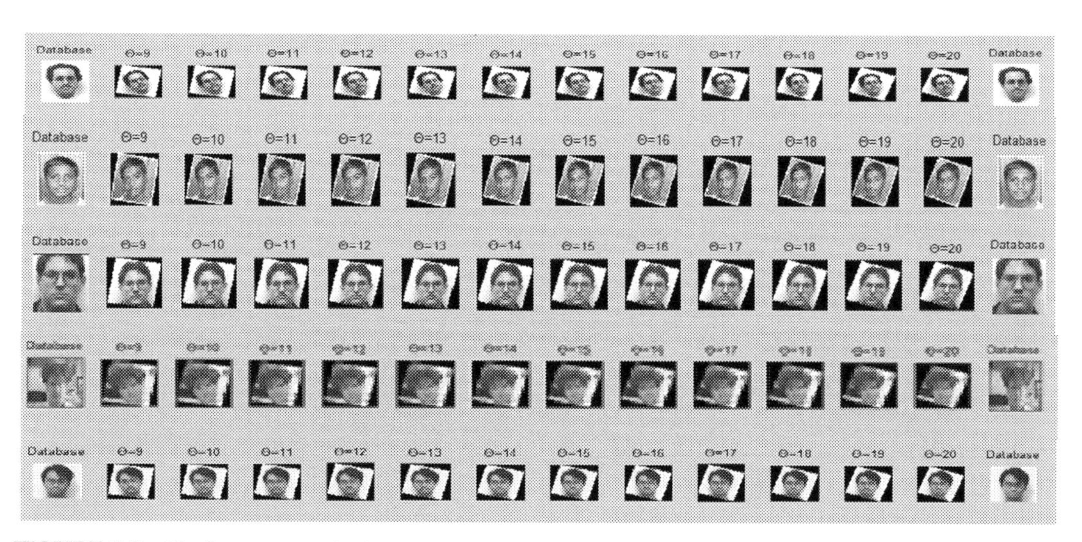

FIGURE 8.8 The first image is the database image, whiles the second and subsequent is the clockwise rotation of images of the same subjects with different scaling and orientation.

FIGURE 8.9 The first row is anticlockwise input images in the test database, while the second row shows database images like MIT, FERET, Yale, and college students.

availability of the image in the database as shown in Fig. 8.7. The resultant data of that person is displayed. The angle orientation technique is used to handle the drawbacks mentioned above. These are represented using two cases (1) distinct angles of input images, i.e., 0°–90° namely clockwise rotation with different scale of the poses. (2) Orientation problem can be solved to increase the reliability of the performance rate and compare with other scale (angle) changes.

10.2 Anticlockwise rotation

The similar process as mentioned in the clockwise rotations in the above description of Fig. 8.8 is again used for the recognition of the images in the anticlockwise rotation, which are rotated in anticlockwise direction to form an input dataset for the angle-oriented recognition system. The major drawbacks in the picture recognition system are to test the multiple scales, orientation. The anticlockwise rotation in angle orientation technique can also be used to solve the above problems.

The anticlockwise rotations with different orientations i.e., 0°–90° are mentioned in Fig. 8.11. The angle-oriented pose recognition system checks user availability in the database as shown in Fig. 8.10. The resultant image of that person is displayed. To handle the drawbacks which are mentioned above, angle orientation technique was used, which are

FIGURE 8.10 Mean values of database images.

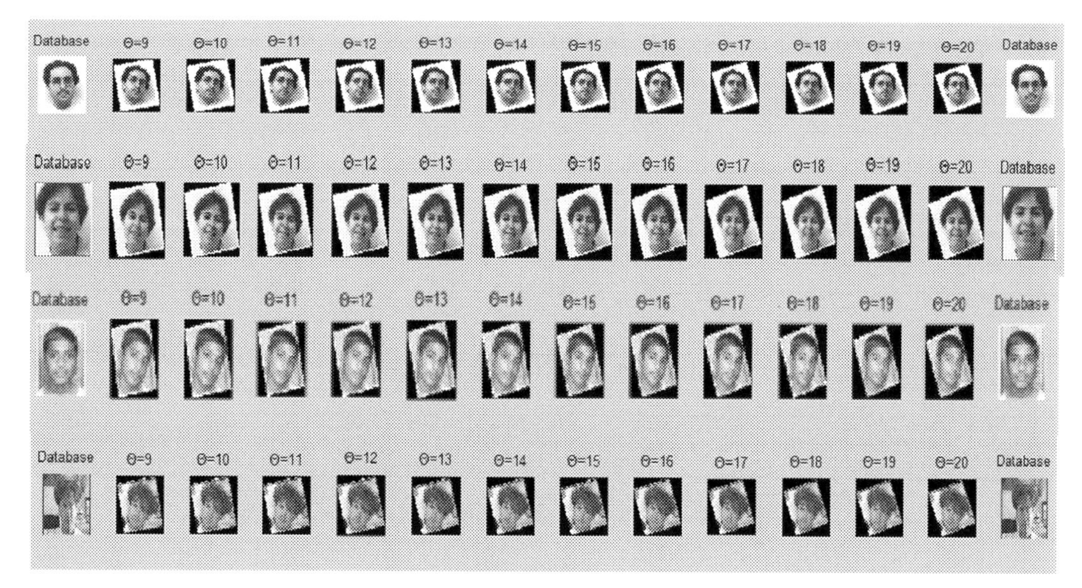

FIGURE 8.11 The first image is the database image, whiles the second and subsequent is the anticlockwise image of the same subjects with different scaling and orientation.

represented using two cases: (1) Different angles of input images i.e., 0°−90° namely anticlockwise rotation with distinct scale of the poses; (2) to enhance the accuracy of performance recognition to higher rate and analyze with other scale (angle) changes, the orientation problem is solved.

10.3 Optimized clockwise angle-oriented images

Optimization is the selection of the best angle or pose image from various database images. This problem comprises of maximizing or reducing original angle-oriented image function by selecting input images systematically from within an allowed set of images and computing the images using image function. The formulation of optimization study and methods used in other formulations constitutes a wide range of applied mathematics and computing operations. In general, optimization involves finding "best available" images of some objective in a specifies domain (or input), including a variety of different types of objectives and domains.

The aforementioned datasets are considered for testing cluster optimization. The recognition rate is obtained using DCT. The input image is taken as variables and it is considered as life random variables. The corresponding probability model fitted to the data is an exponential model. In the process of clockwise or anticlockwise model, take five poses out of 90 poses in input images in all the two clusters, out of which 78 poses are recognized, and 12 poses are unmatched. Deduct the unmatched poses. Out of these 78 recognized poses, only one pose can be like the database pose as per the normalization technique. From the results, it is noticed that the proposed model obtains an optimized rate of 12.8% in recognizing images and deviation of 87.2% with the database poses. The exponential based angle-oriented fuzzy rough cluster images are used as modest model for cluster optimizations when the picture recognition system fails [39]. The exponential distribution is characterized by a constant failure rate, denoted by lambda (λ).

T has an exponential distribution with rate lambda $(\lambda) > 0$ if $P(T \leq t) = 1 - e^{-\lambda t}$ for any nonnegative t. For instance, this function can be used to evaluate with a failure rate of 12.8 along with output 0.9999.

11. Conclusion

In this chapter, an attempt for cluster optimization using angle-oriented fuzzy rough images was presented. Two types of optimization techniques, micro- and macrolevel, are adopted. For microlevel, parallel computation, and for macrolevel, R^2 procedure, are adopted. The software reliability failure rate model for cluster optimization is used by considering the exponential failure rate model. The technique is tested on three types of databases, and experimental results are mentioned. It is also used to optimize redundancy. The future perspective of this work is to compare the different databases with various failure rate models.

References

[1] K. Zhang, Z. Zhang, H. Wang, Z. Li, Y. Qiao, W. Liu, Detecting faces using inside cascaded contextual cnn, in: Proceedings of the IEEE Conference on Computer Vision and Pattern Recognition, 2017, pp. 3171–3179.
[2] X. Zhu, D. Ramanan, Face detection; pose estimation, and landmark localization in the wild, in: Computer Vision and Pattern Recognition (CVPR), 2012 IEEE Conference on, pages 2879-2886, IEEE, 2012.
[3] Y. Wen, K. Zhang, Z. Li, Y. Qiao, A discriminative feature learning approach for deep face recognition, in: European Conference on Computer Vision, Springer, 2016, pp. 499–515.
[4] P. Viola, M.J. Jones, Robust real-time face detection, Int. J. Comput. Vis. 57 (2) (2004) 137–154.
[5] P. Viola, M. Jones, Rapid object detection using a boosted cascade of simple features, in: Computer Vision and Pattern Recognition, CVPR 2001, Proceedings of the 2001, IEEE Computer Society Conference on, volume 1, pages I–I, IEEE, 2001.
[6] B. Yang, J. Yan, Z. Lei, S.Z. Li, Fine-grained evaluation on face detection in the wild, in: Automatic Face and Gesture Recognition (FG), 2015 11th IEEE International Conference and Workshops on, volume 1, pages 1–7, IEEE, 2015.
[7] B. Yang, J. Yan, Z. Lei, S.Z. Li, Aggregate channel features for multi-view face detection, in: Biometrics (IJCB), 2014 IEEE International Joint Conference on, pages 1–8, IEEE, 2014.
[8] M. Turk, A. Pentland, Eigenfaces for recognition, J. Cognit. Neurosci. 3 (1) (1991) 71–86.
[9] S. Yang, P. Luo, C.-C. Loy, X. Tang, Wider face: a face detection benchmark, in: Proceedings of the IEEE Conference on Computer Vision and Pattern Recognition, 2016, pp. 5525–5533.
[10] M.H. Yang, D.J. Kriegman, N. Ahuja, Detecting faces in images: a survey, IEEE Trans. Pattern Anal. Mach. Intell. 24 (1) (2002) 34–58.
[11] N. Ahmed, T. Natarajan, K.R. Rao, IEEE Trans. Comput. C-23 (1) (Jan. 1974) 90–93. Date of Publication: Jan. 1974.
[12] M. Nakazawa, T. Mukasa, B. Stenger, A Photo Booth that Finds Your Sports Player Lookalike, 16th International Conference on Machine Vision Applications, MVA) National Olympics Memorial Youth Center, Tokyo, Japan, May 27–31, 2019.
[13] L. Wolf, T. Hassner, I. Maoz, Face recognition in unconstrained videos with matched background similarity, in: Computer Vision and Pattern Recognition (CVPR), 2011 IEEE Conference on, pages 529–534, IEEE, 2011.

[14] S. Yang, P. Luo, C.-C. Loy, X. Tang, From facial parts responses to face detection: a deep learning approach, in: Proceedings of the IEEE International Conference on Computer Vision, 2015, pp. 3676–3684.
[15] Z. Wu, S. Song, A. Khosla, F. Yu, L. Zhang, X. Tang, J. Xiao, 3D ShapeNets: a deep representation for volumetric shapes, in: Proceedings of the IEEE Conference on Computer Vision and Pattern Recognition, 2015, pp. 1912–1920.
[16] B.J. Ahn, H.-M. Kim, A new family of semicircular models: the semicircular Laplace distributions, Commun. Korean Stat. Soc. 15 (No. 5) (2008) 775–781.
[17] Y. Xiong, K. Zhu, D. Lin, X. Tang, Recognize complex events from static images by fusing deep channels, in: Proceedings of the IEEE Conference on Computer Vision and Pattern Recognition, 2015, pp. 1600–1609.
[18] J. SRao, A. SenGupta, Topics in Circular Statistics, World Scientific Publishing, Singapore, 2001. Series on Multivariate Analysis.
[19] S. Sengupta, J.C. Chen, C. Castillo, V.M. Patel, R. Chellappa, D.W. Jacobs, Frontal to profile face verification in the wild, in: Applications of Computer Vision (WACV), 2016 IEEE Winter Conference on, pages 1–9, IEEE, 2016.
[20] R.N.V. Jagan Mohan, R. Subbarao, International Congress Conference on Productivity, Quality, Reliability, Optimization and Modeling, ICPQROM-2011), 2011.
[21] M. Ziad Hafed, M. Levine, Face recognition using discrete cosine transform||, Int. J. Comput. Vis. 43 (3) (2001) 167–188.
[22] Z. Pawlak, Decision Rules, Bayes' Rule and Rough Sets, New Direction in Rough Sets, Data Mining, and Granular-Soft Computing, 1999, pp. 1–9.
[23] Z. Pawlak, Rough Sets: Theoretical Aspects of Reasoning about Data, Kluwer Academic Publishing, Dordrecht, 1991. ISBN: 978-94-011-3534-4.
[24] Z. Pawlak, S.K.M. Wong, W. Ziarko, Rough sets: probabilistic versus deterministic approach, Int. J. Man Mach. Stud. 29 (1988) 81–95, https://doi.org/10.1016/S0020-7373(88)80032-4.
[25] Z. Pawlak, Rough sets, Int. J. Parallel Program. 11 (5) (1982) 341–356, https://doi.org/10.1007/BF01001956.
[26] Z. Pawlak, Rough Sets Research Report PAS 431, Institute of Computer Science, Polish Academy of Sciences, 1981.
[27] Z. Pawlak, Rough Relations, ICS PAS Reports 435 (1981).
[28] D. Chen, X. Cao, F. Wen, J. Sun, Blessing of dimensionality: high-dimensional feature and its efficient compression for face verification, in: Computer Vision and Pattern Recognition (CVPR), 2013 IEEE Conference on, pages 3025–3032, IEEE, 2013.
[29] Z. Wang, Fast algorithms for the discrete W transform and for the discrete Fourier transform, IEEE Trans. Acoust. Speech Signal Process. 32 (1984) 803–816.
[30] D. Chen, S. Ren, Y. Wei, X. Cao, J. Sun, Joint cascade face detection and alignment, in: European Conference on Computer Vision, Springer, 2014, pp. 109–122.
[31] W.L. Scott, Block-level Discrete Cosine Transform Coefficients for Autonomic Face Recognition, Ph.D. Thesis, Louisiana State University, USA, May 2003.
[32] N. Zhang, M. Paluri, Y. Taigman, R. Fergus, L. Bourdev, Beyond frontal faces: improving person recognition using multiple cues, in: Proceedings of the IEEE Conference on Computer Vision and Pattern Recognition, 2015, pp. 4804–4813.
[33] R.O. Duda, P.E. Hart, Pattern Classification and Scene Analysis, Wiley, New York, NY, 1973.
[34] T. Tanimoto, An Elementary Mathematical Theory of Classification and Prediction, Internal IBM Technical Report, 1957.
[35] D.J. Rogers, T.T. Tanimoto, A computer program for classifying plants, Science 132 (3434) (1960) 1115–1118, https://doi.org/10.1126/science.132.3434.1115.
[36] J. Yan, X. Zhang, Z. Lei, S.Z. Li, Face detection by structural models, Image Vis Comput. 32 (10) (2014) 790–799.
[37] S. Boyd, L. Vandenberghe, In Convex Optimization, Cambridge University Press, 2004, ISBN 978-0-521-83378-3.
[38] K. Rao, P. Yip, Discrete Cosine Transform-Algorithm, Advantages, Applications, Academic, New York, N.Y, 1990.
[39] F. Van Lishout, A. Dubois, M.L. Wang, T. Ewbank, L. Wehenkel, Integrating Facial Detection and Recognition Algorithms into Real-Life Applications, Montefiore Institute – department of EE & CS, University of Liege, Belgium, 2018. Workshop on the Architecture of Smart Cameras – WASC 2018.

CHAPTER

9

Preservation module prediction by weighted differentially coexpressed gene network analysis (WDCGNA) of HIV-1 disease: a case study for cancer

Ria Kanjilal[1], Bandana Barman[1], Mainak Kumar Kundu[2]

[1]Department of Electronics & Communication Engineering, Kalyani Government Engineering College, Kalyani, West Bengal, India; [2]Department of Electronics & Communication Engineering, Brainware Engineering College, Barasat, Kolkata, West Bengal, India

1. Introduction

Genetic analysis is an overall process to identify significant genes among gene expression data set. As these significant genes play important role on disease progression, genetic analysis is used to track the strength of genes and they perform a differential diagnosis of inherited genetic disorder. In this article, a specific gene expression data set is analyzed to diagnose a virulent, chronic and transmissible disease, namely Acquired Immunodeficiency Syndrome (AIDS), caused by the Human Immunodeficiency Virus (HIV) infection. At the beginning of infection, a human usually faces very common symptoms like influenza-fever, enlarged lymph-node, muscle and joint pain, headache, rash, or a sore throat but internally, by damaging the immune system of human body, HIV destroys body's defense against infection, causing varieties of other vulnerable infections and certain malignancies that eventually cause death.

HIV is classified into two classes. Those are mainly HIV-1 and HIV-2 [1]. In worldwide, HIV-1 group M causes the majority of HIV infections. Helper T lymphocytes or helper T cells or CD4+ T cells are the main cellular target of HIV. That is a special class of white

https://doi.org/10.1016/B978-0-12-822260-7.00004-2

blood cells critical to immune system [2]. The CD4+ T cells (Cluster of Differentiation) produces those factors which activate virtually all other cells involve in immune system. So CD4+ T cells play the central role in normal immune system of body. The HIV infection grows with three important stages. Those are Acute Infection, Clinical Latency, and AIDS.

The initial contraction period of HIV is called Acute HIV infection stage or Acute Retroviral Syndrome (ARS). During this initial period, a large amount of virus use CD4+ T cells to replicate and then slowly destroy these cells, causing a sharp fall on CD4 count of body [3]. In this stage, within 2–4 weeks after infection, most of the infected people develop flu-like symptoms. However, duration of symptoms varies usually in 1 or 2 weeks. Then next stage occurs, called clinical latency or asymptomatic HIV or Chronic HIV. In absence of treatment, the Chronic stage exists from 3 years to over 20 years and only a few (~5%) of HIV infected people can restrain the sufficient amount of CD4+ T cells while, most of them turn to AIDS stage. AIDS develops due to a large fall of cluster of differentiation cell count, i.e., <200 cells/μL or due to happening of other specific diseases in connotation with HIV infection [4]. These types of infections may fully or partially affect almost every organ's system and may cause various viral-induced cancers. HIV-1 virus attacks mainly immune system cells, i.e., CD4+ T surface receptors those include macrophages, monocytes, lymphocytes and dendritic cells. This virus also affects uninfected bystander cells e.g., NK cells, neurons, and CD8+ T cells directly or indirectly [5,6]. These cytotoxic T lymphocyte or CD8+ T cells are called killer cells because it kills virus, but it becomes deactivated due to absence of CD4+ T cells. Due to HIV infections, the entire immune system of a human destroys. So, there be a chance to develop cancer disease in patient's body.

Here, Ria Kanjilal (RK) did a conscientious literature survey and developed the idea of the proposed work. RK and Mainak Kumar Kundu (MK) jointly did the data preprocessing and implemented the methods, analyzed the results and drafted the manuscripts. Bandana Barman (BB) did KEGG pathway analysis and supervised the whole work by contributing the intellectual concept of the proposed work. She provided the valuable suggestions to achieve desire result and also revised the final manuscripts.

2. Related work

During last decade, in several literatures, researchers have done a conscientious study on differential expression analysis of microarray genes to understand the intricate network of gene interactions. They also tried to recognize the impact of those genetic interactions on viral replication. Their researches also include Gene Ontology (GO) analysis using three structured networks of defined terms. Those terms describe gene product attributes such as biological processes (BP), cellular components (CC) and molecular functions (MF) as they describe involvement of genes in different metabolic pathways [1], construction of coexpression network using pairwise correlation matrices [7], and preservation pattern of gene coexpression network [8]. Previously, researchers extensively analyzed the preservation pattern of coexpression network for six different brain regions affected by Alzheimer's disease and then they compared the preservation pattern of coexpressed modules. They proposed a novel framework to unveil a substantial association amongst "betweenness centrality" and "degree" of the involved genes which shows a deeper study on the preservation pattern of gene expression of the corresponding brain regions [8]. From an intense literature survey of HIV-1 disease progression, it has been cleared that identification of differentially expressed genes are important to determine the key genes involve in disease mechanism. The research work for detecting differentially expressed genes (DEGs) from samples of wild type HIV-1 Vpr and HIV-1 mutant Vpr explored by B. Barman et al. shows a thorough analysis of wild type HIV-1 Vpr and two HIV-1 mutant Vprs separately using statistical t-test and false discovery rate (FDR). The authors discovered the process to identify the upregulated and down regulated genes from the above-mentioned samples which can be used to form coexpressed networks, which may find the metabolic pathway of a particular disease [9]. The network analysis of weighted coexpressed genes were done before to study endometriosis, oral squamous cell carcinoma tumorigenesis, schizophrenia spectrum disorders, hyperlipidemia, etc. [10–13]. Researchers also predicted diseased gene for molecularly uncharacterized diseases in literature [14].

Here, we did a comprehensive step-by-step analysis using sample data sets of different stages (acute, chronic and nonprogressor stages) of HIV-1 infected patients to determine the Differentially Coexpressed Gene (DCEG) preservation modules. Differentially coexpressed genes are altered coexpression pattern of genes which are observed in different biological conditions. Here, differentially coexpressed genes (DCEGs) play an important role to find out the significant genes which are actively responsible for mechanism of AIDS in human body [15]. The characteristic information of identified DCEGs are defined through eigengene network of coexpression modules, constructed by the differential gene expressions, which are observed in different conditions of three stages of HIV infection.

3. Material and methods

Here details of sample gene microarray dataset is discussed.

3.1 Sample dataset

In this article, we selected HIV-1 microarray dataset from Gene Expression Omnibus (GEO) database. The database was proposed by Hyrcza MD et al. with GEO Series accession no. GSE6740 (available in http://www.ncbi.nlm.nih.gov/geo). The dataset gives a clear overview of human CD4+ and CD8+ T cells from untreated HIV-infected individuals at different clinical stages such as; acute, chronic, nonprogressor and uninfected stages. The dataset shows the rates of disease progression in each stage. These datasets also define gene expression profiles in ex vivo human CD4+ and CD8+ T cells from untreated HIV-infected individuals at different clinical stages. The sample dataset was excerpted from 10 gene chips, among them five gene chips gave the expression of CD4+ T cell and remaining five implied to CD8+ T cell. It consists of total 40 samples and 22283 genes, in which 10 samples are of early HIV-1 infection (acute) stage, 10 samples are of chronic

HIV-1 infection stage, 10 samples of HIV-1 infection nonprogressor (a stage of low or undetectable viral load) stage and other 10 are uninfected samples. For proposed work, 10 uninfected samples are not considered to analyze disease progression in human body.

According to conventional microarray data analysis, mainly interrelationship between pair of all genes are determined, but our proposed technique refines the significant active genes from a complete dataset through the observations on same genes in different stages of HIV infection using statistical T-test method. Then connectivity of pair of genes are identified through coexpression network. The analysis is done on the basis of WGCNA platform maintaining step-by-step manners [16]. The R statistical software with Bioconductor toolbox is used for sample data analysis. A number of R packages those are appropriate to data processing including some issues like data complexity, data size, data evolution etc. are assembled here to get the desired result. KEGG pathway analysis is performed using DAVID webtool (https://david.ncifcrf.gov) on weighted differentially coexpressed genes (WDCG) obtained from Acute-Chronic, Chronic—Nonprogressor, and Nonprogressor—Acute pairs.

3.2 Proposed methods

The sample dataset is a raw data so data preprocessing is very important. After obtaining the experimental dataset the entire research work is done in a step by step manner. Here, the whole method is discussed thoroughly.

3.2.1 Data preprocessing

Preprocessing of any sample data set is quite necessary to obtain a valid experimental result [17]. To examine the gene expression values of samples from GSE6740 data sets, we selected three main stages of HIV-1 infection such as acute, chronic and nonprogressor stages and took the .CEL files of date set to extract the matrices using Affy package of Bioconductor toolbox. In this article, data preprocessing has been done using MAS5.0 algorithm. Here, "mas5()" function is used to remove background noise present in the data set using location-specific background adjustment and correction of Perfect Match (PM) probe intensities depending on Mean Mismatch (MM) probe intensities. This function performs three steps: background correction, PM/MM correction, and finally, finds the expression matrix. After computing expression values, a scaled normalization has been taken into account to adjust each array in the data set [18]. Then to remove noisy probes with low detection level, the "mas5calls()" function has been executed. This function performs Wilcoxon signed rank-based gene expression presence/absence detection algorithm by which the perfectly matched probes are discriminated from mismatched probes. Using MAS5 algorithm, **14869** genes are selected as 'present call' from a data set of 22283 genes.

3.2.2 Differentially expressed genes (DEGs)

Differentially expressed genes (DEGs) identify the number of genes whose expression values are increased (upregulated) and also the number of genes whose expression values are decreased (downregulated) in one sample to another samples. In this article, t-score or t-statistic values of genes are computed between samples of acute and chronic stages, chronic and nonprogressor stages and nonprogressor and acute stages using Welch's statistical t-test method [19]. As the populations of samples of each pair of stages have equal mean and unequal variances, so Welch's t-test is used to determine the "*P*-value" for each pair of stages and accordingly the raw *P*-values are calculated for each gene of these pairs. In Welch's independent-sample *t*-statistic, t is defined in Eq. (9.1).

$$t_s = \frac{\overline{y_1} - \overline{y_2}}{\sqrt{\frac{s_1^2}{n_1} + \frac{s_2^2}{n_2}}} \tag{9.1}$$

Here, $\overline{y_1}$ and $\overline{y_2}$ are first and second sample's mean respectively; n_1 and n_2 are first and second sample's size respectively. s_p^2 is the sample variance (here $P = 1$ or 2). t_s is standardized mean difference using estimated variance. Here, $n_1 = n_2$ and $\overline{y_1} \neq \overline{y_2}$. Estimated variance is associated with degrees of freedom (DoF) and it is approximated using the Welch–Satterthwaite equation as stated in Eq. (9.2).

$$v \approx \frac{\left(\frac{s_1^2}{n_1} + \frac{s_2^2}{n_2}\right)^2}{\left(\frac{s_1^4}{n_1^2 v_1} + \frac{s_2^4}{n_2^2 v_2}\right)} \tag{9.2}$$

Where, $v_1 = (n_1 - 1)$ indicates DoF associated with first variance estimate and $v_2 = (n_2 - 1)$ indicates the DoF associated with second variance estimate. In proposed work, genes expressed in all samples of three pairs of stages are extracted from complete data set, following a significant threshold level of *P*-value less than 0.01. This exhibits total **589 DEGs** in pair of acute and chronic stages, **491 DEGs** in pair of chronic and nonprogressor stages and **1155 DEGs** in pair of nonprogressor and acute stages. Moreover, the connectivity of these differentially expressed genes (DEGs) are analyzed by constructing coexpression networks, which represent the significant potential to identify gene clusters affected by state transitions. To find out number of DEGs, which have high correlation in their expression profile and also to enhance the computation speed, scale-free topology has been implemented [20] by which values of beta, β have been measured. Table 9.1A–C indicate the calculation of scale free topology fitting indices (R2) corresponding to different soft threshold powers beta, β. Through a comprehensive soft power analysis from (Fig. 9.1A–C) and (Tables 9.1A–C), β has been chosen as **β = 16** for acute-chronic stage, **β = 16** for chronic-nonprogressor stage and finally **β = 22** for nonprogressor-acute stage. These values of β define the points where corresponding curves reach at saturation point and we get the maximum values of R2, i.e., 0.88 for acute-chronic stage, 0.95 for chronic-nonprogressor stage and 9.23e-01 for nonprogressor-acute stage.

TABLE 9.1A Table to find out soft power of acute-chronic network.

Sl.	Power	SFT.R.sq	Slope	Truncated.R.sq	Mean.k.	Median.k.	Max.k.
1	1	0.0763	−1.74	0.921	270.000	272.000	373.00
2	2	0.2970	−2.32	0.893	142.000	142.000	243.00
3	3	0.4150	−2.00	0.957	81.400	80.200	162.00
4	4	0.4900	−1.87	0.972	49.500	47.800	110.00
5	5	0.5240	−1.65	0.983	31.500	29.400	76.00
6	6	0.5980	−1.46	0.961	20.900	19.300	54.30
7	7	0.6550	−1.42	0.945	14.200	12.900	39.70
8	8	0.7020	−1.35	0.936	9.990	8.780	29.40
9	9	0.7440	−1.34	0.952	7.170	6.190	22.50
10	10	0.7630	−1.41	0.937	5.240	4.470	17.80
11	12	0.8240	−1.44	0.974	2.960	2.340	11.40
12	14	0.8540	−1.44	0.965	1.760	1.340	7.61
13	16	0.8790	−1.45	0.954	1.100	0.772	5.24
14	18	0.7330	−1.54	0.727	0.717	0.464	3.72
15	20	0.8770	−1.54	0.924	0.482	0.292	2.82

TABLE 9.1B Table to find out soft power of chronic-nonprogressor network.

Sl.	Power	SFT.R.sq	Slope	Truncated.R.sq	Mean.k.	Median.k.	Max.k.
1	1	0.17700	2.000	0.901	233.000	230.000	320.00
2	2	0.00951	0.258	0.897	125.000	121.000	216.00
3	3	0.05020	−0.448	0.891	72.900	68.500	150.00
4	4	0.12800	−0.619	0.838	44.800	41.100	107.00
5	5	0.30100	−0.874	0.810	28.900	25.700	78.00
6	6	0.46000	−1.050	0.819	19.300	16.700	59.20
7	7	0.60200	−1.230	0.815	13.300	11.100	46.10
8	8	0.76100	−1.430	0.839	9.460	7.550	37.00
9	9	0.83500	−1.580	0.864	6.880	5.290	30.40
10	10	0.91900	−1.640	0.936	5.120	3.780	25.50
11	12	0.93300	−1.690	0.938	3.000	1.980	18.80
12	14	0.94300	−1.690	0.928	1.880	1.100	14.70
13	16	0.95300	−1.670	0.941	1.240	0.649	11.90
14	18	0.90400	−1.650	0.879	0.862	0.386	9.96
15	20	0.90800	−1.600	0.885	0.624	0.241	8.51

TABLE 9.1C Table to find out soft power of nonprogressor-acute network.

Sl.	Power	SFT.R.sq	Slope	Truncated.R.sq	Mean.k.	Median.k.	Max.k.
1	1	2.02e-01	1.92	0.861	579.000	581.000	783.00
2	2	1.91e-05	0.0112	0.913	326.000	324.000	550.00
3	3	7.75e-02	−0.5730	0.960	197.000	193.000	400.00
4	4	2.17e-01	−0.8900	0.976	126.000	120.000	298.00
5	5	3.78e-01	−1.1300	0.986	83.800	77.800	227.00
6	6	4.87e-01	−1.2900	0.989	57.700	52.100	176.00
7	7	6.06e-01	−1.5300	0.987	40.800	35.600	139.00
8	8	6.75e-01	−1.6100	0.988	29.600	25.100	111.00
9	9	7.23e-01	−1.6400	0.996	21.900	18.000	89.90
10	10	7.53e-01	−1.7200	0.994	16.500	13.200	73.50
11	12	8.11e-01	−1.7600	0.992	9.800	7.420	50.50
12	14	8.50e-01	−1.7500	0.969	6.130	4.280	35.90
13	16	8.81e-01	−1.7200	0.968	4.010	2.560	26.10
14	18	9.00e-01	−1.7300	0.970	2.710	1.610	19.50
15	20	9.06e-01	−1.6900	0.959	1.890	1.020	14.80
16	22	9.23e-01	−1.6700	0.970	1.350	0.663	11.70
17	24	9.13e-01	−1.6800	0.960	0.994	0.436	9.36
18	26	9.30e-01	−1.6300	0.966	0.744	0.297	7.59
19	28	9.62e-01	−1.5600	0.977	0.567	0.205	6.21
20	30	9.61e-01	−15,500	0.970	0.440	0.143	5.35

3.2.3 *Concept of adjacency matrix*

Each coexpression network is represented by adjacency matrix $A = [a_{ij}]$, which encodes connection strength of each pair of nodes. Adjacency matrix is a symmetric matrix and for unweighted networks, entries of this matrix is 0 or 1 which indicate if the pair of nodes are connected or not. In this work, the weighted networks have been constructed and entries of the adjacency matrix indicate the connection strength between gene pairs. In next step, the adjacency matrix is used to measure dissimilarity between nodes which is an important parameter for cluster analysis and module detection. Since, the similarity between genes should be reflected at expression and network topology level, the coexpression adjacency matrix is analyzed further to compute Topological Overlap Measure (TOM) matrix which builds another adjacency matrix considering topological similarity.

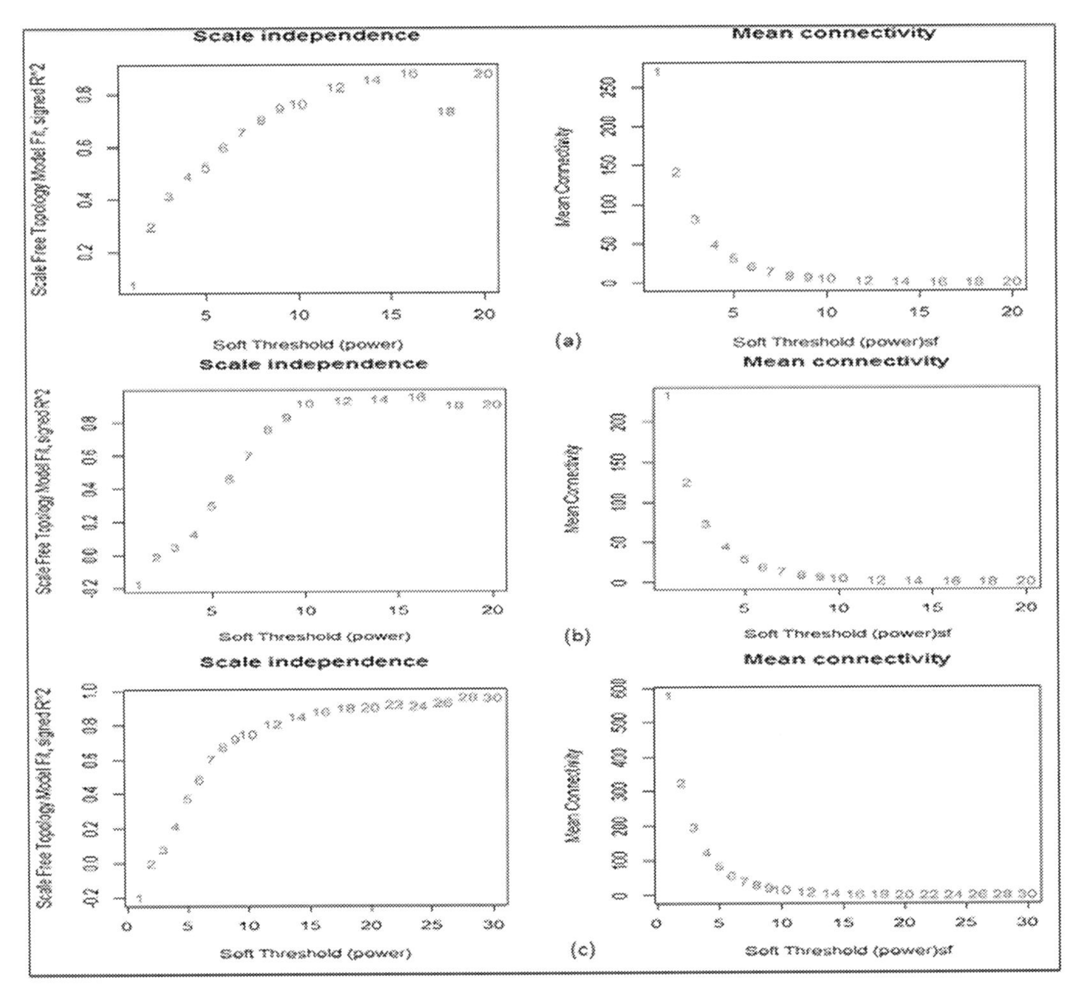

FIGURE 9.1 (A) Mean connectivity plot to find soft threshold of acute-chronic coexpressed network; (B) Mean connectivity plot to find soft threshold of chronic-nonprogressor coexpressed network; (C) Mean connectivity plot to find soft threshold of chronic-nonprogressor coexpressed network.

The adjacency matrix helps to generate WDCGN for differentially coexpressed genes obtained from acute-chronic, chronic-nonprogressor, and nonprogressor-acute pair.

3.3 TOM similarity

TOM has been introduced for less sensitive networks and it not only computes direct connection strength, but also mediates connection strengths by shared neighbors. TOM measurement includes signed and unsigned network analysis. According to signed correlation network, nodes representing negative correlation are considered unconnected (i.e., connection strength ≅ zero). The nodes with strong negative correlations hold high connection strengths for unsigned correlation networks. Also the adjacency is based on absolute value of correlation (consider equal +ve and −ve correlations) in unsigned network [21]. The TOM is defined by Eq. (9.3). Eqs. (9.4) and (9.5) explain the values of parameters used in Eq. (9.3).

$$TOM_{ij}(a_{ij}) = \frac{a_{ij} + \sum_{u \neq i,j} \widetilde{a}_{iu} \widetilde{a}_{uj}}{\min(k_i, k_j) + 1 - a_{ij}} \tag{9.3}$$

Where k_i and k_j denote the connectivity of nodes i and j:

$$k_i = \sum_{u \neq i,j} |\widetilde{a}_{ui}| \tag{9.4}$$

And the modified adjacency matrix is

$$\widetilde{a}_{ij} = a_{ij} \times sign(cor(x_i, x_j)) \tag{9.5}$$

3.4 TOM based dissimilarity measure

Eq. (9.3) can be rewritten by TOM based dissimilarity matrix. That is shown in Eq. (9.6).

$$Diss_{ij} = 1 - TOM_{ij}(a_{ij}) \tag{9.6}$$

At the time of clustering of gene expression profile, TOM-based dissimilarity $Diss_{ij}$ leads to more distinct gene modules than any standard measurement [20]. By assuming value of soft power using soft threshold method, noise of correlation matrix has been reduced and thus TOM based dissimilarity results 151 genes for acute-chronic differential coexpression network, 133 genes for chronic-nonprogressor differential coexpression network and finally 220 genes for nonprogressor-acute differential coexpression network. Fig. 9.2A–C show the corresponding networks.

As a whole TOM analysis explains the strength of connectivity between paired genes in a network so that noise in network can be minimized by calculating the TOM dissimilarity.

3.5 Module detection using dynamic tree cutting

The hierarchical clustering of obtained DCEGs signifies the objects into a dendogram whose branches are desired clusters. Among different methods of hierarchical gene clustering, the "Dynamic Tree" cut top-down algorithm is used to implement an adaptive, iterative process of cluster decomposition and combination. It stops while the number of clusters becomes stable [22]. From Fig. 9.3A–C, it is clear that the dynamic tree cut method provides flexible and stabilized gene clustering and also generate the modules shown in Tables 9.2A–C, for the mentioned differential coexpression networks.

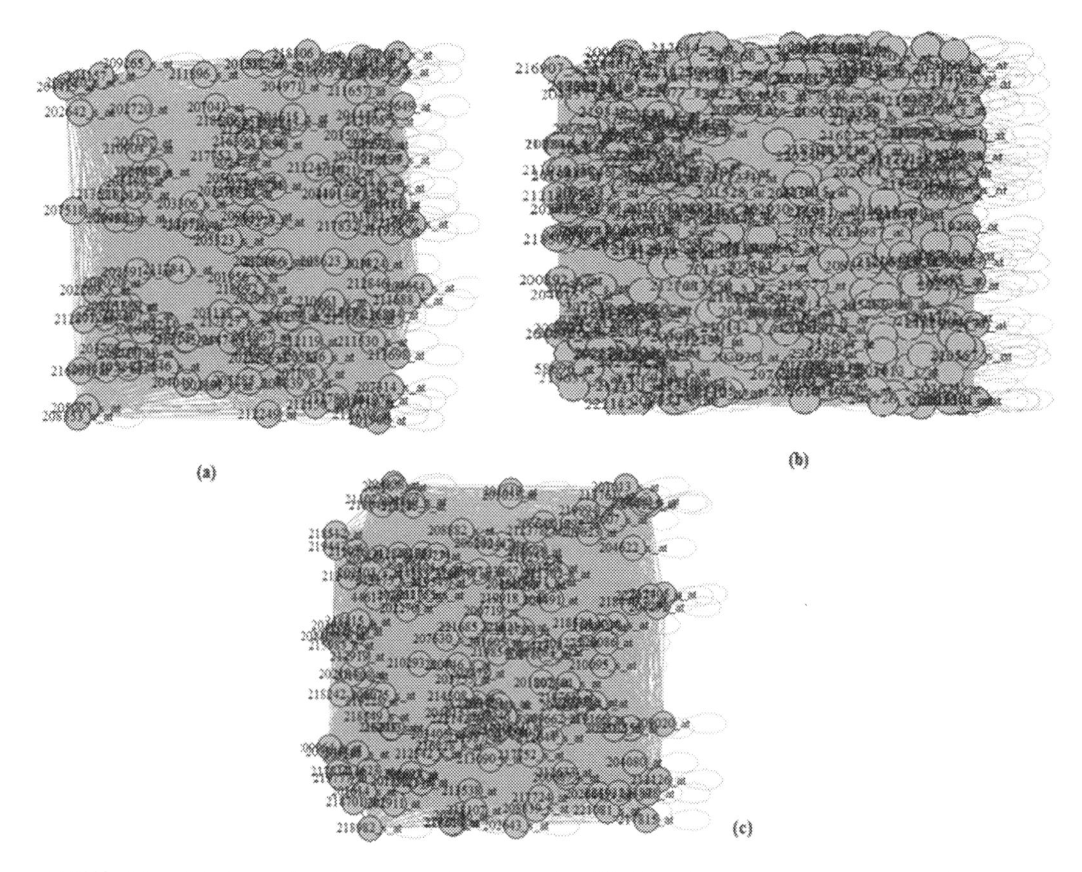

FIGURE 9.2 (A) Acute-chronic coexpression network; (B) Chronic-nonprogressor coexpression network; (C) Nonprogressor-acute coexpression network.

In TOM similarity and dissimilarity measure, strength of differentially coexpressed gene network connectivity is analyzed. This analysis also identifies the gene modules which contain strongly interconnected gene patterns so that this research work able to generate the robust relationships in terms of weighted differentially coexpressed gene network (WDCGN). TOM is used as the input of hierarchical clustering where average linkage process is applied by TOM dissimilarity measure. Then all modules were plotted as dendogram using dynamic hybrid tree cutting approach. As results highly correlated eigengenes' module were merged.

3.6 Connectivity plot or TOM plot

To represent connectivity patterns of adjacency matrix or to measure interconnectedness of pair-wise nodes, the TOM plot has been realized considering TOM-based dissimilarity as input. Fig. 9.4A–C show the TOM plots which represent color-coded depiction of the values of TOM measure including its sorted rows and columns using hierarchical clustering tree.

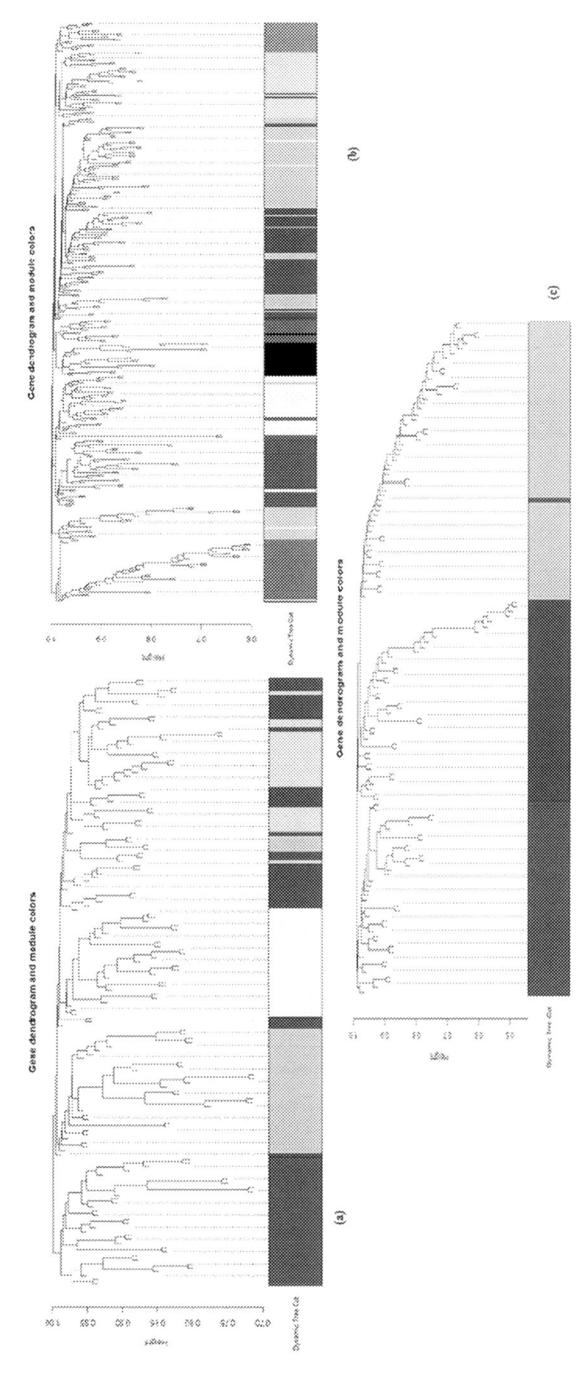

FIGURE 9.3 (A) Gene Dendogram and module colors of acute-chronic coexpression network; (B) Gene Dendogram and module colors of nonprogressor-acute network; (C) Gene Dendogram of chronic-nonprogressor network.

TABLE 9.2A Table for module detection in acute-chronic network.

Modules	Blue	Brown	Green	Gray	Turquoise	Yellow
No. of genes	33	32	24	438	35	27

TABLE 9.2B Table for module detection in chronic-nonprogressor network.

Modules	Blue	Brown	Gray	Turquoise
No. of genes	40	39	358	54

TABLE 9.2C Table for module detection in nonprogressor-acute network.

Modules	Black	Blue	Brown	Green	Gray	Magenta	Pink	Purple	Red	Turquoise	Yellow
No. of genes	24	60	45	43	760	21	21	20	42	76	43

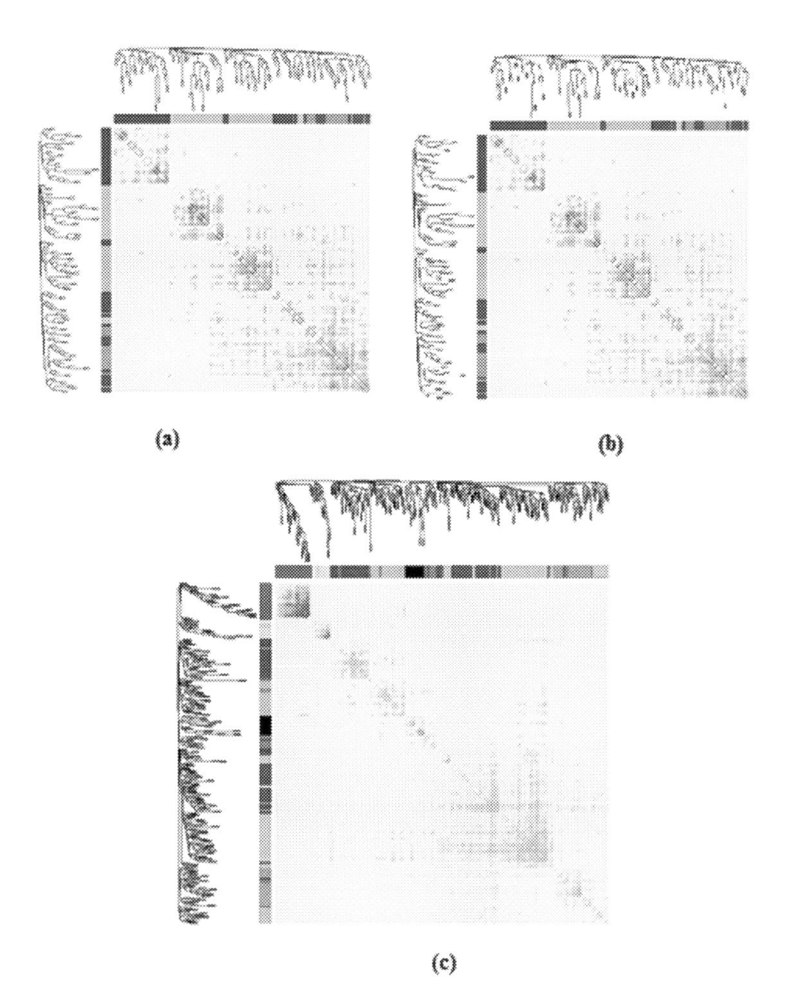

FIGURE 9.4 (A) TOM plot of acute-chronic coexpression network, (B) TOM plot of chronic-nonprogressor coexpression network, (C) TOM plot of nonprogressor-acute coexpression network.

4. Result and analysis

In this section, result of the experiments with their analysis is mentioned.

4.1 Eigengene network

The main aim to build an eigengene network is that, eigengenes of different modules carry the information of correlation between corresponding modules [23,24]. The module eigengenes of different modules can be highly correlated and the correlations between two eigengenes can be used to express eigengene coexpression networks. A weighted eigengene coexpression network can be computed followed by the expression [25] mentioned in Eq. (9.7).

$$a_{qp} = |cor(E^q, E^p)|^{\beta} \quad (9.7)$$

Where E^q and E^p represent eigengenes of two distinct modules. Moreover, it is a coexpression network reduction-process which involves thousands of genes, maintaining an order of magnitude smaller meta-network and representing of module by one eigengene per module. Fig. 9.5A–C show the acute-chronic, chronic-nonprogressor and nonprogressor-acute eigengene networks. The tables corresponding to Fig. 9.6A–C are color-coded by correlation according to color legend. Each cell of these tables represents the correlation value (P-value) between two modules. Within each cell, upper values represent correlation coefficients between module eigengenes and lower values represent corresponding P-value.

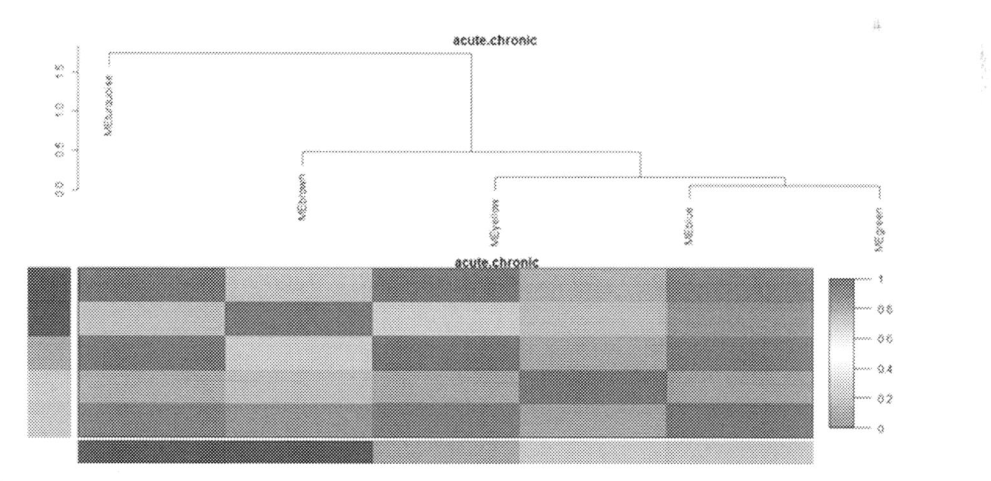

FIGURE 9.5 (A) Hierarchical clustering of module colors and eigengene network of acute-chronic; (B) Hierarchical clustering of module colors and eigengene network of chronic-nonprogressor; (C) Hierarchical clustering of module colors and eigengene network of nonprogressor-acute.

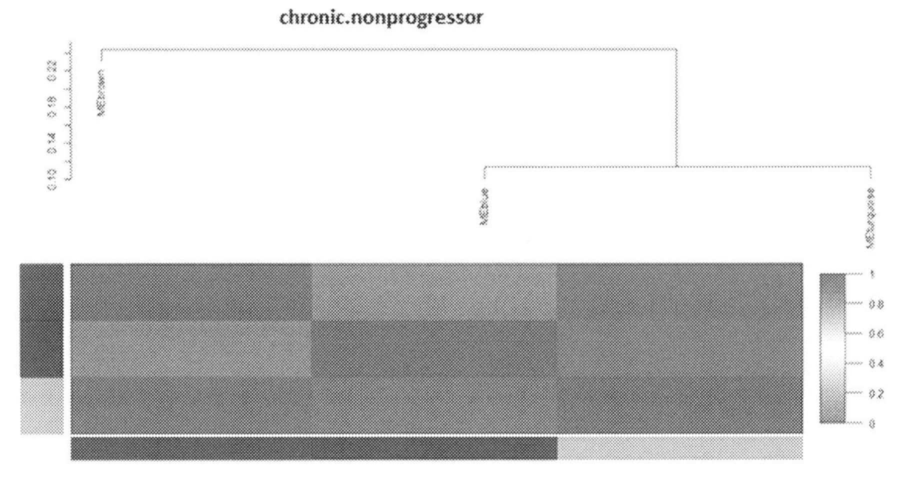

FIGURE 9.5

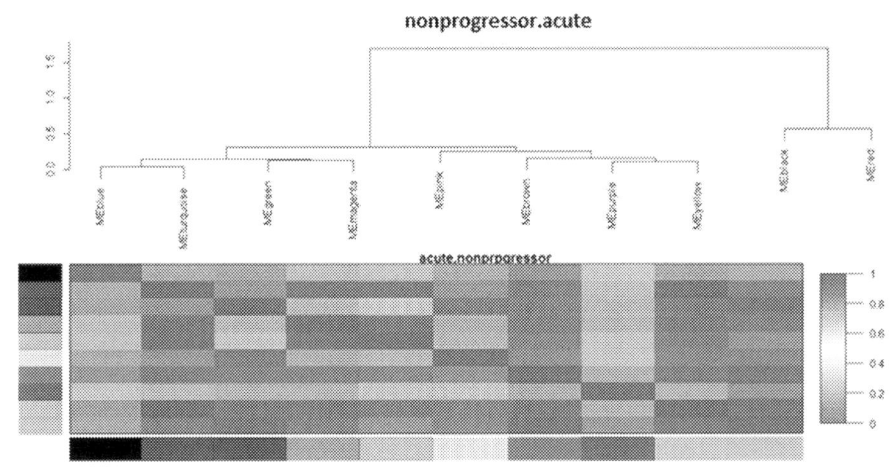

FIGURE 9.5 cont'd

4.2 Module preservation pattern

The significance of module preservation pattern is that, it defines whether a module is preserved in another data set. To find out module preservation, the reference and target datasets are presumed and then through module preservation technique, an important statistical analysis done to identify invalid, nonreproducible modules due to array outliers.

Basic steps include the computation of median rank preservation and Z summary preservation.

The statistical observations of both preservations in three pairs of coexpression networks are shown in Tables 9.3A, B, 9.4A, B, and 9.5A, B.

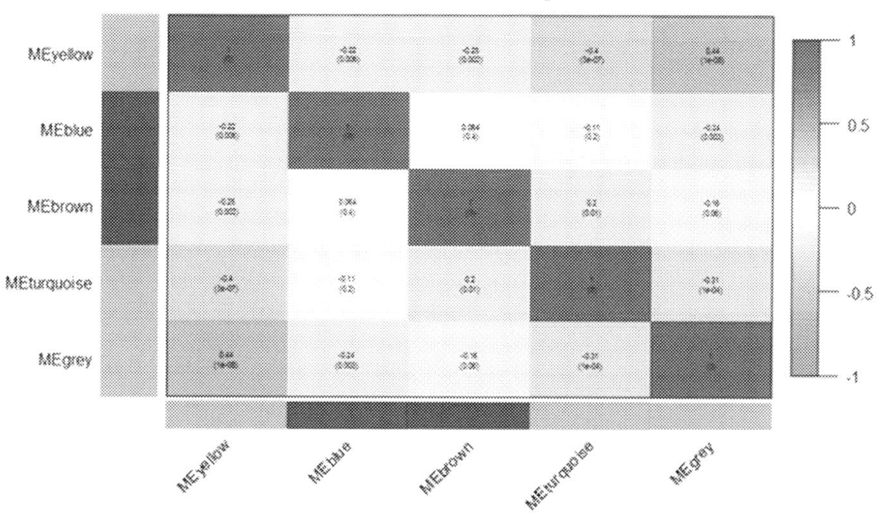

FIGURE 9.6 (A) Eigengene network with *P*-values of acute-chronic network; (B) Eigengene network with *P*-values of chronic-nonprogressor network; (C) Eigengene network with *P*-values of nonprogressor-acute network.

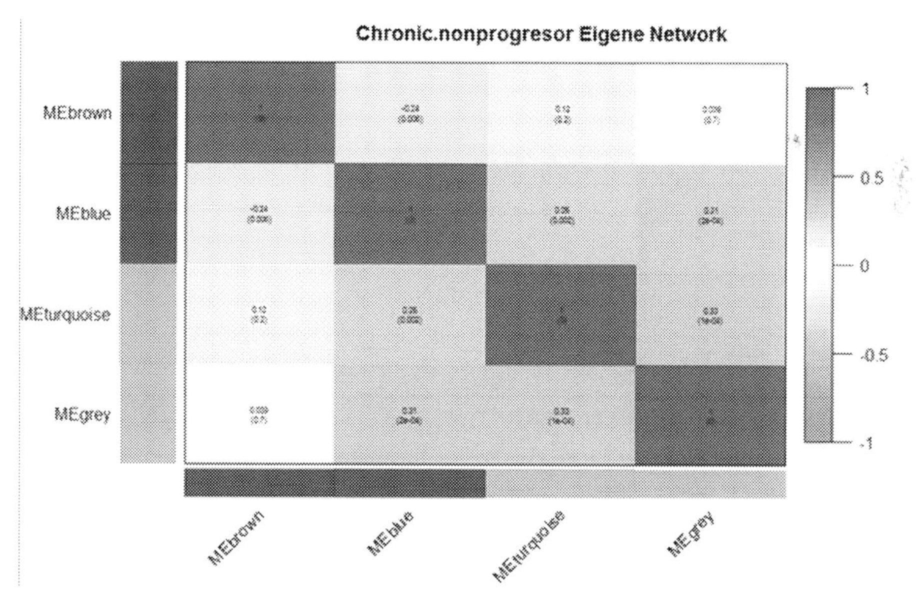

FIGURE 9.6 cont'd

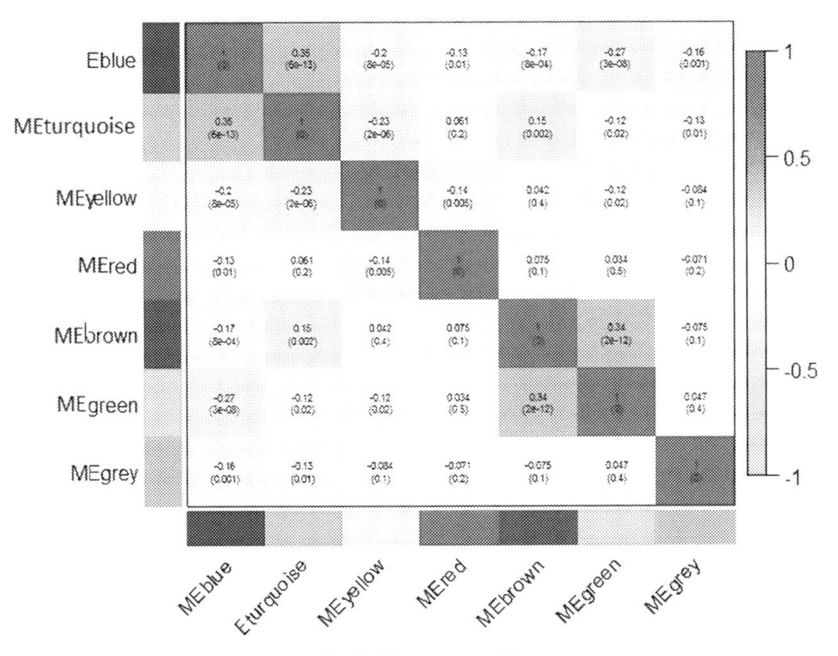

FIGURE 9.6 cont'd

The median rank versus module size plots indicate that, the module preservation gets lower with increase of median rank. Since median rank preservation contrasts Z summary preservation, the modules which show higher Z summary (i.e., >0.10), have strong preservation characteristics. However, modules with Z summary < -2 show no evidence of preservation. The preservation of acute-chronic modules in chronic-nonprogressor network has been analyzed and result shown in Fig. 9.7A indicates that "blue" among other modules has no evidence of preservation. Fig. 9.8A shows the heat map representation of "blue" module gene expressions and panel defines the rows correspond to genes, columns correspond to acute-chronic samples. The color code of heat map reveals that red gene expressions are overexpressed and greens are underexpressed in the corresponding test dataset. In preservation pattern of above pair of coexpression network, "turquoise" and "brown" show strong preservation in chronic-nonprogressor samples. Similarly, the preservation pattern analysis of chronic-nonprogressor modules in nonprogressor-acute coexpression network shown in Fig. 9.7B implies the strongly preserve modules are blue and turquoise and brown module has no evidence of preservation. Fig. 9.8B shows the heat map of brown module. Also preservation pattern of nonprogressor-acute modules in acute-chronic samples shown in Fig. 9.7C has been discussed in acute-chronic samples, which defines that turquoise, blue and red modules are strongly preserved and brown module has no preservation in Fig. 9.8C.

The median Rank and Z summary statistics evaluates module preservation as it correlates connectivity and density of WDCGN. When Zsummary is more than 0.10, it is considered

TABLE 9.3A Median rank preservation of acute-chronic coexpression network.

Median rank preservation

	Module Size	medianRank.pres	medianRankDensity.pres
Blue	9	4	4.0
Brown	5	2	1.0
Gold	18	2	3.5
Turquoise	18	2	2.5
Yellow	4	3	3.5

	medianRankConnectivity.pres	propVarExplained.pres
Blue	4	0.4727879
Brown	3	0.6903681
Gold	2	0.6038882
Turquoise	1	0.4960642
Yellow	3	0.5384513

	meanSignAwareKME.pres	meanSignAwareCorDat.pres	meanAdj.pres
Blue	0.3808592	0.24603008	0.1003166
Brown	0.6769647	0.40124020	0.1973806
Gold	0.3793134	0.02564339	0.2145083
Turquoise	0.6101949	0.37175921	0.1317934
Yellow	0.5747782	0.22082298	0.0980482

	meanMAR.pres	Cor.kIM	Cor.kME	Cor.kMEall	Cor.cor	Cor.MAR
Blue	0.4122538	−0.3108420	0.3411721	0.095718364	−0.13850076	−0.30040154
Brown	0.3378528	−0.1869667	0.3505761	0.057211212	−0.10901822	−0.25972178
Gold	0.4917474	0.1272168	0.1211767	0.002324142	−0.08758612	0.09162292
Turquoise	0.4018293	0.1507930	0.1450592	0.060892511	0.10269509	0.35673561
Yellow	0.2910416	−0.1602369	0.5902263	0.053864344	−0.34535378	−0.07245456

that strong evidence of a module is preserved. If both modules are preserved then lower medianRank exhibits stronger preservation than higher medianRankmodule.

Heat map reveals a graphical comparison between one preservation module with other preservation module. It also allows to define and to characterize a preservation modules by its higher expression in a lesser expression module.

TABLE 9.3B Z summary preservation of acute-chronic coexpression network.

Z summary preservation				
	Module size	**Zsummary.pres**	**Zdensity.pres**	**Zconnectivity.pres**
Blue	9	−0.6238116	−1.038235847	−0.2093874
Brown	5	0.1222110	0.609038709	−0.3646167
Gold	18	0.3726854	1.038623169	−0.2932523
Turquoise	18	0.3531688	0.006694934	0.6996427
Yellow	4	−0.4221132	−0.530581008	−0.3136454
	Z.propVarExplained.pres		**Z.meanSignAwareKME.pres**	
Blue	−0.73545185		−2.6581798	
Brown	1.15607943		0.4807498	
Gold	1.74038839		0.5807851	
Turquoise	−0.08341789		0.2711604	
Yellow	−0.66170449		−0.1223349	
	Z.meanSignAwareCorDat.pres	**Z.meanAdj.pres**	**Z.meanMAR.pres**	**Z.cor.kIM**
Blue	−1.28390831	−0.7925634	0.37568585	−0.5939115
Brown	0.47077398	0.7373276	0.25794080	−0.4519754
Gold	−0.20552598	1.4964613	1.26277353	0.3871941
Turquoise	0.09680776	−0.2194289	−0.39956734	0.8385407
Yellow	−0.63185766	−0.4293044	0.01727506	−0.3136454
	Z.cor.kME	**Z.cor.kMEall**	**Z.cor.cor**	**Z.cor.MAR**
Blue	0.15093727	−0.08456563	−0.2093874	−0.6888737
Brown	−0.01163741	−0.48915874	−0.3646167	−0.5597711
Gold	−0.29325234	−1.12885986	−0.9918220	0.2297870
Turquoise	0.41556165	−0.55693204	0.6996427	2.3283256
Yellow	0.04297466	−0.80470676	−0.7456837	−0.1240446

5. KEGG pathway analysis

Kyoto Encyclopedia of Genes and Genomes (KEGG) is a resource to understand the functionalities and utilities of different cells, organisms and ecosystems generated by genome sequencing and other high-throughput experimental technologies. KEGG PATHWAY analysis represents molecular interaction, reaction and relation networks in metabolism, cellular

TABLE 9.4A Median rank preservation of chronic-nonprogressor coexpression network.

Median rank preservation

	Module size	medianRank.pres	medianRankDensity.pres
Blue	11	1	1.0
Brown	12	3	3.5
Gold	25	3	3.5
Turquoise	27	2	2.0

	medianRankConnectivity.pres	propVarExplained.pres
Blue	4	0.4197922
Brown	3	0.2614451
Gold	1	0.2948744
Turquoise	2	0.3065462

	meanSignAwareKME.pres	meanSignAwareCorDat.pres	meanAdj.pres
Blue	0.4677956	0.22826262	0.07276209
Brown	0.1592020	0.04454638	0.01867521
Gold	0.1480857	0.02418015	0.04081909
Turquoise	0.3228983	0.12543889	0.03296258

	meanMAR.pres	Cor.kIM	Cor.kME	Cor.kMEall	Cor.cor	Cor.MAR
Blue	0.2161275	0.26202320	0.09706958	0.24245197	−0.04587115	0.14716665
Brown	0.1667972	0.50307901	0.12513691	0.15160349	0.01963549	0.57532945
Gold	0.2462872	−0.06509092	0.37221288	0.15831598	0.02875392	0.14373066
Turquoise	0.2465503	0.08399607	0.13679869	0.08080586	0.02205132	0.09229416

process, human disease, drug discovery, etc. It is a reference database for metabolic pathway analysis also. So, we performed KEGG pathway analysis for weighted differentially coexpressed gene (WDCG) obtained from gene expression pairs (Acute−Chronic, Chronic−Nonprogressor, and Nonprogressor−Acute) to obtain whether any gene is significant in cancer disease or not. As results, we found that some genes are significant for the growth of different cancer in human body. The list of those genes which are significant both in HIV infection and cancer disease are listed in Table 9.6.

TABLE 9.4B Z summary preservation of chronic-nonprogressor coexpression network.

Z summary preservation				
	moduleSize	**Zsummary.pres**	**Zdensity.pres**	**Zconnectivity.pres**
Blue	11	0.81454513	2.1700003	−0.54091001
Brown	12	−0.34396000	−1.3023159	0.61439590
Gold	25	−0.39419625	−0.7697458	−0.01864669
Turquoise	27	0.04344028	−0.2090797	0.29596029
	Z.propVarExplained.pres		**Z.meanSignAwareKME.pres**	
Blue	2.0734670		1.35292508	
Brown	−1.4448381		−1.20297791	
Gold	−0.4596777		−1.07981397	
Turquoise	−0.1983461		0.09251999	
	Z.meanSignAwareCorDat.pres	**Z.meanAdj.pres**	**Z.meanMAR.pres**	**Z.cor.kIM**
Blue	2.2665335	2.8110606	0.35846249	0.6522697
Brown	−1.4016539	−1.0112367	−0.24213197	1.8120641
Gold	−1.7136055	0.7951525	−0.03171544	−0.5596116
Turquoise	−0.2449743	−0.2198134	0.08692995	0.2959603
	Z.cor.kME	**Z.cor.kMEall**	**Z.cor.cor**	**Z.cor.MAR**
Blue	−1.23065711	3.0467913	−0.54091001	0.49212657
Brown	−0.70688260	1.3001715	0.61439590	2.21007210
Gold	1.35766410	0.8848753	−0.01864669	0.40887241
Turquoise	−0.07973247	−0.2121010	0.35569576	0.09283485

In Table 9.6, KEGG Pathway analysis is performed on WDCG obtained from pair of Acute−Chronic gene expression data set (i.e., AC−CH−WDCG-KEGG PATHWAY label in the Table), Chronic−Nonprogressor gene expression data set (i.e., CH-NONPRO-WDCG-KEGG PATHWAY label in the Table) and Nonprogressor−Acute gene expression data set (i.e., NONPRO-AC-WDCG-KEGG PATHWAY in the table). In Table 9.6, it is noticed that the genes which are significant in different types of cancers are listed.

TABLE 9.5A Median rank preservation of nonprogressor-acute coexpression.

Median rank preservation			
	ModuleSize	**medianRank.pres**	**medianRankDensity.pres**
Blue	10	4	5.5
Brown	5	6	6.0
Gold	27	4	4.5
Green	3	5	6.0
Gray	1	NA	NA
Red	3	1	1.0
Turquoise	25	3	2.5
Yellow	8	4	3.0
	medianRankConnectivity.pres		**propVarExplained.pres**
Blue	3		0.2920665
Brown	5		0.3607329
Gold	3		0.2892397
Green	3		0.4230382
Gray	NA		NA
Red	7		0.6460400
Turquoise	4		0.3613779
Yellow	5		0.3456505
	meanSignAwareKME.pres	**meanSignAwareCorDat.pres**	**meanAdj.pres**
Blue	0.09762589	0.017860038	0.0241921340
Brown	0.10076153	−0.006336579	0.0013307094
Gold	0.17327768	0.025191924	0.0150194528
Green	0.11211812	−0.115865610	0.0000442355
Gray	NA	NA	NA
Red	0.78635930	0.443100867	0.0715218377
Turquoise	0.47970122	0.223461416	0.0336942607
Yellow	0.21553175	0.046310586	0.0427845907

(Continued)

TABLE 9.5A Median rank preservation of nonprogressor-acute coexpression.—cont'd

Median rank preservation						
	ModuleSize		medianRank.pres		medianRankDensity.pres	
	meanMAR.pres	**Cor.kIM**	**Cor.kME**	**Cor.kMEall**	**Cor.cor**	**Cor.MAR**
Blue	1.639891e-01	0.4370209	0.29882792	0.12022962	0.16161126	0.05850731
Brown	5.152633e-03	−0.6226024	0.67893053	0.04312264	−0.08943769	−0.47572098
Gold	1.676059e-01	0.2879673	0.36794285	0.33947752	0.15332737	0.29314270
Green	8.389989e-01	0.9957398	0.62381095	0.13758229	−0.94966384	0.99901003
Gray	NA	NA	NA	NA	NA	NA
Red	1.422736e-01	−0.9949295	0.99982745	0.47326285	−0.99990938	−0.05384727
Turquoise	2.145766e-01	0.1528412	0.05455824	0.41527232	0.10248446	0.30323667
Yellow	1.855901e-01	−0.4893360	0.27978615	0.18410651	−0.03140195	−0.56669438

TABLE 9.5B Z summary preservation of nonprogressor-acute coexpression network.

Z summary preservation				
	Module size	**Zsummary.pres**	**Zdensity.pres**	**Zconnectivity.pres**
Blue	10	0.35407302	−0.8140705	1.52221658
Brown	5	−0.56967446	−0.8664558	−0.27289308
Gold	27	0.87732728	0.2224452	1.53220935
Green	3	−0.56866532	−1.0720062	−0.06532425
Gray	1	NaN	NA	NA
Red	3	0.06077257	1.5486736	−1.42712845
Turquoise	25	2.35512751	3.6404818	1.06977323
Yellow	8	−25986653	0.1189195	−0.63865252
	Z.propVarExplained.pres		**Z.meanSignAwareKME.pres**	
Blue	−6.332019e-01		−0.99493915	
Brown	−6.122583e-01		−1.13871296	
Gold	1.730736e-05		0.44487311	
Green	−1.048043e+00		1.30566833	

TABLE 9.5B Z summary preservation of nonprogressor-acute coexpression network.—cont'd

Z summary preservation				
	Module size	**Zsummary.pres**	**Zdensity.pres**	**Zconnectivity.pres**
Gray	NA			NA
Red	1.458926e+00			1.63842175
Turquoise	3.810469e+00			3.47049491
Yellow	2.738897e-01			−0.03605074
	Z.meanSignAwareCorDat.pres	**Z.meanAdj.pres**	**Z.meanMAR.pres**	**Z.cor.kIM**
Blue	−1.0483781	0.3548895	0.40883586	1.9258544
Brown	−0.8355380	−0.8973737	−0.90811086	−1.6196080
Gold	0.5638332	−1.5514136	−1.45703445	1.5322093
Green	−1.0959691	−0.6338833	−0.69847321	0.9821886
Gray	NA	NA	NA	NA
Red	2.0049484	1.4560975	1.45749992	−1.4271285
Turquoise	7.0430347	3.1518770	0.05901294	1.0697732
Yellow	−0.2612488	2.2456842	1.36396684	−1.1551872
	Z.cor.kME	**Z.cor.kMEall**	**Z.cor.cor**	b
Blue	0.02027550	0.08781364	1.52221658	0.2435460
Brown	1.02355849	−0.78678662	−0.27289308	−1.1833050
Gold	0.47974658	3.37517150	1.61010657	1.9038293
Green	−0.06532425	−0.50517180	−1.18939339	0.9847020
Gray	NA	NA	NA	NA
Red	1.55115052	2.80816251	−1.57928666	−0.2047462
Turquoise	−0.72798263	5.1110196	1.61780409	1.6338998
Yellow	−0.63865252	1.70884570	−0.09780969	−1.2585380

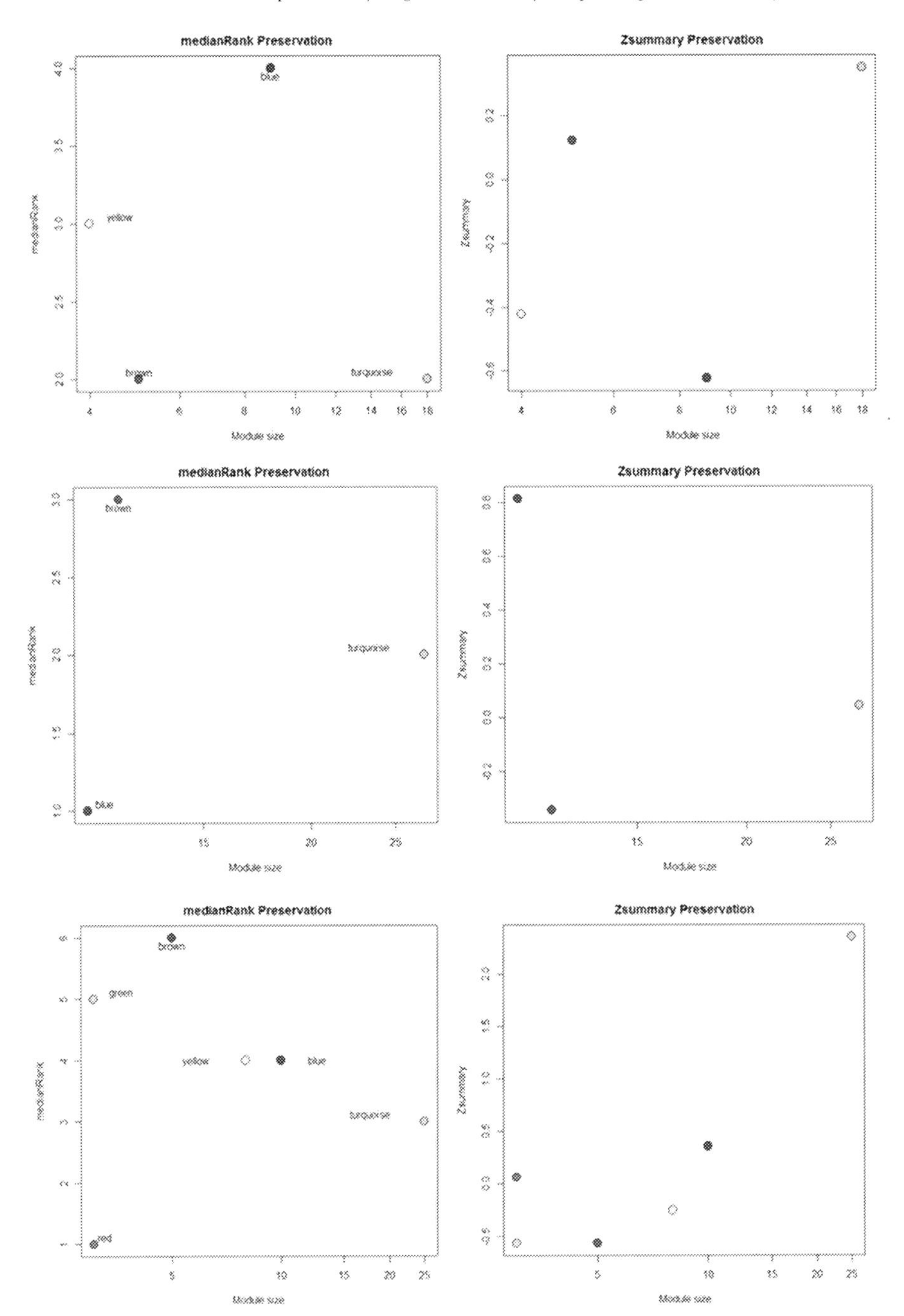

FIGURE 9.7 (A) Median Rank and Z summary preservation of acute-chronic coexpression network; (B) Median Rank and Z summary preservation of chronic-nonprogressor coexpression network; (C) Median Rank and Z summary preservation of nonprogressor-acute coexpression network.

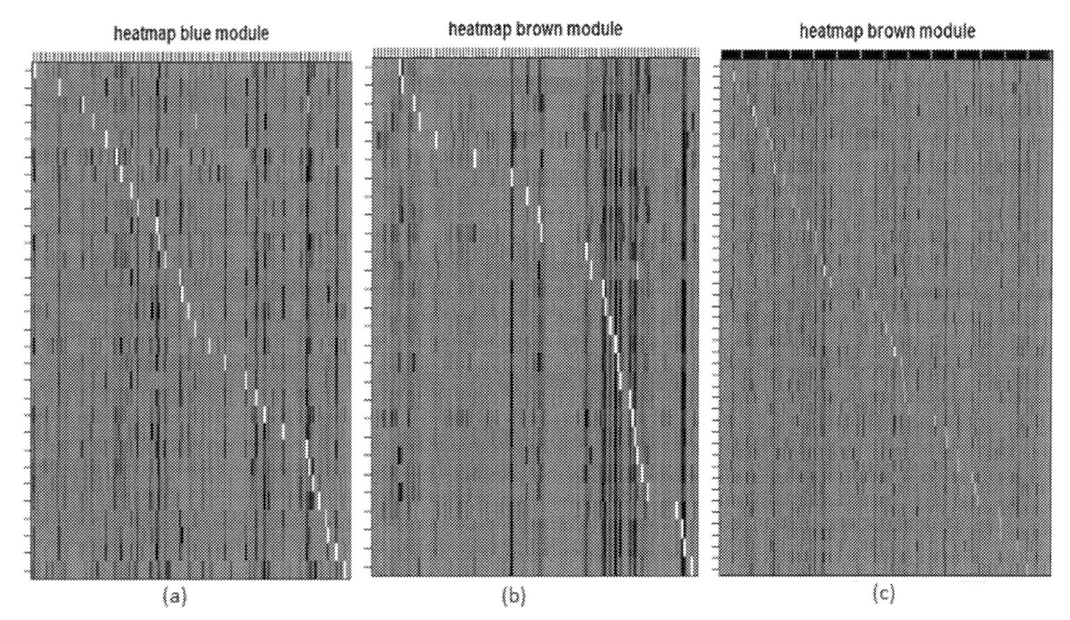

FIGURE 9.8 (A) Heatmap representation of blue module (preservation of acute-chronic modules in chronic-nonprogressor samples), (B) Heatmap representation of brown module (preservation of chronic-nonprogressor modules in nonprogressor-acute samples), (C) Heatmap representation of brown module (preservation of nonprogressor-acute modules in acute-chronic samples).

TABLE 9.6 KEGG PATHWAY analysis.

AC–CH–WDCG-KEGG PATHWAY		
ID	**Gene name**	**KEGG PATHWAY**
216951_at	Fc fragment of IgG receptor Ia (FCGR1A)	hsa04145:Phagosome, hsa04380:Osteoclast differentiation, hsa04640:Hematopoietic cell lineage, hsa04666:Fc gamma R-mediated phagocytosis, hsa05140:Leishmaniasis, hsa05150: *Staphylococcus aureus* infection, hsa05152: Tuberculosis, hsa05202: Transcriptional misregulation in cancer, hsa05322: Systemic lupus erythematosus,
209054_s_at	Wolf–Hirschhorn syndrome candidate 1(WHSC1)	hsa00310: Lysine degradation, hsa05202: Transcriptional misregulation in cancer,
202246_s_at	Cyclin dependent kinase 4 (CDK4)	hsa04110: Cell cycle, hsa04115: p53 signaling pathway, hsa04151: PI3K-Akt signaling pathway, hsa04530: Tight junction, hsa04660: T cell receptor signaling pathway, hsa05161: Hepatitis B, hsa05162: Measles, hsa05166: HTLV-I infection, hsa05200: Pathways in cancer, hsa05203: Viral carcinogenesis, hsa05212: Pancreatic cancer, hsa05214: Glioma, hsa05218: Melanoma, hsa05219: Bladder cancer, hsa05220: Chronic myeloid leukemia, hsa05222: Small cell lung cancer, hsa05223: Nonsmall cell lung cancer,
210095_s_at	Insulin like growth factor binding protein 3(IGFBP3)	hsa04115: p53 signaling pathway, hsa05202: Transcriptional misregulation in cancer,
202911_at	mutS homolog 6 (MSH6)	hsa03430: Mismatch repair, hsa05200: Pathways in cancer, hsa05210: Colorectal cancer,
212610_at	Protein tyrosine phosphatase, non-receptor type 11 (PTPN11)	hsa04014: Ras signaling pathway, hsa04630: Jak-STAT signaling pathway, hsa04650: Natural killer cell mediated cytotoxicity, hsa04670: Leukocyte transendothelial migration, hsa04722: Neurotrophin signaling pathway, hsa04920: Adipocytokine signaling pathway, hsa04931: Insulin resistance, hsa05120: Epithelial cell signaling in Helicobacter pylori infection, hsa05168: Herpes simplex infection, hsa05205: Proteoglycans in cancer, hsa05211: Renal cell carcinoma, hsa05220: Chronic myeloid leukemia,
211023_at	Pyruvate dehydrogenase (lipoamide) beta(PDHB)	hsa00010: Glycolysis/Gluconeogenesis, hsa00020: Citrate cycle (TCA cycle), hsa00620: Pyruvate metabolism, hsa01100: Metabolic pathways, hsa01130: Biosynthesis of antibiotics, hsa01200: Carbon metabolism, hsa04066: HIF-1 signaling pathway, hsa04922: Glucagon signaling pathway, hsa05230:Central carbon metabolism in cancer,

CH-NONPRO-WDCG-KEGG PATHWAY		
ID	**Gene name**	**KEGG PATHWAY**
204039_at	CCAAT/enhancer binding protein alpha(CEBPA)	hsa04932: Non-alcoholic fatty liver disease (NAFLD), hsa05200: Pathways in cancer, hsa05202: Transcriptional misregulation in cancer, hsa05221: Acute myeloid leukemia
216951_at	Fc fragment of IgG receptor Ia (FCGR1A)	hsa04145: Phagosome, hsa04380: Osteoclast differentiation, hsa04640: Hematopoietic cell lineage, hsa04666: Fc gamma R-mediated phagocytosis, hsa05140: Leishmaniasis, hsa05150: *Staphylococcus aureus* infection, hsa05152: Tuberculosis, hsa05202: Transcriptional misregulation in cancer, hsa05322: Systemic lupus erythematosus,
209054_s_at	Wolf-Hirschhorn syndrome candidate 1(WHSC1)	hsa00310: Lysine degradation, hsa05202: Transcriptional misregulation in cancer,
208353_x_at	Ankyrin 1(ANK1)	hsa05205: Proteoglycans in cancer,
211896_s_at	decorin(DCN)	hsa04350: TGF-beta signaling pathway, hsa05205: Proteoglycans in cancer,
207518_at	Diacylglycerol kinase epsilon (DGKE)	hsa00561: Glycerolipid metabolism, hsa00564: Glycerophospholipid metabolism, hsa01100: Metabolic pathways, hsa04070: Phosphatidy linositol signaling system, hsa05231: Choline metabolism in cancer,
208623_s_at	ezrin(EZR)	hsa04670: Leukocyte transendothelial migration, hsa04810: Regulation of actin cytoskeleton, hsa04971: Gastric acid secretion, hsa05130: Pathogenic Escherichia coli infection, hsa05205: Proteoglycans in cancer, hsa05206: MicroRNAs in cancer,
205247_at	notch 4(NOTCH4)	hsa04320: Dorso-ventral axis formation, hsa04330: Notch signaling pathway, hsa04919: Thyroid hormone signaling pathway, hsa05206: MicroRNAs in cancer,

(*Continued*)

TABLE 9.6 KEGG PATHWAY analysis.—cont'd

CH-NONPRO-WDCG-KEGG PATHWAY		
ID	**Gene name**	**KEGG PATHWAY**
212249_at	Phosphoinositide-3-kinase regulatory subunit 1(PIK3R1)	hsa04012: ErbB signaling pathway, hsa04014: Ras signaling pathway, hsa04015: Rap1 signaling pathway, hsa04024: cAMP signaling pathway, hsa04062: Chemokine signaling pathway, hsa04066: HIF-1 signaling pathway, hsa04068: FoxO signaling pathway, hsa04070: Phosphatidylinositol signaling system, hsa04071: Sphingolipid signaling pathway, hsa04150: mTOR signaling pathway, hsa04151: PI3K-Akt signaling pathway, hsa04152: AMPK signaling pathway, hsa04210: Apoptosis, hsa04370: VEGF signaling pathway, hsa04380: Osteoclast differentiation, hsa04510: Focal adhesion, hsa04550: Signaling pathways regulating pluripotency of stem cells, hsa04611: Platelet activation, hsa04620: Toll-like receptor signaling pathway, hsa04630: Jak-STAT signaling pathway, hsa04650: Natural killer cell mediated cytotoxicity, hsa04660: T cell receptor signaling pathway, hsa04662: B cell receptor signaling pathway, hsa04664: Fc epsilon RI signaling pathway, hsa04666: Fc gamma R-mediated phagocytosis, hsa04668: TNF signaling pathway, hsa04670: Leukocyte transendothelial migration, hsa04722: Neurotrophin signaling pathway, hsa04725: Cholinergic synapse, hsa04750: Inflammatory mediator regulation of TRP channels, hsa04810: Regulation of actin cytoskeleton, hsa04910: Insulin signaling pathway, hsa04914: Progesterone-mediated oocyte maturation, hsa04915: Estrogen signaling pathway, hsa04917: Prolactin signaling pathway, hsa04919: Thyroid hormone signaling pathway, hsa04923: Regulation of lipolysis in adipocytes, hsa04930: Type II diabetes mellitus, hsa04931: Insulin resistance, hsa04932: Nonalcoholic fatty liver disease (NAFLD), hsa04960: Aldosterone-regulated sodium reabsorption, hsa04973: Carbohydrate digestion and absorption, hsa05100: Bacterial invasion of epithelial cells, hsa05142: Chagas disease (American trypanosomiasis), hsa05146: Amebiasis, hsa05160: Hepatitis C, hsa05161: Hepatitis B, hsa05162: Measles, hsa05164: Influenza A, hsa05166: HTLV-I infection, hsa05169: Epstein–Barr virus infection, hsa05200: Pathways in cancer, hsa05203: Viral carcinogenesis, hsa05205: Proteoglycans in cancer, hsa05210: Colorectal cancer, hsa05211: Renal cell carcinoma, hsa05212: Pancreatic cancer, hsa05213: Endometrial cancer, hsa05214:Glioma, hsa05215: Prostate cancer, hsa05218: Melanoma, hsa05220: Chronic myeloid leukemia, hsa05221: Acute myeloid leukemia, hsa05222: Small cell lung cancer, hsa05223: Nonsmall cell lung cancer, hsa05230: Central carbon metabolism in cancer, hsa05231: Choline metabolism in cancer,
208067_x_at	Ubiquitously transcribed tetratricopeptide repeat containing, Y-linked (UTY)	hsa05202: Transcriptional misregulation in cancer,

NONPRO-AC-WDCG-KEGG PATHWAY		
ID	**Gene name**	**KEGG PATHWAY**
204039_at	CCAAT/enhancer binding protein alpha(CEBPA)	hsa04932: Nonalcoholic fatty liver disease (NAFLD), hsa05200: Pathways in cancer, hsa05202: Transcriptional misregulation in cancer, hsa05221: Acute myeloid leukemia,
212180_at	CRK like proto-oncogene, adaptor protein(CRKL)	hsa04010: MAPK signaling pathway, hsa04012: ErbB signaling pathway, hsa04015: Rap1 signaling pathway, hsa04062: Chemokine signaling pathway, hsa04510: Focal adhesion, hsa04666: Fc gamma R-mediated phagocytosis, hsa04722: Neurotrophin signaling pathway, hsa04810: Regulation of actin cytoskeleton, hsa04910: Insulin signaling pathway, hsa05100: Bacterial invasion of epithelial cells, hsa05131: Shigellosis, hsa05200: Pathways in cancer, hsa05206: MicroRNAs in cancer, hsa05211: Renal cell carcinoma, hsa05220: Chronic myeloid leukemia,
216951_at	Fc fragment of IgG receptor Ia (FCGR1A)	hsa04145: Phagosome, hsa04380: Osteoclast differentiation, hsa04640: Hematopoietic cell lineage, hsa04666: Fc gamma R-mediated phagocytosis, hsa05140: Leishmaniasis, hsa05150: *Staphylococcus aureus* infection, hsa05152: Tuberculosis, hsa05202: Transcriptional misregulation in cancer, hsa05322: Systemic lupus erythematosus,
219657_s_at	Kruppel like factor 3 (KLF3)	hsa05202: Transcriptional misregulation in cancer,
35160_at	LIM domain binding 1(LDB1)	hsa05202: Transcriptional misregulation in cancer,
210567_s_at	S-phase kinase associated protein 2 (SKP2)	hsa04068: FoxO signaling pathway, hsa04110: Cell cycle, hsa04120: Ubiquitin mediated proteolysis, hsa05168: Herpes simplex infection, hsa05169: Epstein–Barr virus infection, hsa05200: Pathways in cancer, hsa05203: Viral carcinogenesis, hsa05222: Small cell lung cancer,
205558_at	TNF receptor associated factor 6 (TRAF6)	hsa04010: MAPK signaling pathway, hsa04064: NF-kappa B signaling pathway, hsa04120: Ubiquitin mediated proteolysis, hsa04144: Endocytosis, hsa04380: Osteoclast differentiation, hsa04620: Toll-like receptor signaling pathway, hsa04621: NOD-like receptor signaling pathway, hsa04622: RIG-I-like receptor signaling pathway, hsa04722: Neurotrophin signaling pathway, hsa05133: Pertussis, hsa05140: Leishmaniasis, hsa05142: Chagas disease (American trypanosomiasis), hsa05145: Toxoplasmosis, hsa05152: Tuberculosis, hsa05160: Hepatitis C, hsa05162: Measles, hsa05168: Herpes simplex infection, hsa05169: Epstein–Barr virus infection, hsa05200: Pathways in cancer, hsa05222: Small cell lung cancer,
209054_s_at	Wolf-Hirschhorn syndrome candidate 1(WHSC1)	hsa00310: Lysine degradation, hsa05202: Transcriptional misregulation in cancer,
208353_x_at	Ankyrin 1(ANK1)	hsa05205: Proteoglycans in cancer,

(*Continued*)

TABLE 9.6 KEGG PATHWAY analysis.—cont'd

NONPRO-AC-WDCG-KEGG PATHWAY		
ID	**Gene name**	**KEGG PATHWAY**
202246_s_at	cyclin dependent kinase 4(CDK4)	hsa04110: Cell cycle, hsa04115: p53 signaling pathway, hsa04151: PI3K-Akt signaling pathway, hsa04530: Tight junction, hsa04660: T cell receptor signaling pathway, hsa05161: Hepatitis B, hsa05162: Measles, hsa05166: HTLV-I infection, hsa05200: Pathways in cancer, hsa05203: Viral carcinogenesis, hsa05212: Pancreatic cancer, hsa05214: Glioma, hsa05218: Melanoma, hsa05219: Bladder cancer, hsa05220: Chronic myeloid leukemia, hsa05222: Small cell lung cancer, hsa05223: Non-small cell lung cancer,
207518_at	Diacylglycerol kinase epsilon(DGKE)	hsa00561: Glycerolipid metabolism, hsa00564: Glycerophospholipid metabolism, hsa01100: Metabolic pathways, hsa04070: Phosphatidy linositol signaling system, hsa05231: Choline metabolism in cancer,
208623_s_at	ezrin(EZR)	hsa04670: Leukocyte transendothelial migration, hsa04810: Regulation of actin cytoskeleton, hsa04971: Gastric acid secretion, hsa05130: Pathogenic Escherichia coli infection, hsa05205: Proteoglycans in cancer, hsa05206: MicroRNAs in cancer,
204689_at	Hematopoietically expressed homeobox(HHEX)	hsa04950: Maturity onset diabetes of the young, hsa05202: Transcriptional misregulation in cancer,
203710_at	Inositol 1,4,5-trisphosphate receptor type 1 (ITPR1)	hsa04020: Calcium signaling pathway, hsa04022: cGMP-PKG signaling pathway, hsa04070: Phosphatidylinositol signaling system, hsa04114: Oocyte meiosis, hsa04270: Vascular smooth muscle contraction, hsa04540: Gap junction, hsa04611: Platelet activation, hsa04713: Circadian entrainment, hsa04720: Long-term potentiation, hsa04723: Retrograde endocannabinoid signaling, hsa04724: Glutamatergic synapse, hsa04725: Cholinergic synapse, hsa04726: Serotonergic synapse, hsa04728: Dopaminergic synapse, hsa04730: Long-term depression, hsa04750: Inflammatory mediator regulation of TRP channels, hsa04912: GnRH signaling pathway, hsa04915: Estrogen signaling pathway, hsa04918: Thyroid hormone synthesis, hsa04921: Oxytocin signaling pathway, hsa04922: Glucagon signaling pathway, hsa04924: Renin secretion, hsa04925: Aldosterone synthesis and secretion, hsa04970: Salivary secretion, hsa04971: Gastric acid secretion, hsa04972: Pancreatic secretion, hsa05010: Alzheimer's disease, hsa05016: Huntington's disease, hsa05205: Proteoglycans in cancer,

TABLE 9.6 KEGG PATHWAY analysis.—cont'd

NONPRO-AC-WDCG-KEGG PATHWAY		
ID	**Gene name**	**KEGG PATHWAY**
202911_at	mutS homolog 6 (MSH6)	hsa03430:Mismatch repair, hsa05200:Pathways in cancer, hsa05210: Colorectal cancer,
209959_at	Nuclear receptor subfamily 4 group A member 3(NR4A3)	hsa05202: Transcriptional misregulation in cancer,
204209_at	Phosphate cytidylyltransferase 1, choline, alpha (PCYT1A)	hsa00564: Glycerophospholipid metabolism, hsa01100: Metabolic pathways, hsa05231: Choline metabolism in cancer,
212240_s_at	Phosphoinositide-3-kinase regulatory subunit 1(PIK3R1)	hsa04012: ErbB signaling pathway, hsa04014: Ras signaling pathway, hsa04015: Rap1 signaling pathway, hsa04024: cAMP signaling pathway, hsa04062: Chemokine signaling pathway, hsa04066: HIF-1 signaling pathway, hsa04068: FoxO signaling pathway, hsa04070: Phosphatidylinositol signaling system, hsa04071: Sphingolipid signaling pathway, hsa04150: mTOR signaling pathway, hsa04151: PI3K-Akt signaling pathway, hsa04152: AMPK signaling pathway, hsa04210: Apoptosis, hsa04370: VEGF signaling pathway, hsa04380: Osteoclast differentiation, hsa04510: Focal adhesion, hsa04550: Signaling pathways regulating pluripotency of stem cells, hsa04611: Platelet activation, hsa04620: Toll-like receptor signaling pathway, hsa04630: Jak-STAT signaling pathway, hsa04650: Natural killer cell mediated cytotoxicity, hsa04660: T cell receptor signaling pathway, hsa04662: B cell receptor signaling pathway, hsa04664: Fc epsilon RI signaling pathway, hsa04666: Fc gamma R-mediated phagocytosis, hsa04668: TNF signaling pathway, hsa04670: Leukocyte transendothelial migration, hsa04722: Neurotrophin signaling pathway, hsa04725: Cholinergic synapse, hsa04750: Inflammatory mediator regulation of TRP channels, hsa04810: Regulation of actin cytoskeleton, hsa04910: Insulin signaling pathway, hsa04914: Progesterone-mediated oocyte maturation, hsa04915: Estrogen signaling pathway, hsa04917: Prolactin signaling pathway, hsa04919: Thyroid hormone signaling pathway, hsa04923: Regulation of lipolysis in adipocytes, hsa04930: Type II diabetes mellitus, hsa04931: Insulin resistance, hsa04932: Non-alcoholic fatty liver disease (NAFLD), hsa04960: Aldosterone-regulated sodium reabsorption, hsa04973: Carbohydrate digestion and absorption, hsa05100: Bacterial invasion of epithelial cells, hsa05142: Chagas disease (American trypanosomiasis), hsa05146: Amebiasis, hsa05160: Hepatitis C, hsa05161: Hepatitis B, hsa05162: Measles, hsa05164: Influenza A, hsa05166: HTLV-I

(*Continued*)

TABLE 9.6 KEGG PATHWAY analysis.—cont'd

NONPRO-AC-WDCG-KEGG PATHWAY		
ID	**Gene name**	**KEGG PATHWAY**
		infection, hsa05169: Epstein–Barr virus infection, hsa05200: Pathways in cancer, hsa05203: Viral carcinogenesis, hsa05205: Proteoglycans in cancer, hsa05210: Colorectal cancer, hsa05211: Renal cell carcinoma, hsa05212: Pancreatic cancer, hsa05213: Endometrial cancer, hsa05214: Glioma, hsa05215: Prostate cancer, hsa05218: Melanoma, hsa05220: Chronic myeloid leukemia, hsa05221: Acute myeloid leukemia, hsa05222: Small cell lung cancer, hsa05223: Nonsmall cell lung cancer, hsa05230: Central carbon metabolism in cancer, hsa05231: Choline metabolism in cancer,
212610_at	Protein tyrosine phosphatase, nonreceptor type 11 (PTPN11)	hsa04014: Ras signaling pathway, hsa04630: Jak-STAT signaling pathway, hsa04650: Natural killer cell mediated cytotoxicity, hsa04670: Leukocyte transendothelial migration, hsa04722: Neurotrophin signaling pathway, hsa04920: Adipocytokine signaling pathway, hsa04931: Insulin resistance, hsa05120: Epithelial cell signaling in Helicobacter pylori infection, hsa05168: Herpes simplex infection, hsa05205: Proteoglycans in cancer, hsa05211: Renal cell carcinoma, hsa05220: Chronic myeloid leukemia,
209050_s_at	ral guanine nucleotide dissociation stimulator (RALGDS)	hsa04014: Ras signaling pathway, hsa04015: Rap1 signaling pathway, hsa05200: Pathways in cancer, hsa05210: Colorectal cancer, hsa05212: Pancreatic cancer, hsa05231: Choline metabolism in cancer,
209152_s_at	Transcription factor 3(TCF3)	hsa04550: Signaling pathways regulating pluripotency of stem cells, hsa05166: HTLV-I infection, hsa05202: Transcriptional misregulation in cancer,
208067_x_at	Ubiquitously transcribed tetratricopeptide repeat containing, Y-linked(UTY)	hsa05202: Transcriptional misregulation in cancer,

6. Conclusion

Here, we did a comprehensive analysis of GEO dataset of HIV-1 infected samples and KEGG pathway analysis for weighted differentially coexpressed gene to get significant genes responsible for cancer disease. We ultimately found sets of DCEGs, which are significantly responsible for disease progression in human body. Moreover, we have analyzed the preservation patterns of differentially coexpressed genes of each coexpression networks in other pairs of stages which result the modules for strong preservation and the modules which have no preservation evidences. The preservation patterns decipher that the most significant preservation can be found for nonprogressor-actue coexpression network in acute-chronic coexpression network. Also, the correlation between each module of each coexpression network and corresponding *P* values are computed by determining the eigengene networks.

Preservation analysis indicates the preservation of each modules across weighted differentially coexpressed gene networks i.e., WDCGN. To understand this, *Z-summary* measure of preservation of each module over another module is estimated.

After KEGG Pathway analysis, the genes significant in cancer disease also obtained. As a future work, drug target genes and drug discovery for the diseases can be done. However, the further analysis can be done to find out the integrated consensus genes or marker genes. The integrated consensus modules which will show the high expression values in case of disease progression. This analysis will highlight the key genes which will be significantly responsible for disease mechanism. Furthermore, this can direct one to find an avenue of designing of new drugs to build up a strong immune system despite of HIV-1 infection.

From the step-by-step analysis of preservation module obtained from HIV prognosis dataset, the name of genes which are significant in different cancer disease are also found. So, in future gene marker as well as drug target gene of these diseases can be identified and furthermore the drug design may be possible for those disease.

References

[1] K. Devadas, S. Biswas, et al., Analysis of host gene expression profile in HIV-1 and HIV-2 infected T-cells, PLoS One 11 (1) (2016) e0147421, https://doi.org/10.1371/journal.pone.0147421.

[2] M. Catalfamo, C. Wilhelm, et al., CD4 and CD8 T cell immune activation during chronic HIV infection: roles of homeostasis, HIV, type I IFN, and IL-7, J. Immunol. 186 (2011) 2106–2116.

[3] G.M.D. Pantaleo, et al., Studies in subjects with long-term nonprogressive human immunodeficiency virus infection, N. Engl. J. Med. 332 (4) (1995) 209–216.

[4] J. Blankson, Control of HIV-1 replication in elite suppressors, Discov. Med. 9 (46) (2010) 261–266.

[5] S.L. Heath, et al., CD8 T-cell proliferative capacity is compromised in primary HIV-1 infection in, J. Acquir. Immune Defic. Syndr. 56 (2011) 213–221.

[6] S. Kottilil, K. Shin, et al., Innate immune dysfunction in HIV infection: effect of HIV envelope-NK cell interactions1,2, J. Immunol. 176 (2006) 1107–1114.

[7] S.k. Md, M. Hossain, et al., Preservation affinity in consensus modules among stages of HIV-1 progression, BMC Bioinformatics 18 (2017) 181.

[8] S. Ray, et al., A comprehensive analysis on preservation patterns of gene co-expression networks during Alzheimer's disease progression, BMC Bioinformatics 18 (2017) 579.

[9] B. Barman, A. Mukhopadhyay, Detection of differentially expressed genes in wild type HIV-1 Vpr and two HIV-1 mutant Vprs, Adv. Intell. Syst. Comput. 327 (2014) 597–604.

[10] M.R. Bakhtiarizadeh, B. Hosseinpour, et al., Weighted gene co-expression network analysis of endometriosis and identification of functional modules associated with its main hallmarks, Front. Genet. 9 (2018) 453, https://doi.org/10.3389/fgene.2018.00453.
[11] X. Zhang, H. Feng, et al., Application of weighted gene co-expression network analysis to identify key modules and hub genes in oral squamous cell carcinoma tumorigenesis, Onco Targets Ther. 2018 (11) (2018) 6001–6021.
[12] A.S. Feltrin, A.C. Tahira, et al., Assessment of complementarity of WGCNA and NERI results for identification of modules associated to schizophrenia spectrum disorders, PLoS One 14 (1) (2019) e0210431, https://doi.org/10.1371/journal.pone.0210431.
[13] L. Miaoa, R. Xing, et al., Weighted gene Co-expression network analysis identifies specific modules and hub genes related to hyperlipidemia, Cell. Physiol. Biochem. 48 (2018) 1151–1163, https://doi.org/10.1159/000491982.
[14] J.J. Cáceres, A. Paccanaro, Disease gene prediction for molecularly uncharacterized diseases, PLoS Comput. Biol. 15 (7) (2019) e1007078, https://doi.org/10.1371/journal.pcbi.1007078.
[15] B.M. Tesson, R. Breitling, et al., DiffCoEx: a simple and sensitive method to find differentially coexpressed gene modules, BMC Bioinformatics 11 (2010) 497.
[16] P. Langfelder, S. Horvath, WGCNA: an R package for weighted correlation network analysis, BMC Bioinformatics 9 (2008) 559.
[17] F. Rafii, et al., Data preprocessing and reducing for microarray data exploration and analysis, Int. J. Comput. Appl. 132 (16) (2015) 0975–8887.
[18] P. Stafford, Methods in Microarray Normalization, CRC Press, 2008, p. 304. ISBN: 978-1-42005-278-7; hardback.
[19] L.L. Zhenqiu, Y. Ke-Hai, Welch's t test, in: N.J. Salkind (Ed.), Book: Encyclopedia of Research Design, Sage, Thousand Oaks, CA, 2010, pp. 1620–1623, https://doi.org/10.13140/RG.2.1.3057.9607. ISBN: 9781412961271.
[20] B. Zhang, S. Horvath, A general framework for weighted gene co-expression network analysis, Stat. Appl. Genet. Mol. Biol. 4 (1) (2005) 17. Produced by The Berkeley Electronic Press.
[21] P. Langfelder, Signed vs. Unsigned topological overlap matrix, Tech. Rep. (2013).
[22] P. Langfelder, B. Zhang, et al., Defining clusters from a hierarchical cluster tree: the dynamic tree cut package for R, Bioinform. Appl. Note 24 (5) (2008) 719–720.
[23] P. Langfelder, S. Horvath, Eigengene networks for studying the relationships between co-expression modules, BMC Syst. Biol. 1 (2007) 54.
[24] S. Ray, et al., Discovering preservation pattern from Co-expression modules in progression of HIV-1 disease: an eigengene based approach, in: Intl. Conference on Advances in Computing, Communications and Informatics (ICACCI), Jaipur, India, 2016.
[25] S. Horvath, J. Dong, Geometric interpretation of gene Co-expression network analysis, PLoS Comput. Biol. 4 (8) (2008) e1000117.

CHAPTER

10

Computational intelligence for genomic data: a network biology approach

Parameswar Sahu[1], *Fahmida Khan*[2], *Subrat Kumar Pattanayak*[2]

[1]Indian Council of Agricultural Research (ICAR-CIFRI), Barrackpore, West Bengal, India;
[2]Department of Chemistry, National Institute of Technology, Raipur, Chhattisgarh, India

1. Introduction

Biosciences is the front end of current research arena, as it contributes to healthcare systems and health-related community data at a depth of proteome and genome level understanding. Today, the healthcare system generates a huge amount of highly complicated data, which is overwhelming throughout the Internet for research purposes. Big data usage is under rapid growth these days [1]. It is an advantage for the scientific community that the Internet allows us to obtain every type of information in need. Current population and its impact on the expansion of "big data" for healthcare management is on an accelerated hike. It is a better scope for the people working with big data development and management. Adaptations of automated healthcare record systems are highly common in nonfederal acute care hospitals in the United States [2,3]. Advancement in the field of the Internet-of-things with context to the healthcare system gives access to generate and collect personal health data from consumers. Big data basically refers to three sections, such as structured unstructured and semistructured datasets, which demands a need for a high-end storage facility [4]. Massive amount of datasets need an architecture that can afford parallel processing of these complicated attributes. Between centralized storage and distributed storage systems, distributed storage systems are highly preferable as they are cost effective and allow parallel processing of data than centralized storage facility. Invention of cloud server platforms provides high processing resources to users [4]. Cloud environments provide domain-based platform to its users for up gradation of existing databases along with enriched

Handbook of Computational Intelligence in Biomedical Engineering and Healthcare
https://doi.org/10.1016/B978-0-12-822260-7.00005-4

portfolio for healthcare systems. Cloudlets function in a circular circuit pattern, which enables it to work with the genomic and proteomic data, sending them to the web. This is the mainstream of the healthcare sector generating a large amount of electronic datasets. Traditional software and database management systems are not enough for big data management and analysis. Therefore, an advanced computation system is the requirement for such amount of data manipulation and optimization, which can provide a platform to clinicians for better treatment for human health. Such a system is based on the features of the patient's genomic/proteomic dataset, existing information available on genomic/proteomic datasets, and advanced tools and techniques for optimization and manipulation for decision support.

2. Next generation sequencing overview

Next Generation Sequencing (NGS), alias deep sequencing, is the advanced method of DNA sequencing that has revolutionized genomic research. Healthcare sectors generate large volume of genomic datasets by virtue of NGS. Thus, NGS is also called as "Massively parallel sequencing." NGS has transformed modern biology to such an extent that it takes only a single day to sequence an entire genome. Previously used Sanger sequencing method required a decade to publish the final draft of human genome [5]. Development of advanced sequencing technology made it cost-effective and time saving. Small-sized whole genomes can be sequenced within a day. NGS facilitates via its high-throughput method to decrypt the human genome and exploration of novel genes and proteins along with metabolites associated with diseases. Targeted sequencing method is an advantage for the identification of mutation positions responsible for disease. An overview of emerging DNA sequencing technology and applications associated with it has been shown in Table 10.1.

This table describes the development and advances that occur in the DNA sequencing method and associated technologies with different time scale. This process started with the discovery of the double helical structure of DNA by James Watson and Francis Crick in the 1950s. The structure of DNA raised questions in researchers to unravel the genetic information stored in it. This led to the development of two different sequencing methods in the mid-1970s by Frederick Sanger and Allan Maxam and Walter Gilbert. Sanger sequencing is the method that incorporates chain terminating dideoxynucleotides to determine the sequence of a DNA.

Afterward different sequencing methods like shotgun sequencing, paired-end sequencing, pyrosequencing, single-molecule massively parallel sequencing, and next-generation sequencing were developed. Random DNA strands are sequenced by using shotgun sequencing method. The pair-end sequencing method is used to sequence ends of fragments to generate high quality sequence data which is very helpful in detection of genomic rearrangement, repetitive sequences, gene fusion, and novel transcripts. Pyrosequencing is based on the method which relies on the detection of pyrophosphate release and the generation of light on nucleotide incorporation. Massive parallel sequencing is another approach for sequencing which is also known as next-generation sequencing or second-generation sequencing. With this method it is now possible to analyze genomes, exomes, a defined subset of genes, transcriptomes, and even methylation across the genome. These technologies have and will continue to completely transform the clinical practice.

TABLE 10.1 A brief history of advancement of DNA sequencing and associated technologies.

Year	Application milestones	References
1953	Double helical structure of DNA	[6]
1977	Maxam-Gilbert sequencing	[7]
	Genome sequencing	[8]
1982	Shotgun sequencing	[9]
1990	Paired-end sequencing	[10]
1983,1991	Expressed sequence tags	[11,12]
1995	Gene expression analysis	[13]
1996	Pyrosequencing	[14]
1998	Large-scale human SNP discovery	[15]
2000	Massively parallel sequencing by ligation	[16]
2003	Single-molecule massively parallel sequencing	[17,18]
2004	Metagenome assembly	[19]
2005	Bacterial genome resequencing with NGS	[20,21]
2007	(ChIP–seq) using NGS	[22]
	Large-scale targeted sequencing	[23–26]
2007–08	NGS resequencing of human genome and cancer genome	[27–30]
2008	RNA-Seq using NGS	[31–35]
2009	Exome resequencing using NGS	[36]
	Ribosome profiling using NGS	[37]
2010	Completion of phase I of the 1000 genomes project	[38]
	Large genome de novo assembly from short reads	[39]
2011	NGS resequencing of Haplotype-resolved human genome	[40,41]
2016	De novo assembly of human genome with PacBio	[42]
2017	Human genome de novo assembly with nanopore	[43]

Between the time period 2 years from 2005, some industries announced various distinct sequencing platforms viz., Roche (454 sequencing) [44], Illumina (Solexa technology) [45], and Life Technologies (ABI SOLiD sequencing) [46]. All these techs have their own integrity and deficiency. Characteristics features of different sequencing platforms have been given in Table 10.2.

ABI Sanger Sequencer uses dideoxy terminator chemistry and generates 96 single end reads per run whereas Illumina uses reversible terminators and generates both single end and paired end reads. Illumina also generates variable reads per run for different versions

TABLE 10.2 Characteristics features of different sequencing platforms.

Platform\instrument	Introduced in the year	Throughput range (Gb)	Read length (bp)	Error rate (%)	Chemistry
Sanger sequencing ABI 3500/3730	2002	0.0003	Up to 1 kb	0.3	Dideoxy terminator
Illumina MiSeq	2011	0.3–15	1 × 36 to 2 × 300	0.1	Reversible terminators
IonTorrent PGM	2011	0.08–2	Up to 400	1	Proton detection
IonTorrent proton	2012	10–15	Up to 200	1	Proton detection
Illumina MiniSeq	2013	1.7–7.5	1 × 75 to 2 × 150	1	Reversible terminators
Pacific BioSciences PacBio RSII	2013	0.5–1	Up to 60 kb (Average 10 kb, N50 20 kb)	15	Real-time SMS
Illumina NextSeq	2014	10–120	1 × 75 to 2 × 150	1	Reversible terminators
Illumina HiSeq (2500)	2014	10–1000	1 × 50 to 2 × 250	0.1	Reversible terminators
IonTorrent S5	2015	0.6–15	Up to 400	1	Proton detection
Oxford Nanopore MInION	2015	0.1–1	Up to 100 kb	12	Real-time SMS
Pacific BioSciences Sequel	2016	5–10	Up to 60 kb (Average 10 kb, N50 20 k)	NA	Real-time SMS

of sequencers, i.e., MiSeq and MiniSeq produces 25 million reads per run, NextSeq and HiSeq generates 400 million and five billion reads per run. Ion Torrent uses proton detection method and generates single end reads ranging from 500 reads per run to 80 million reads per run for different sequencing platforms. Pacific Biosciences uses real-time single molecule sequencing and generates single end reads ranging between 432 and 660 reads per run for various sequencers. Oxford Nanopore MinION is the only portable sequencer for DNA and RNA sequencing which can produce 30 gigabytes of raw DNA sequence or 7–12 million reads for RNA sequence.

3. Different sequencing platforms

Different sequencing platform like Roche/453 technology, array-based Illumina sequencing platform, Supported Oligonucleotide Ligation and Detection (SOLiD) System are discussed in the following subsections.

3.1 Roche/454

The Roche/453 technology (http://www.454.com) is an incorporation of multiple technologies to influence sequencing technology platforms. The assimilation of emulsion PCR and pyrosequencing technology consolidates its ability to sequence long reads. Nyren and coworker communicated in their work about an advanced sequencing approach that relies on luminometric inorganic pyrophosphate detection method [47]. The advanced 454 GS FLX system involves Titanium chemistry which produces about one million reads per instrument run with lengths of up to 1000 base pairs. Albeit the higher costs as compared to other platforms, the Titanium chemistry gives an advantage to be convenient for various applications, in addition to de novo assembly [48] and metagenomics [49]. Despite of all, 454 sequencing program has a comparably higher error rate for calling indels (deletion) and insertions in homopolymers [50].

3.2 Illumina/Solexa

The array-based Illumina sequencing platform is a DNA sequencing-by-synthesis high tech, that works alongside reversible terminator chemistry [51]. Illumina developed the Genome Analyzer (GA), which is the first sequencing system that generates more than 1 gigabase (Gb) of high-throughput sequences per run within 2–3 days. Furthermore, Illumina upgraded the sequencing platform to GA IIx and HiSeq 2000 systems that produces higher sequence with expanded read lengths. Despite of ultrahigh-throughput sequence generation and cost-effective favors, the Illumina systems' adequacy is limited to short-read sequencing. The dephasing effects are the prime component to limitation for read lengths [52].

3.3 Life Technologies/SOLiD

The SOLiD System is a short-read sequencing method that relies on ligation, distributed by Applied Biosystems (http://www.solid.appliedbiosystems.com). The SOLiD platform was flourished in the George Church laboratory and drafted in 2005 in addition to the *Escherichia coli* genome resequencing [21]. The sequencing library of SOLiD is based on emulsion PCR alike the 454 protocol. The SOLiD system thrives with lowest error rate among the current NGS systems.

3.4 De-novo assembly, sequence alignment and variant calling

NGS Technology works with different file formats, such as; binary base call format (BCL), Sequence Read Format (SRF), FASTQ, FASTA, SAM, BAM, BED, etc. The raw output of all Illumina-based NGS machines like NextSeq, HiSeq, and NovaSeq Sequencing Systems is the binary base call (.bcl) format. These files are named after, and represent base calls per cycle, which is a binary file that contains both the base call and the quality of that base call for every "tile" in every cycle. The binary base call formatted files needs conversion to FASTQ format for further analysis. The FASTQ file format is universally used to represent raw sequencing data in the bioinformatics community. Generated FASTQ data undergoes alignment with a reference genome for further investigation. Numerous mapping softwares/software suites exist such as; BWA, Bowtie2 [53,54], etc. If the sequenced data does not have a reference genome, then de novo method is required to implied on the reads. This method includes the examination of reads against each other to check overlap between them to build a large contig file containing contiguous sequences. Alignment of sequenced data of FASTQ format result in SAM files which is universal for mapped sequence. After aligned to a reference genome, the sequenced reads undergoes variant calling to determine the existence of SNPs, de novo SNVs, and INDELs. In variant calling method, the mapped data and the reference were examined side by side. SAM tools mpileup and the Genome Analysis Toolkit (GATK) are two openly available source for variant calling programs that are based on Bayesian algorithm. In this contemplate the aim is to build a transit between NGS data with network biology to obtain the "virtual targets" for diseases. From sequenced data genes of interest can be obtained. Various NGS data analysis tools and their application have been given in Table 10.3.

TABLE 10.3 NGS data analysis tools and their application.

Tool [source]	Web address	Application
FASTX Toolkit	http://hannonlab.cshl.edu/fastx_toolkit/	An assembly of tools for short-reads preprocessing
The Genome Analysis Toolkit (GATK)	http://bioops.info/2011/05/gatk-the-genome-analysis-toolkit/	An organized software library suite. Available tools include; ➢ Depth of coverage analyzers ➢ Quality score recalibrator ➢ SNP/indel caller ➢ Local realigner
Genome-wide Complex Trait Analysis (GCTA)	http://complextraitgenomics.com/	Evaluates the percentage of phenotypic variance by detection of single nucleotide polymorphisms (SNP) for complex traits
Burrows Wheeler Aligner	http://bio-bwa.sourceforge.net/	Maps low-divergent sequences by three different algorithms; ➢ BWA-backtrack (designed for Illumina sequence reads up to 100 bp) ➢ BWA-SW (longer sequences ranged from 70 bp to 1 Mbp) ➢ BWA-MEM(longer sequences ranged from 70 bp to 1 Mbp)
SAM Tools	http://samtools.sourceforge.net/	SAM tools enables the analyst to manipulate sequence alignments available in SAM format. SAM tools features for ➢ Generating alignments ➢ Indexing ➢ Sorting ➢ Merging in a per-position format
BFAST (Blat-like Fast Accurate Search Tool)	https://sourceforge.net/apps/mediawiki/bfast/	Used for fast and accurate mapping of short reads to reference sequences

TABLE 10.3 NGS data analysis tools and their application.—cont'd

Tool [source]	Web address	Application
Bowtie	http://bowtie-bio.sourceforge.net/index.shtml	Memory-efficient and ultraswift short read aligner
VariationHunter	http://compbio.cs.sfu.ca/software-variation-hunter	High throughput technologies for the detection of structural variation in one or manifold simultaneously
mrFAST	http://mrfast.sourceforge.net/	Microread fast alignment search tool. This tool is "designed to map short reads generated with the Illumina platform to reference genome assemblies; in a fast and memory-efficient manner"
SOAP Aligner	http://soap.genomics.org.cn/soapaligner.html	High-speed and authentic alignment for large amounts of short read datasets generated by Illumina/Solexa genome analyzer
SOAPdenovo	http://soap.genomics.org.cn/soapdenovo.html	Unique short-read assembler to build a de novo draft assembly Especially developed for Illumina GA short reads
SOAPsnp	http://soap.genomics.org.cn/soapsnp.html	Resequencing utility to assemble consensus sequence for the newly sequenced genome
Celera Assembler/CABOG	https://sourceforge.net/projects/breakdancer/	*De novo* whole-genome shotgun DNA sequence assembler.
BreakDancer	https://sourceforge.net/projects/breakdancer/	A Perl package that enables structural variants detection from paired-end sequencing reads
CNVnator	http://sv.gersteinlab.org/cnvnator/	Tool for genotyping and copy number variation (CNV) detection
PEMer	http://sv.gersteinlab.org/pemer/	Tool for construction and analysis of structural variants (SV)
VarScan	http://varscan.sourceforge.net/	This command line based tool is used for "variant detection in massively parallel sequencing data"
Velvet	http://www.ebi.ac.uk/zerbino/velvet/	It is a "sequence assembler for very short reads," maintained by EBI. "Velvet is a de novo genomic assembler specially designed for short read sequencing technologies, such as Solexa or 454, developed by Daniel Zerbino and Ewan Birney at the European Bioinformatics Institute (EMBL-EBI)"
SSAHA2 (Sequence Search and Alignment by Hashing Algorithm)	http://www.sanger.ac.uk/science/tools	A pair wise sequence alignment tool. "SSAHA2 reads of most sequencing platforms (ABI-Sanger, Roche 454, Illumina-Solexa) and a range of output formats (SAM, CIGAR, PSL etc.) are supported. A pile-up pipeline for analysis and genotype calling is available as a separate package"
String Graph Assembler (SGA)	https://github.com/jts/sga	De novo genome assembler based on the string graphs concept. Memory efficient

4. Different scores and parameters involved in biological network

Heuristic progress of science has given access to understanding the complex mechanisms of various diseases such as nephrological diseases, cardiovascular diseases, cancer, and neurological disorders. Previously anticipated theory about diseases has changed, as they are generally caused by multiple icky molecular mutations (multiple noisy/complex molecular mutations) in lieu the outcome of a single deficit. Thus, studies involved with diseases have been using network-centric approaches [55–57]. Network models are the representation of correlation among the entities present in a system. These models are set for the study of complex systems in many disciplines such as computer-aid learning, biological sciences including mathematics. Inside a biological system, molecular complexes are the result of intermolecular interactions. These molecular complexes are also known as biological modules or pathways that execute various biological functions. The paradigm is shifting from one-drug one-target concept toward multiple targets. This is the necessity of time, as in case of complex diseases it is not possible to get deep insight into the pathology and manifestation of the disease. Here the network pharmacology plays an important role to construct novel networks and predict the novel targets and drugs. The whole concept is based on system biology and network pharmacology. The nodes of a network are the basic components which can be metabolites, genes, or proteins. The edges represent the connection between different nodes. These connections between the nodes can be physical, regulatory, or genetic interactions. Nodes having a large number of connections or interactions are known as hubs. Nodes which are connecting two different nodes in the network are called as bridging nodes. In biological studies we are mostly using the scale-free and complex networks. The biological network can be analyzed through different scores and different parameters such as degree centrality, clustering coefficient, network density, node-degree distribution, between centrality, closeness centrality, and eccentricity [58].

(1) Number of nodes: The number of nodes present in the cluster.
(2) Degree centrality: It is defined as the number of links incident upon a node. For a directed network there are two versions for measure: (1) in-degree and (2) out-degree. In-degree: the number of in-coming links or predecessor nodes
(3) Out-degree: the number of out-going links or successor nodes.
(4) Clustering coefficient: The clustering coefficient corresponds to the interconnectivity of a node with its neighbors.
(5) Network density: Estimation of the ratio of probable number of ties between the entities of a network.
(6) Node degree distribution: It is the degree of a node and can be defined as the number of edges connecting to a node, and the degree distribution corresponds to the probability distribution of the degrees in a network.
(7) Between centrality: It determines the frequency for the shortest paths among every pair of nodes pass through that node.
(8) Closeness centrality: Closeness of a node within a connected graph or network is measured as the total of the length of the shortest paths among the node and all other nodes present in the graph.
(9) Eccentricity: Eccentricity is the maximum distance between any pair of nodes.

Obtained genomic information from NGS data can used to generate biological networks using various tools like; Cytoscape (http://www.cytoscape.org/) [59], STRING (https://string-db.org/) [60], DisGeNET (http://www.disgenet.org/) [61], and many more. Cytoscape is an open-source software package for visualization of biological networks comprised of proteins, genes and various interactions. Prediction of gene function and construction of pathways are the tasks can be performed by exploring complex biological interactions by the help of diverse annotation and experimental data. "ReactomeFI" plugin available in Cytoscape software package is based on a functional interaction network of genes unified with human curated pathways obtained from Reactome database and other pathway databases [62]. The plugin lets us access the Reactome Functional Interaction (FI) network out of which FI subnetwork for selected genes can be constructed. The plugin lets us access the Reactome FI network database out of which FI subnetwork for selected genes can be constructed. Reactome FI also facilitates the user to build and analyze network modules for closely interacting set of genes along with allowing this platform to perform functional enrichment to annotate generated modules.

5. Genomic data mining and biological network analysis: a case study

To study the infrastructure of Parkinson's disease, genomic information were availed from the Gene Cards database (https://www.genecards.org/) and the GWAS Catalog (https://www.ebi.ac.uk/gwas/). More than 5000 genes were explored and further validated via literature evidence and using different online databases such as UniProt, GeneAtlas, and MalaCards. Finally, total 70 genes were identified on the basis of functional values obtained from PANTHER [63] database and these encoded protein coding gene sequences of selected genes were retrieved from UniProt database and subjected for network analysis. STRING (https://string-db.org/) web server was used to establish protein networks between selected Parkinson's disease associated genes, with consideration of medium confidence parameter, i.e., 0.4. Highest confidence and lowest confidence parameters for STRING server are 0.9 and 0.1, respectively. The Cluster ONE [64] algorithm of the Cytoscape tool was used to explore interconnections between overlapping proteins within a network (Fig. 10.1).

The Cluster ONE algorithm resulted in weighted clusters of proteins within the network with confidence value. Further, reliable gene functional network was derived from Reactome and other pathway databases using the ReactomeFI plugin. The ReactomeFI plugin constructed an FI subnetwork based on FI data set to depict the underlying evidence of immensely interacting set of genes and to annotate the modules of functionally enriched groups. Network built using Reactome FI has been shown in Fig. 10.2.

Cluster ONE generated six clusters, out of which we choose four clusters to highlight in the network, based on the number of nodes in a cluster. The clusters are made on the basis of density of the genes present in a cluster, quality of the cluster, and the *P*-value. Furthermore, discovered four clusters within the network were labeled as cluster A (20 nodes), cluster B (8 nodes), cluster C (5 nodes), and cluster D (5 nodes). Cluster ONE uses a different color scheme to distinguish significant nodes from insignificant ones. Nodes with *P*-value less than 0.05 (highly significant) were depicted using red (dark gray in printed version) color whereas nodes with *P*-values between 0.05 and 0.1 (significant) were represented using yellow

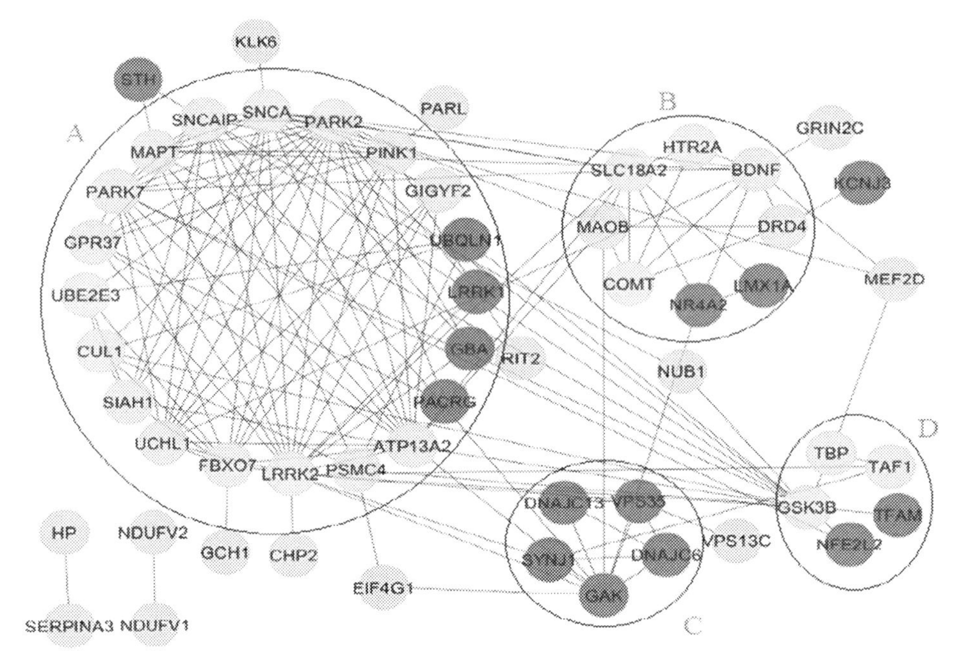

FIGURE 10.1 Schematic representation of clustered genes of network using Cluster ONE plugin in Cytoscape software suite. Different colors (online) represent various groups of genes that are shown in modules.

(light gray in printed version). Similarly, gray was used to decipher less significant nodes within the network. The nodes that are directly or indirectly involved with a disease can be a potential target for drug binding based on the influence on different pathways, molecular functions and cellular components along with expression studies.

6. Summary and conclusions

Big data basically refers to three sections, such as structured, unstructured, and semistructured datasets. Cloud environments provide domain-based platform to users for upgrading existing databases along with enriched portfolio for healthcare systems. Cloudlets function in a circular circuit pattern which enables it to work with the genomic and proteomic data, making them up to the web. This is a mainstream of healthcare sector generating large amount of electronic datasets. Traditional software's and database management systems are not enough for big data management and analysis. Therefore, an advanced computation system is the requirement for such amount of data manipulation and optimization which can provide a platform to clinicians for better treatment for human health. A total of 70 genes were identified on the basis of functional values obtained from PANTHER database and these encoded protein coding gene sequences of selected genes were retrieved from UniProt database and subjected for network analysis. STRING web server was used to establish protein networks between selected Parkinson's disease associated genes, with consideration of medium

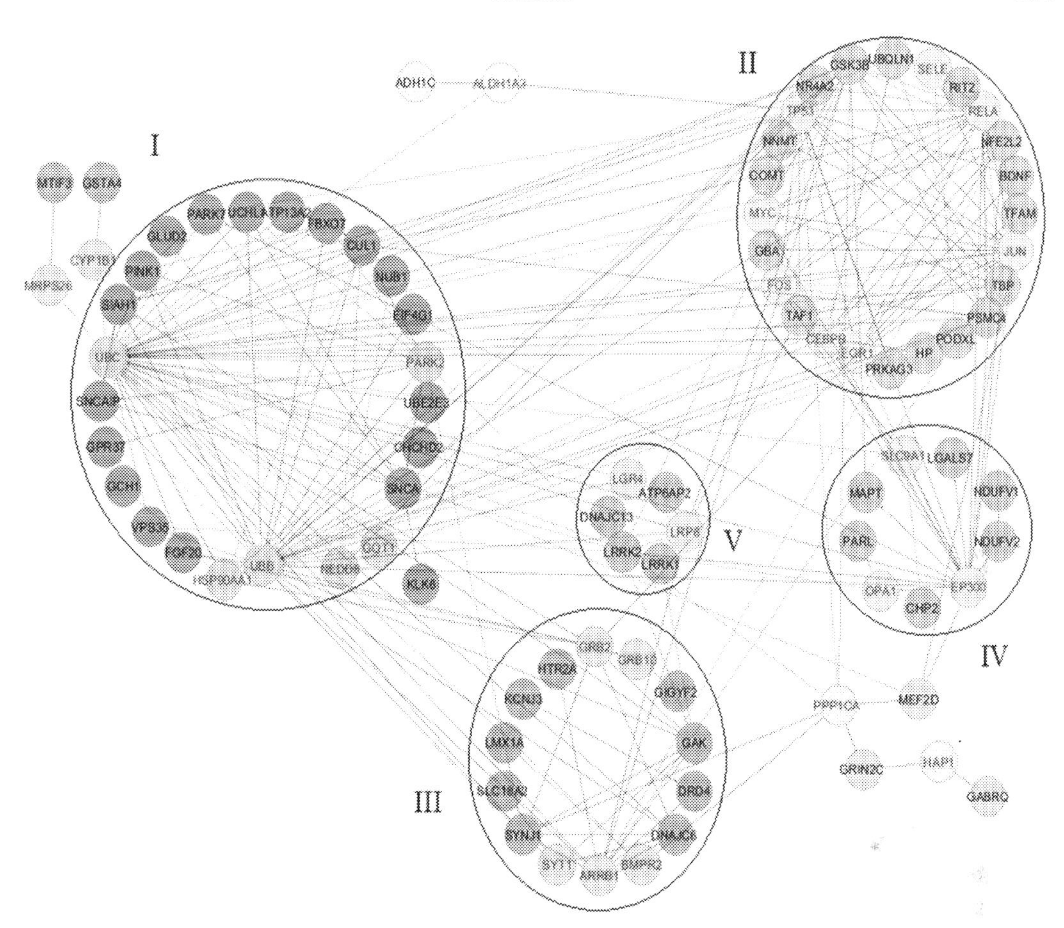

FIGURE 10.2 Schematic representation of constructed network of PD associated genes from pathway enrichment of 70 genes using Reactome FI plugin.

confidence parameter i.e., 0.4. Highest confidence and lowest confidence parameters for STRING server are 0.9 and 0.1, respectively. Employing this concept, novel "virtual targets" for various diseases can be attained based on biological networks constructed from the genomic information of raw sequences, which can revolutionize the era of theranostics in a numerous way. Virtual target or virtual lead is a potential drug target to prevent the irregular mutation [65]. These "virtual targets" may code for proteins. Further, structural and functional characteristics of these novel proteins can be analyzed by virtue of in silico study [66].

References

[1] W. Raghupathi, V. Raghupathi, Big data analytics in healthcare: promise and potential, Health Inf. Sci. Syst. 2 (1) (2014) 3.

[2] R. Zhang, H. Wang, R. Tewari, G. Schmidt, D. Kakrania, May. Big data for medical image analysis: a performance study, in: 2016 IEEE International Parallel and Distributed Processing Symposium Workshops (IPDPSW), IEEE, 2016, pp. 1660–1664.

[3] D. Charles, M. Gabriel, T. Searcy, Adoption of electronic health record systems among US non-federal acute care hospitals: 2008-2012, ONC Data Brief 9 (2013) 1–9.

[4] C.R. Panigrahi, M. Tiwary, B. Pati, H. Das, Big data and cyber foraging: future scope and challenges, in: Techniques and Environments for Big Data Analysis, Springer, Cham, 2016, pp. 75–100.

[5] S. Behjati, P.S. Tarpey, What is next generation sequencing? Arch. Dis. Child. Educ. Pract. 98 (6) (2013) 236–238.

[6] J.D. Watson, F.H.C. Crick, Macmillan publishers ltd molecular structure of nucleic acids, Nature 171 (1953) 737–738.

[7] A.M. Maxam, W. Gilbert, A new method for sequencing DNA, Proc. Natl. Acad. Sci. U.S.A. 74 (2) (1977) 560–564.

[8] F. Sanger, G.M. Air, B.G. Barrell, N.L. Brown, A.R. Coulson, J.C. Fiddes, C.A. Hutchison, P.M. Slocombe, M. Smith, Nucleotide sequence of bacteriophage ϕX174 DNA, Nature 265 (5596) (1977) 687.

[9] F. Sanger, A.R. Coulson, G.F. Hong, D.F. Hill, G.D. Petersen, Nucleotide sequence of bacteriophage λ DNA, J. Mol. Biol. 162 (4) (1982) 729–773.

[10] A. Edwards, H. Voss, P. Rice, A. Civitello, J. Stegemann, C. Schwager, J. Zimmermann, H. Erfle, C.T. Caskey, W. Ansorge, Automated DNA sequencing of the human HPRT locus, Genomics 6 (4) (1990) 593–608.

[11] M.D. Adams, J.M. Kelley, J.D. Gocayne, M. Dubnick, M.H. Polymeropoulos, H. Xiao, C.R. Merril, A. Wu, B. Olde, R.F. Moreno, Complementary DNA sequencing: expressed sequence tags and human genome project, Science 252 (5013) (1991) 1651–1656.

[12] S.D. Putney, W.C. Herlihy, P. Schimmel, A new troponin T and cDNA clones for 13 different muscle proteins, found by shotgun sequencing, Nature 302 (5910) (1983) 718.

[13] V.E. Velculescu, L. Zhang, B. Vogelstein, K.W. Kinzler, Serial analysis of gene expression, Science 270 (5235) (1995) 484–487.

[14] M. Ronaghi, S. Karamohamed, B. Pettersson, M. Uhlén, P. Nyrén, Real-time DNA sequencing using detection of pyrophosphate release, Anal. Biochem. 242 (1) (1996) 84–89.

[15] D.G. Wang, J.B. Fan, C.J. Siao, A. Berno, P. Young, R. Sapolsky, G. Ghandour, N. Perkins, E. Winchester, J. Spencer, L. Kruglyak, Large-scale identification, mapping, and genotyping of single-nucleotide polymorphisms in the human genome, Science 280 (5366) (1998) 1077–1082.

[16] S. Brenner, M. Johnson, J. Bridgham, G. Golda, D.H. Lloyd, D. Johnson, S. Luo, S. McCurdy, M. Foy, M. Ewan, R. Roth, Gene expression analysis by massively parallel signature sequencing (MPSS) on microbead arrays, Nat. Biotechnol. 18 (6) (2000) 630.

[17] J. Shendure, S. Balasubramanian, G.M. Church, W. Gilbert, J. Rogers, J.A. Schloss, R.H. Waterston, DNA sequencing at 40: past, present and future, Nature 550 (7676) (2017) 345–353.

[18] I. Braslavsky, B. Hebert, E. Kartalov, S.R. Quake, Sequence information can be obtained from single DNA molecules, Proc. Natl. Acad. Sci. U.S.A. 100 (7) (2003) 3960–3964.

[19] G.W. Tyson, J. Chapman, P. Hugenholtz, E.E. Allen, R.J. Ram, P.M. Richardson, V.V. Solovyev, E.M. Rubin, D.S. Rokhsar, J.F. Banfield, Community structure and metabolism through reconstruction of microbial genomes from the environment, Nature 428 (6978) (2004) 37.

[20] M. Margulies, M. Egholm, W.E. Altman, S. Attiya, J.S. Bader, L.A. Bemben, J. Berka, M.S. Braverman, Y.J. Chen, Z. Chen, S.B. Dewell, Genome sequencing in microfabricated high-density picolitre reactors, Nature 437 (7057) (2005) 376.

[21] J. Shendure, G.J. Porreca, N.B. Reppas, X. Lin, J.P. McCutcheon, A.M. Rosenbaum, M.D. Wang, K. Zhang, R.D. Mitra, G.M. Church, Accurate multiplex polony sequencing of an evolved bacterial genome, Science 309 (5741) (2005) 1728–1732.

[22] D.S. Johnson, A. Mortazavi, R.M. Myers, B. Wold, Genome-wide mapping of in vivo protein-DNA interactions, Science 316 (5830) (2007) 1497–1502.

[23] T.J. Albert, M.N. Molla, D.M. Muzny, L. Nazareth, D. Wheeler, X. Song, T.A. Richmond, C.M. Middle, M.J. Rodesch, C.J. Packard, G.M. Weinstock, Direct selection of human genomic loci by microarray hybridization, Nat. Methods 4 (11) (2007) 903.

[24] D.T. Okou, K.M. Steinberg, C. Middle, D.J. Cutler, T.J. Albert, M.E. Zwick, Microarray-based genomic selection for high-throughput resequencing, Nat. Methods 4 (11) (2007) 907.

[25] G.J. Porreca, K. Zhang, J.B. Li, B. Xie, D. Austin, S.L. Vassallo, E.M. LeProust, B.J. Peck, C.J. Emig, F. Dahl, Y. Gao, Multiplex amplification of large sets of human exons, Nat. Methods 4 (11) (2007) 931.

[26] E. Hodges, Z. Xuan, V. Balija, M. Kramer, M.N. Molla, S.W. Smith, C.M. Middle, M.J. Rodesch, T.J. Albert, G.J. Hannon, W.R. McCombie, Genome-wide in situ exon capture for selective resequencing, Nat. Genet. 39 (12) (2007) 1522.

[27] D.R. Bentley, S. Balasubramanian, H.P. Swerdlow, G.P. Smith, J. Milton, C.G. Brown, K.P. Hall, D.J. Evers, C.L. Barnes, H.R. Bignell, J.M. Boutell, Accurate whole human genome sequencing using reversible terminator chemistry, Nature 456 (7218) (2008) 53.

[28] D.A. Wheeler, M. Srinivasan, M. Egholm, Y. Shen, L. Chen, A. McGuire, W. He, Y.J. Chen, V. Makhijani, G.T. Roth, X. Gomes, The complete genome of an individual by massively parallel DNA sequencing, Nature 452 (7189) (2008) 872.

[29] J. Wang, W. Wang, R. Li, Y. Li, G. Tian, L. Goodman, W. Fan, J. Zhang, J. Li, J. Zhang, Y. Guo, The diploid genome sequence of an Asian individual, Nature 456 (7218) (2008) 60.

[30] T.J. Ley, E.R. Mardis, L. Ding, B. Fulton, M.D. McLellan, K. Chen, D. Dooling, B.H. Dunford-Shore, S. McGrath, M. Hickenbotham, L. Cook, DNA sequencing of a cytogenetically normal acute myeloid leukaemia genome, Nature 456 (7218) (2008) 66.

[31] N. Cloonan, A.R. Forrest, G. Kolle, B.B. Gardiner, G.J. Faulkner, M.K. Brown, D.F. Taylor, A.L. Steptoe, S. Wani, G. Bethel, A.J. Robertson, Stem cell transcriptome profiling via massive-scale mRNA sequencing, Nat. Methods 5 (7) (2008) 613.

[32] R. Lister, R.C. O'Malley, J. Tonti-Filippini, B.D. Gregory, C.C. Berry, A.H. Millar, J.R. Ecker, Highly integrated single-base resolution maps of the epigenome in Arabidopsis, Cell 133 (3) (2008) 523–536.

[33] A. Mortazavi, B.A. Williams, K. McCue, L. Schaeffer, B. Wold, Mapping and quantifying mammalian transcriptomes by RNA-Seq, Nat. Methods 5 (7) (2008) 621.

[34] U. Nagalakshmi, Z. Wang, K. Waern, C. Shou, D. Raha, M. Gerstein, M. Snyder, The transcriptional landscape of the yeast genome defined by RNA sequencing, Science 320 (5881) (2008) 1344–1349.

[35] B.T. Wilhelm, S. Marguerat, S. Watt, F. Schubert, V. Wood, I. Goodhead, C.J. Penkett, J. Rogers, J. Bähler, Dynamic repertoire of a eukaryotic transcriptome surveyed at single-nucleotide resolution, Nature 453 (7199) (2008) 1239.

[36] S.B. Ng, E.H. Turner, P.D. Robertson, S.D. Flygare, A.W. Bigham, C. Lee, T. Shaffer, M. Wong, A. Bhattacharjee, E.E. Eichler, M. Bamshad, Targeted capture and massively parallel sequencing of 12 human exomes, Nature 461 (7261) (2009) 272.

[37] N.T. Ingolia, S. Ghaemmaghami, J.R. Newman, J.S. Weissman, Genome-wide analysis in vivo of translation with nucleotide resolution using ribosome profiling, Science 324 (5924) (2009) 218–223.

[38] 1000 Genomes Project Consortium, A map of human genome variation from population-scale sequencing, Nature 467 (7319) (2010) 1061.

[39] R. Li, W. Fan, G. Tian, H. Zhu, L. He, J. Cai, Q. Huang, Q. Cai, B. Li, Y. Bai, Z. Zhang, The sequence and de novo assembly of the giant panda genome, Nature 463 (7279) (2010) 311.

[40] J.O. Kitzman, A.P. MacKenzie, A. Adey, J.B. Hiatt, R.P. Patwardhan, P.H. Sudmant, S.B. Ng, C. Alkan, R. Qiu, E.E. Eichler, J. Shendure, Haplotype-resolved genome sequencing of a Gujarati Indian individual, Nat. Biotechnol. 29 (1) (2011) 59.

[41] H.C. Fan, J. Wang, A. Potanina, S.R. Quake, Whole-genome molecular haplotyping of single cells, Nat. Biotechnol. 29 (1) (2011) 51.

[42] J.S. Seo, A. Rhie, J. Kim, S. Lee, M.H. Sohn, C.U. Kim, A. Hastie, H. Cao, J.Y. Yun, J. Kim, J. Kuk, De novo assembly and phasing of a Korean human genome, Nature 538 (7624) (2016) 243.

[43] M. Jain, S. Koren, K.H. Miga, J. Quick, A.C. Rand, T.A. Sasani, J.R. Tyson, A.D. Beggs, A.T. Dilthey, I.T. Fiddes, S. Malla, Nanopore sequencing and assembly of a human genome with ultra-long reads, Nat. Biotechnol. 36 (4) (2018) 338.

[44] B. Steuernagel, S. Taudien, H. Gundlach, M. Seidel, R. Ariyadasa, D. Schulte, A. Petzold, M. Felder, A. Graner, U. Scholz, K.F. Mayer, De novo 454 sequencing of barcoded BAC pools for comprehensive gene survey and genome analysis in the complex genome of barley, BMC Genom. 10 (1) (2009) 547.

[45] O. Morozova, M.A. Marra, Applications of next-generation sequencing technologies in functional genomics, Genomics 92 (5) (2008) 255–264.

[46] K.J. McKernan, H.E. Peckham, G.L. Costa, S.F. McLaughlin, Y. Fu, E.F. Tsung, C.R. Clouser, C. Duncan, J.K. Ichikawa, C.C. Lee, Z. Zhang, Sequence and structural variation in a human genome uncovered by short-read, massively parallel ligation sequencing using two-base encoding, Genome Res. 19 (9) (2009) 1527–1541.

[47] P. Nyrén, B. Pettersson, M. Uhlén, Solid phase DNA minisequencing by an enzymatic luminometric inorganic pyrophosphate detection assay, Anal. Biochem. 208 (1) (1993) 171–175.
[48] A. Gilles, E. Meglécz, N. Pech, S. Ferreira, T. Malausa, J.F. Martin, Accuracy and quality assessment of 454 GS-FLX Titanium pyrosequencing, BMC Genom. 12 (1) (2011) 245.
[49] N.J. Loman, R.V. Misra, T.J. Dallman, C. Constantinidou, S.E. Gharbia, J. Wain, M.J. Pallen, Performance comparison of benchtop high-throughput sequencing platforms, Nat. Biotechnol. 30 (5) (2012) 434.
[50] K.E. Wommack, J. Bhavsar, J. Ravel, Metagenomics: read length matters, Appl. Environ. Microbiol. 74 (5) (2008) 1453–1463.
[51] D.P. Smith, K.G. Peay, Sequence depth, not PCR replication, improves ecological inference from next generation DNA sequencing, PLoS One 9 (2) (2014) e90234.
[52] M.L. Metzker, Sequencing technologies—the next generation, Nat. Rev. Genet. 11 (1) (2010) 31.
[53] H. Li, R. Durbin, Fast and accurate short read alignment with Burrows–Wheeler transform, Bioinformatics 25 (14) (2009) 1754–1760.
[54] B. Langmead, S.L. Salzberg, Fast gapped-read alignment with Bowtie 2, Nat. Methods 9 (4) (2012) 357.
[55] X. Wang, N. Gulbahce, H. Yu, Network-based methods for human disease gene prediction, Briefings Funct. Genom. 10 (5) (2011) 280–293.
[56] S.A. Pendergrass, K. Brown-Gentry, S.M. Dudek, E.S. Torstenson, J.L. Ambite, C.L. Avery, S. Buyske, C. Cai, M.D. Fesinmeyer, C. Haiman, G. Heiss, The use of phenome-wide association studies (PheWAS) for exploration of novel genotype-phenotype relationships and pleiotropy discovery, Genet. Epidemiol. 35 (5) (2011) 410–422.
[57] Y.A. Kim, T.M. Przytycka, Bridging the gap between genotype and phenotype via network approaches, Front. Genet. 3 (2013) 227.
[58] B. Boezio, K. Audouze, P. Ducrot, O. Taboureau, Network-based approaches in pharmacology, Mol. Inform. 36 (10) (2017) 1700048, https://doi.org/10.1002/minf.201700048.
[59] P. Shannon, A. Markiel, O. Ozier, N.S. Baliga, J.T. Wang, D. Ramage, N. Amin, B. Schwikowski, T. Ideker, Cytoscape: a software environment for integrated models of biomolecular interaction networks, Genome Res. 13 (11) (2003) 2498–2504.
[60] D. Szklarczyk, A.L. Gable, D. Lyon, A. Junge, S. Wyder, J. Huerta-Cepas, M. Simonovic, N.T. Doncheva, J.H. Morris, P. Bork, L.J. Jensen, STRING v11: protein–protein association networks with increased coverage, supporting functional discovery in genome-wide experimental datasets, Nucleic Acids Res. 47 (D1) (2018) D607–D613.
[61] J. Piñero, À. Bravo, N. Queralt-Rosinach, A. Gutiérrez-Sacristán, J. Deu-Pons, E. Centeno, J. García-García, F. Sanz, L.I. Furlong, DisGeNET: a comprehensive platform integrating information on human disease-associated genes and variants, Nucleic Acids Res. (2016) gkw943.
[62] G. Wu, E. Dawson, A. Duong, R. Haw, L. Stein, ReactomeFIViz: a Cytoscape app for pathway and network-based data analysis, F1000Research (2014) 3.
[63] P.D. Thomas, A. Kejariwal, N. Guo, H. Mi, M.J. Campbell, A. Muruganujan, B. Lazareva-Ulitsky, Applications for protein sequence–function evolution data: mRNA/protein expression analysis and coding SNP scoring tools, Nucleic Acids Res. 34 (Suppl. l_2) (2006) W645–W650.
[64] T. Nepusz, H. Yu, A. Paccanaro, Detecting overlapping protein complexes in protein-protein interaction networks, Nat. Methods 9 (5) (2012) 471.
[65] A. Gangopadhyay, H.J. Chakraborty, A. Datta, Targeting the dengue β-OG with serotype-specific alkaloid virtual leads, J. Mol. Graph. Model. 73 (2017) 129–142.
[66] H.J. Chakraborty, A. Gangopadhyay, S. Ganguli, A. Datta, Protein structure prediction, in: Applying Big Data Analytics in Bioinformatics and Medicine, IGI Global, 2018, pp. 48–79.

CHAPTER

11

A Kinect-based motor rehabilitation system for stroke recovery

Sriparna Saha, Neha Das

Department of Computer Science and Engineering, Maulana Abul Kalam Azad University of Technology, Kolkata, West Bengal, India

1. Introduction

Physical retardation is one such challenge that needs constant exercise and rehabilitation to improve its condition, and it is found that over a billion world populations are affected with disabilities in some way or the other [1]. Such impairments are caused by patients' motor malfunctions and these disabilities cause restrictions to their normal movements requiring accurate motor control, angle and range of motion with proper strength resulting in their long term dependency. Such people often lack the basic requirements in performing their day-to-day activities like brushing teeth, or getting dressed or even bathing.

One of the most dreadful health problems that lead to quite a number of deaths across the globe every year is "stroke." Statistics in Sweden showed that the number of deaths per year crosses as many as 25,000 [2]. The main cause of stroke lies in the deficiency of oxygen in brain leading to a brain-malfunction or permanent memory loss or a severe brain-injury. A stroke is a result of insufficient reception of oxygen to the brain cells. Motor rehabilitation is very crucial for stroke-affected patients even after they are cured. And, researchers have motivated to find new ideas how doing exercises can be made easy and systematic as well as more acceptable to help engage patients [3].

The initial stage of stroke recovery is to reduce the brain impairment as much as possible and increase normal blood flow to the brain to its optimum level. The more quickly the stroke can be detected, the better the improvement in the patient can be possible. Emergency medical team is needed for maximizing the recovery process. The patients should be admitted to the appropriate stroke recovery center at the earliest. These medical units are equipped with high end medical aids to stabilize the patient's condition. After the recovery of the patient from stroke affected condition, the doctors usually prescribe them some free hand exercises

Handbook of Computational Intelligence in Biomedical Engineering and Healthcare
https://doi.org/10.1016/B978-0-12-822260-7.00007-8

to aid in recovery process. These bodily movements of patients fall into the domain of motor rehabilitation.

By providing motor rehabilitative measures to a disabled person, the person could overcome the physical weakness. This type of rehabilitation [4] could be possible through some training or by some therapeutic measures after proper consultation with a physician. With the new-age improvements in sensor technology [5], the physician can present virtually. This inspired the authors of the current work to delve into human gesture recognition domain for stroke recovery using motor rehabilitation. The exercises, act as rehabilitative tool [6], should be entertaining as well as fruitful. Thus the use of video games for the purpose of providing remedial measures to ailing individuals become famous among medical persons during last decade. The interactive way of doing body movements motivates one to be physically fit without constant intervention of a trainer physically present over there [7,8].

In this chapter, we have proposed a Kinect-based system where the patient does the upper extremities function exercises. After consulting with the doctor, the severity of brain stroke can be gazed. The patients are asked to perform some exercises prescribed by the doctors. As it is inconvenient for the patients to visit the hospital frequently, so we have proposed this system where the performance of the exercises by the patients can be monitored from their homes. This is done using Kinect sensor kindred with the Microsoft's software development kit (SDK) [9]. Here, multilayer perceptron (MLP) [10] along with back propagation [10] algorithm is used to compute the extent of correctness of the performed exercise to measure the improvement of the patient on a daily basis.

Section 2 explains the existing literatures present in the current research area. Section 3 gives the details of the work proposed in this paper followed by the block diagram of the working principle and their descriptions with the exercises considered along with their respective features extracted. Section 4 gives test results and findings of the experiment, and Section 5 draws conclusions from it.

2. Literature survey

There are many strategies that have been developed to realize the concept of computerized vision-based system to help the physically disabled patients with their dedicated set of exercises. Examples of such works are the development of virtual reality (VR) and motion-based games developed for rehabilitation as in Jack et al. [11] where VR based PC has been efficiently used to help in the rehabilitation process in patients of stroke. Table 11.1 is dedicated to give idea about some existing works in the field of motor rehabilitation.

3. Proposed work

The proposed work is implemented to provide a home monitoring system for the stroke affected patients (Fig. 11.1). After affected by the stroke, the patients suffer from muscle stiffness; to overcome this, the doctors suggest them to do specific set of exercises. It is very

TABLE 11.1 Analysis of other methods in literature survey.

Sl. No.	Author names	Factors involved
1.	B. Lange et al. [12]	Authors introduced a support vector machine (SVM) based strategy where features are measured from space-time domain, such that complicated gesture patterns can be identified. But the work has limited applicability to certain actions only; on the contrary our proposed work finds ample scope for a larger set of gestures.
2.	M. Pedraza-Hueso, S. Martín-Calzón, F. J. Díaz-Pernas, and M. Martínez-Zarzuela [13]	Gaming has been done as the mean of entertainment, but the authors invested their time to design a serious game to provide improvement through physical exercises as well as mental functionalities for aged individuals. For ensuring of proper body movements for the participants, the players' are asked to so some tasks and by which their physical functionalities are measured. And for cognitive aspect, the persons were told to memorize the hurdles faced by them during playing. Kinect-based exercise-based games, commonly called "exergames" have been proposed to improve mobility, flexibilities, aerobic capacities, strength, and coordination in patients. The mental and physical activities in the games help the patient with their cognitive as well as motor disabilities simultaneously. Also, another helpful feature of the approach is the remote access of the therapists in configuring the therapies and exercises. Thus, the system intuitively monitors the patients performing their activities and at certain instants, also triggers feedbacks as and when necessary.
3.	V. Nenonen, A. Lindblad, V. Häkkinen, T. Laitinen, M. Jouhtio, and P. Hämäläinen [14]	On that particular paper, the authors changed the way traditional computer games are being played. The authors described a concept of gaming where constant heart rate of the players are monitored and based on that speed of the game changes. So, the players are supposed to maintain a steady heart rate to accomplish a decent score. This game has been developed in lines with skiing and shooting.
4.	C. Schuldt, I. Laptev, and B. Caputo [15]	A rule based approach has been used for the rehabilitation process but the pre-defined dimensional values taken as the initial conditions for the rule may not be a suitable standard for all types of participants since everyone's movement capabilities are not identical. The effectiveness of the system lies in its area of usage which is the patient's home itself where the patient can try out his designated sets of exercises at the comfort of his/her own home and the correctness of his exercise lies in the rule-based approach. The approach assesses the results in real time. Here, for each set of exercises there is a rule-based correctness of exercise prescribed which is independent of the physical appearance of the users. A set of basic rule elements have been designed to express how efficiently and accurately the exercises have been performed by the patients with such motor disabilities. The system has the advantage of multi-mode operations. The system can also study the patients' movements besides providing them with proper guidance and real time assistance.
5.	B. Lange, S. Flynn, R. Proffitt, C.-Y. Chang, and A. Rizzo [16]	The kinect-based system shows a success percentage in monitoring process, however it is meant for patients with cognitive impairments.

(Continued)

TABLE 11.1 Analysis of other methods in literature survey.—cont'd

Sl. No.	Author names	Factors involved
6.	A. Askın, E. Atar, H. Koçyiğit, and A. Tosun [17]	Consequences of virtual reality (VR) training have been investigated for recovery of upper extremity in stroke patients. But, the authors have only investigated a limited number of disordered people, so the work lacks in investigating the efficacy of neuro-rehabilitative measures for a larger mass.
7.	R. Kayyalil, S. Shirmohammadil, A. El Saddik, and E. Lemaire [18]	A 3D environment-based device has been used to facilitate the rehabilitation process in stoke-affected patients for motor rehabilitation. The approach of rehabilitation is effective in its distinct way of interaction with the patient, thus improving the rehabilitation process. It uses a "haptic-based virtual rehabilitation system" with their innovative tactile feedback to help in both upper extremity (UE) rehabilitation and lower extremity (LE) rehabilitation. The paper also presents a dedicated set of exercises especially meant for this kind of virtual environment set up to provide rehabilitative measures. It uses therapists-verified "CyberGlove/ CyberGrasp" and other tactile sensitive hardware devices to assist the process with improvements incorporating into more realistic exercises related to their daily chores or activities like that of eating or drinking actions and many more.
8.	T. Engineering [19]	Although physical or cognitive rehabilitation are complicated processes requiring the omnipresence of clinical experts and proper tools throughout the time of cure and training of the disease through the rehabilitation process, there are certain computerized vision-support systems also which are present currently to provide such proper assistance to these experts in carrying out their duties pretty well thus easing their tasks and improving the quality of rehabilitation training for the patients thereby providing them with an ever increasing motivation. Therefore, such kinds of computerized vision-based systems for physical rehabilitation are of much importance and beneficial if can be implemented successfully.
9.	M. Shaughnessy, C. B. M. Resnick, and C. R. F. Macko [20]	These computerized systems demand disciplined and timely participation in certain set of repetitive exercises that help to train the people with their motor disabilities. These trainings are sufficiently good enough to help the patients overcome their disabilities if constructed with proper sets of exercises and coupled with a routine practice of these exercises. But it is observed in a study that only a certain number of patients with motor disabilities actually perform such recommended physical activities.

crucial to investigate whether the ailed persons are doing these physical activities properly or not. So they need to visit the doctors frequently. The present work comes to rescue from the tedious job of recurrent visit to hospitals and can calculate the day-to-day improvement in the performance of the patient from the ease of the patient's home itself.

The work consists of the following steps:

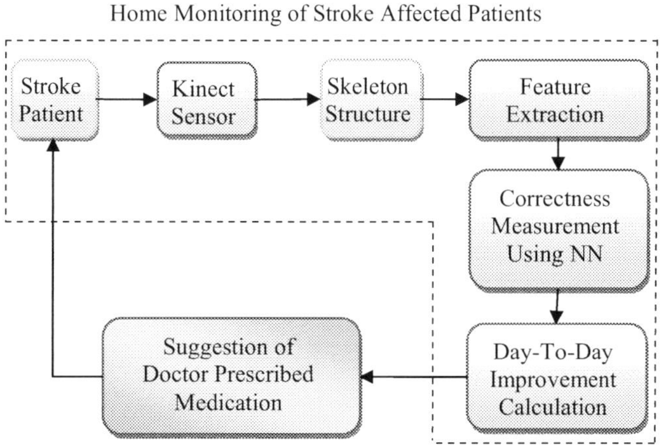

Home Monitoring of Stroke Affected Patients

Stroke Patient → Kinect Sensor → Skeleton Structure → Feature Extraction → Correctness Measurement Using NN → Day-To-Day Improvement Calculation → Suggestion of Doctor Prescribed Medication → Stroke Patient

FIGURE 11.1 Exercise monitoring of stroke-affected patients.

3.1 Kinect sensor

Microsoft's Kinect sensor [21] is used as the data acquisition tool in this work where the major component of Kinect is the VGA RGB camera to capture the image with a depth sensor to calculate the depth map of the user at a certain distance from the Kinect sensor. Here, Kinect helps in recording the skeleton structure of stroke patients by detecting 20 joints at 30 frames per second in three dimensional spaces. Humans present within a range of 12 to 3.5 m can be tracked by this sensory architecture as provided in Fig. 11.2. Not only the physical structure of a human can be traced, but if needed voice gestures can also be monitored by the multi-array microphone assortment. The Kinect is able to tilt along the vertical axis of its own to focus on moving subject up to ±27 degrees. One sample data containing 20 body joins captured by Kinect Xbox 360 is presented in Fig. 11.3 and how the tracking is possible in real-time by capturing humans present in front of it is shown in Fig. 11.4. Kinect finds its applications in wide range of domains from rehabilitative ones to virtual-reality interactions.

Using this compact device we are going to track the patient's response in return of a visual or an audio stimulus and see how effective this device is in training the patient and how capable it is in projecting the patient's actual state of the disease-affected body. Initially, the patient will be asked to perform a series of tasks with the level of fitness being an increasing function of execution of the tasks with possible accuracy, by the patient, in correct order and form. Gradually and steadily the degree of difficulty of the training will increase to reflect the patient's improving condition. First, a set of tasks will be given to the patient and

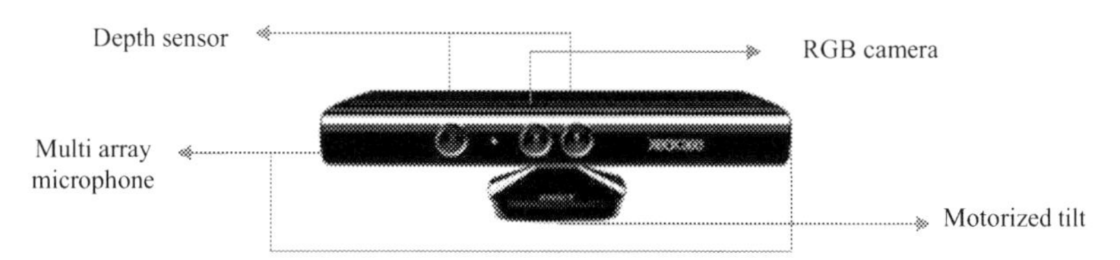

FIGURE 11.2 Different parts of Kinect sensor.

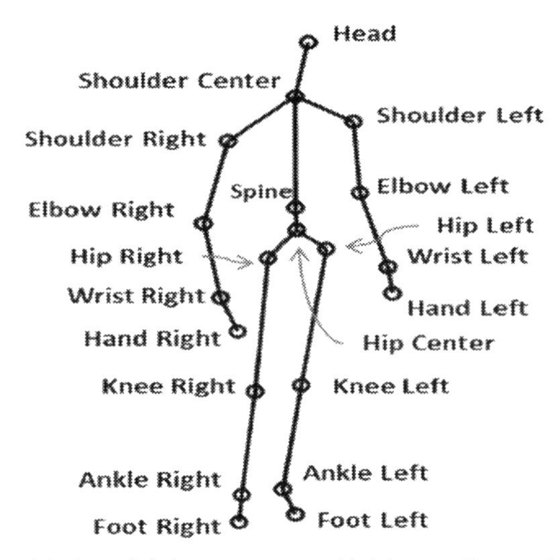

FIGURE 11.3 Pictorial view of skeleton structure with joint coordinates along with their names.

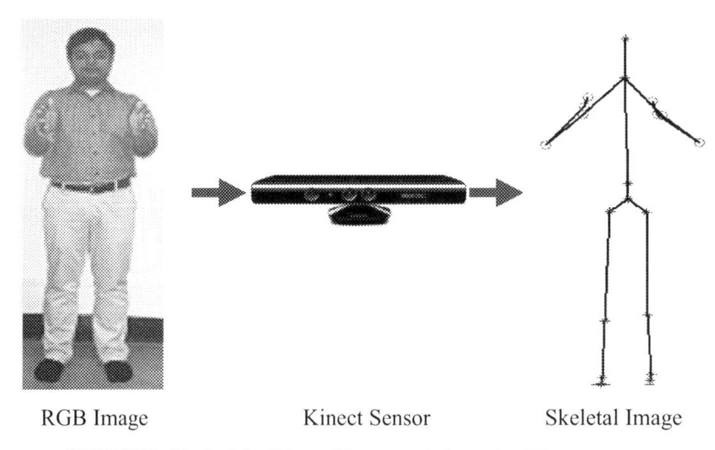

FIGURE 11.4 Tracking of human skeleton by Kinect sensor.

once she/he completes that with a permitted level or zero level of mistakes, the patient will move to the next higher level of the training procedure, that is, training with more difficult set of tasks to be executed.

3.2 Feature extraction

We have prepared a training dataset to measure the closeness of the performed exercise done by the stroke affected patient with the healthy subjects. Thus, for training purpose of neural network (NN), only healthy subjects' data are considered. For a particular exercise, N subjects have been considered. For any n-th subject $(1 \leq n \leq N)$, let total M number of frames be taken at equal intervals. Thus for any m-th frame $(1 \leq m \leq M)$ of the s-th exercise

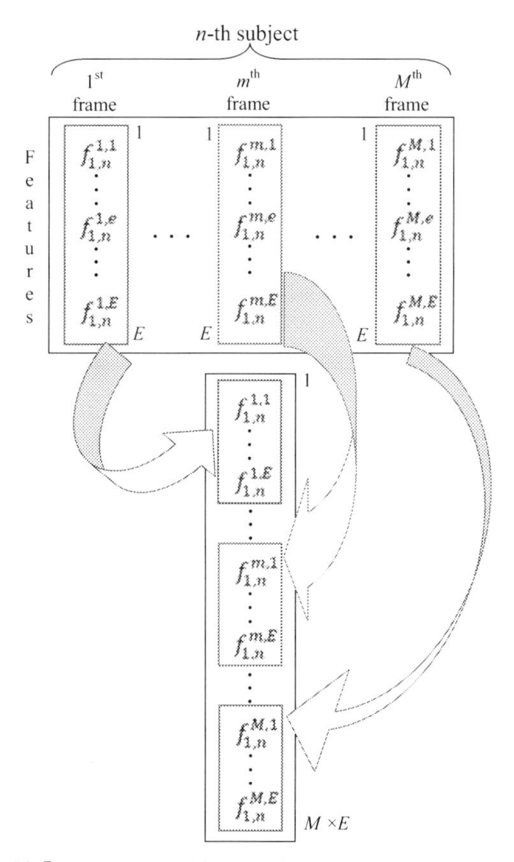

FIGURE 11.5 Extraction of features for training dataset fot n-th subject.

performed by the n-th subject, E ($1 \leq e \leq E$) number of total features have been extracted. The extracted features from all frames are arranged in the manner as depicted by Fig. 11.5, such the features can be fed to the NN. For mapping purpose, for that particular exercise s, the features extracted [22] from stroke patients are depicted.

3.3 Correctness measurement of exercise for stroke recovery

Artificial neural network (ANN) is a simulation based on the biological neural network in human brain that is responsible for processing information. ANN is widely used in various fields of machine learning including speech recognition, pattern recognition, computer aid and vision and text processing. ANN works like a human brain in two ways: (1) acquisition of knowledge through the process of learning, and (2) knowledge thus obtained is stored in the interneuron gap called synapse as different values of weights. ANN is therefore used to perform decision-making by using the knowledge stored using different training sets. ANN can thus also be used in diagnosing diseases by pattern recognition and matching and also, as a classification or a prediction tool in various fields.

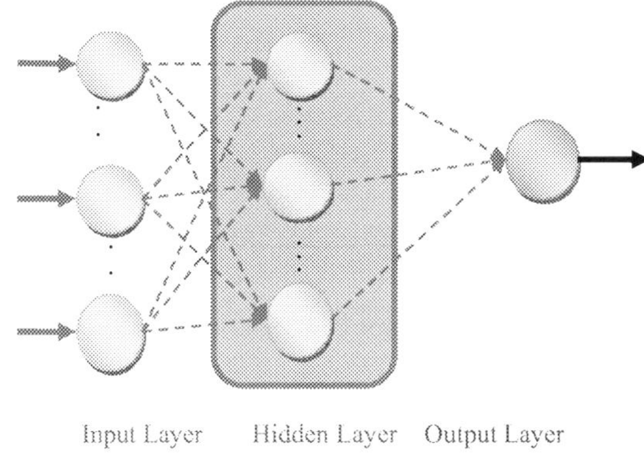

FIGURE 11.6 Neuron configuration for MLP network.

The aim of the work is to dictate the patient to perform a specific exercise and check its correctness with the training dataset. To achieve this, an MLP [10] with back propagation neural network (BPNN) [10] is used. The input layer contains equal number of neurons as that of the size of feature vector for any subject n, that is, $M \times E$. While measuring the correctness of that exercise, we already knew which exercise the patient was performing, so there is only one neuron is present in the output layer (Fig. 11.6).

(1) *Training procedure*: Here, the MLP is fed with random weights during the starting stage of the algorithm for the associated neurons in the first iteration $w^{(k)}$ with the bias equal to 1. The error is calculated by Eq. (11.1).

$$E = ||p - \widehat{p}|| \tag{11.1}$$

And the iterative gradient for the weight space is calculated using the cost function $J(w)$ by given in Eq. (11.2).

$$J(w) = \sum_{n} ||p_{n'} - \widehat{p}_{n'}|| \tag{11.2}$$

Then, the result is obtained in terms of gradient descent as $\frac{\partial J(w)}{\partial w^{(k)}}$ [23]. The value multiplied by a learning rate α by which slope or gradient of loss function is modified. Hence, the new weight is obtained following Eq. (11.3).

$$W^{(k+1)} = w^{(k)} - \alpha \frac{\partial J(w)}{\partial w^{(k)}} \tag{11.3}$$

(2) *Mapping procedure*: The output is a single exercise whose expected value is 1. This output is the exercise that is being performed by the patient. Mapping involves how accurate the exercise is with respect to the specific exercise (done by healthy subjects) the patient needs to perform. If this mapping score is increased after performing the exercises after a certain interval of weeks, then we could say that the patient is recovering from the muscle stiffness generated by the unfortunate stroke.

4. Experimental results

This section is dedicated to provide detailed outline of different stages of the experiment carried out for the proposed work.

4.1 Dataset description

In this work, we have collected data from different age group persons based on their stage of stroke severity. The detailed outline of subjects' description is depicted in Table 11.2.

4.2 Exercises considered for motor rehabilitation

Even after a few months poststroke, many patients still keep on facing troubles in moving their affected upper limbs properly and with ease. This decreased UE function inhibits the patients in doing their daily chores. Since more advanced motor skills are needed to perform this UE related chores, thus its recovery is more crucial as well as more difficult to achieve. Thus, patients having stroke are at a greater risk of not being able to maintain the quality of life. These are the eight easy and very important movements of upper limbs, the exercises ("shoulder abduction," "shoulder adduction," "shoulder extension," "shoulder extension," "shoulder flexion," "elbow flexion," "elbow extension" and "outward rotation") help them to achieve the improved UE functions. Due to shortage of space, only some of the prescribed exercises for right hand are shown pictorially in Fig. 11.7. Left-hand side exercises are also done in the same manner.

Even after three months poststroke, around 37% patients [20] still keep on facing troubles in moving their affected upper limbs properly and with ease. This decreased UE function inhibits the patients in doing their daily chores. Since more advanced motor skills are needed to perform this UE related chores, thus its recovery is more crucial as well as more difficult to

TABLE 11.2 Subject description.

Training dataset			Testing dataset		
Health Condition	**Number of Subjects (N)**	**Age**	**Stroke Severity**	**Number of Subjects**	**Age**
Normal	54	30–39	Chronic	09	55–65 yrs.
			Medium	12	60–65 yrs.
			Acute	16	50–60 yrs.

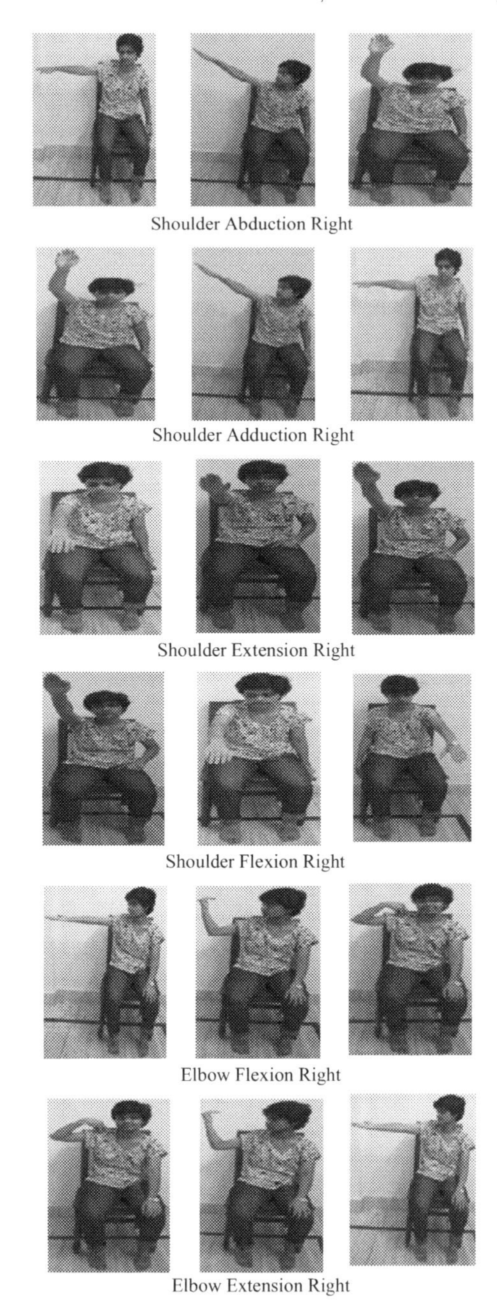

FIGURE 11.7 Exercises considered after consultation with doctors for motor rehabilitation.

achieve. Thus, patients having stroke are at a greater risk of not being able to maintain the quality of life.

There are many ways to achieve the recovery of motor skills to help recover the UE function. The exercises to practice to achieve the improved UE functions are as follows:

1. Shoulder Abduction:
   Step 1: Initially, the arm is in 90 degree angle from the body.
   Step 2: Move the arm upward as much as possible to make a 180 degree from the normal position of the arm.
2. Shoulder Adduction
   Step 1: Initially, the arm in 90 degree angle from the body.
   Step 2: Move the arm downward as much as possible to make a 0 degree i.e., the normal position of the arm.
3. Shoulder Extension:
   Step 1: The arm is flexed along the shoulder joint.
   Step 2: Relax the arm by moving the arm down toward the normal position and extend backward as much as possible.
4. Shoulder Flexion:
   Step 1: Initially, the arm from the shoulder is in the normal or relaxed position.
   Step 2: Move the arm from the shoulder upward along the body as much as possible to flex the shoulder joint as shown.
5. Elbow Flexion:
   Step 1: Initially, the arm stretched out from the elbow.
   Step 2: Move the arm along the elbow to bring it as close to the shoulder as possible to flex the arm from elbow.
6. Elbow Extension:
   Step 1: The arm is initially in the elbow-flexed position.
   Step 2: Extend or stretch out the arm from the elbow.

4.3 Extracted features

All total $E = 12$ features are measured for the current work using MLP NN for motor rehabilitation using Kinect. They are as follows:

1. Distance between shoulder left and elbow left
2. Distance between shoulder right and elbow right
3. Distance between elbow left and wrist left
4. Distance between elbow right and wrist right
5. Distance between spine and elbow left
6. Distance between spine and elbow right
7. Distance between spine and wrist left
8. Distance between spine and wrist right
9. Angle between shoulder left, elbow left and elbow left, wrist left

TABLE 11.3 Skeletal features obtained for a specific subject.

		Frame number			
		1	47	93	139
Skeletal image					
Feature values	1	0.40	0.41	0.40	0.39
	2	0.56	0.56	0.50	0.56
	3	0.30	0.31	0.33	0.30
	4	0.12	0.15	0.20	0.13
	5	0.79	0.79	0.76	0.74
	6	1.25	1.23	1.17	1.24
	7	1.14	1.15	1.13	1.12
	8	1.66	1.71	1.34	1.68
	9	166.22	167.86	169.19	170.15
	10	104.29	104.22	102.61	100.68
	11	149.24	148.98	148.80	146.71
	12	139.47	142.02	148.28	146.28

10. Angle between shoulder right, elbow right and elbow right, wrist right
11. Angle between shoulder center and shoulder left
12. Angle between shoulder center and shoulder right

The features obtained for 10-th subject belonging to "chronic stage" are provided in Table 11.3. The joints that contribute to the feature space are highlighted in the skeletal images.

4.4 Design parameters for multilayer perceptron

In this work an MLP has been fed $M \times E$ number of input features for a particular sample for S number of sample space where $M = 100$ denotes the number of frames at equal interval per sample. $S = 16$ denotes the total number of exercises considering both the left and right side. The total number of computational units or neurons taken in the hidden layer is denoted by H and is equal to 20 and the output is generated using one neuron.

4.5 Performance analysis

We have asked the doctor to differentiate the stroke severity into three stages for each testing dataset. When a patient is diagnosed with "chronic stage," its matching score should be between 0 and 0.33. For "medium stage" and "acute stage," the score values should be in the range of 0.34–0.66 and 0.67 to one respectively. Due to space limitation, results are summarized in Table 11.4. The accuracy (in %) for the proposed work is calculated using Eq. (11.4).

$$\text{Accuracy} = \frac{\text{No. of correctly identified stages}}{\text{Total no. of samples(/subjects)}} \times 100\% \tag{11.4}$$

The Error Histogram plot obtained for the proposed work is represented in Fig. 11.8. The mean square error (MSE) noted for the proposed work is 0.0338. The ROC plot for the proposed work is depicted in Fig. 11.9.

The accuracy of multilayer perceptron (MLP) along with back propagation neural network (BPNN) gradient descent (GD) is compared with three other algorithms namely: Levenberg–Marquardt optimization (LM) [24], radial basis function (RBF) [25] and feed forward neural network (FF) [26]. For BPNN with GD, the parameter set is: maximum number of epochs to train as 1000, learning rate as 0.01, maximum validation failures as six and minimum performance gradient as $1e^{-5}$. For BPNN with LM, the parameter set is: maximum number of epochs to train as 1000, maximum validation failures as 6, minimum performance gradient as $1e^{-7}$, initial mu as 0.001, mu decrease factor as 0.1, mu increase factor as 10 and maximum mu as $1e^{10}$. For RBF, the parameter set is: sum squared error goal as 0.02 and spread constant as 1. The weight adaptation for FF is similarly done like BPNN and size of hidden layer is taken as 10. The accuracies of the competitive algorithms are given in Fig. 11.10.

TABLE 11.4 Results obtained for proposed work.

Testing dataset	Sample number	Matching score	Stage predicted by proposed work	Stage predicted by doctor	Accuracy 12 (%)
Chronic	1	0.24	1	1	
	2	0.19	1	1	
	...	...	...	...	78.9
	11	0.46	2	1	
	12	0.30	1	1	
Medium	1	0.51	2	2	
	...	...	...	...	76.6
	12	0.49	2	2	
Acute	1	1	3	3	
	...	...	...	...	79.5
	16	0.69	3	2	

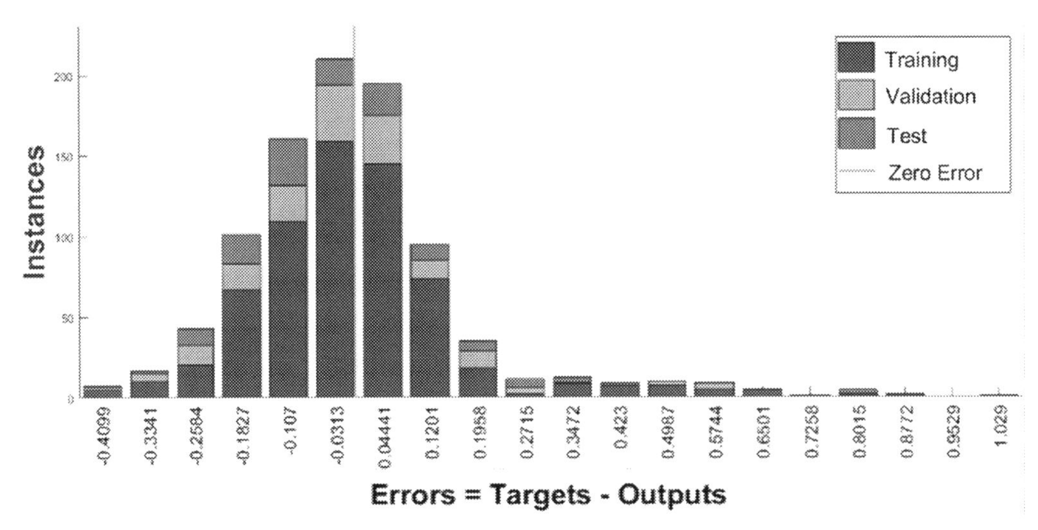

FIGURE 11.8 Error histogram plot obtained in proposed work.

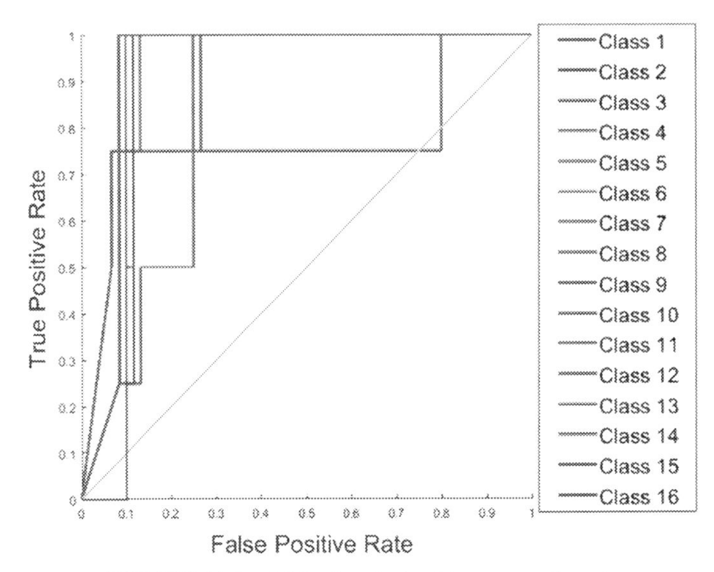

FIGURE 11.9 ROC plot obtained in proposed work.

4.6 Statistical analysis

To statistically justify the outcomes achieved for the proposed work, the authors have carried out two statistical algorithms. For McNemar's statistical method, the chosen algorithm for the current work, i.e., BPNN (GD) is considered as a reference algorithm *A*, while the

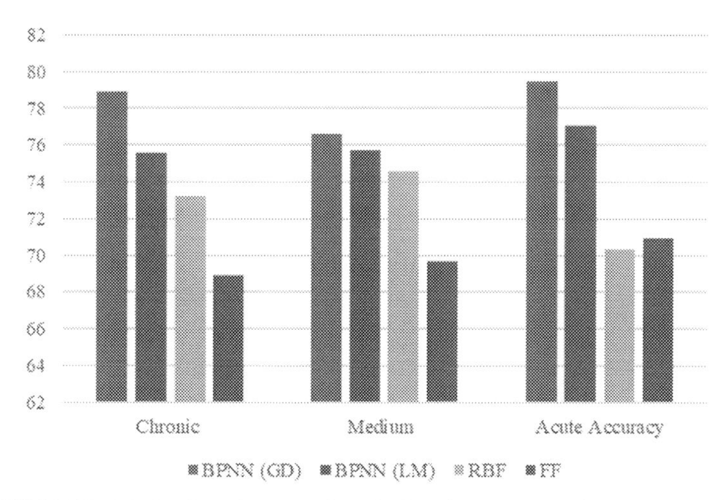

FIGURE 11.10 Evaluation of proposed method with respect to other existing techniques.

other three existing ones from already present literatures are taken as algorithm *B*. Now, how many numbers of times one algorithm is able to detect correctly while the competitor one fails are calculated. n_{01} is that many cases and n_{10} is the number of cases misclassified by *B* but not by *A*. The McNemar's statistic *Z* is measured by Eq. (11.5) [27].

$$Z = \frac{(|n_{01} - n_{10}| - 1)^2}{n_{01} + n_{10}} \tag{11.5}$$

The statistical analysis is carried out for dataset three and the results are shown in Table 11.5. It can be noticed that the null hypothesis is rejected when $Z > 3.84$, where 3.84 is the critical value of χ^2 for 1 degree of freedom at probability of 0.05 [28]. From this table, conclusion can be drawn that the second best method in comparison with the proposed algorithm is BPNN (LM) algorithm and for all the other cases the null hypothesis is rejected.

Another statistical test, the Friedman test, is also performed. The metric (χ^2) for Friedman statistical test is calculated using Eq. 11.6 [29].

$$\chi^2 = \frac{12D}{C(C+1)}\left[\sum_{c=1}^{C} R_c^2 - \frac{C(C+1)^2}{4}\right] \tag{11.6}$$

Here, $D =$ total number of datasets $= 3$ and $C =$ total number of methods used for comparison $= 4$. The values in Table 11.4 clearly shows that χ^2 is greater than the critical value 7.82 for $(C-1) = 3$ degrees of freedom at probability of 0.05, clearly rejecting the null hypothesis (Table 11.6).

TABLE 11.5 McNemar's statistical test.

	A = BPNN (GD)			
B	n_{01}	n_{10}	Z	Comment
BPNN (LM)	6	8	0.0714	Accept
RBF	6	25	10.4516	Reject
FF	4	16	6.0500	Reject

TABLE 11.6 Friedman test.

	Ranking of algorithms				
Algorithm	Chronic	Medium	Acute	Average ranking	Statistics 22 (χ^2)
BPNN(GD)	1	1	1	1.00	8.20
BPNN(LM)	2	2	2	2.00	
RBF	3	3	4	3.33	
FF	4	4	3	3.67	

5. Conclusion and future work

The proposed work deals with rehabilitation of stroke affected patients, performing a specified exercise for upper limbs, using Kinect sensor. Twelve features have been extracted to identify each of the 16 exercises performed by the patient. A multilayer perceptron with GD optimization has been then tested to provide motor rehabilitation for stroke recovery by constant monitoring of the patients while performing the exercises. The proposed work outperforms other standard algorithms with 78.3% accuracy when compared with doctor suggested data given for that specific patient.

The chapter can be brief by the following topics:

(1) *Robustness analysis*: The proposed method is applicable for different sorts of persons having wide variation in their body structures in respect to height, weight and sexuality. Also, the work has been verified for diverse datasets and each time outperforms its competitors.

(2) *Convenience*: The cost associated with the current study is negligible due to its minimum hardware requirements, which is Kinect sensor. Not to forget, this device can work throughout a day without the need for refreshing and is sufficient for a normal size room. As for the proposed study, the images are not captured maintaining the privacy of the users, thus it makes the system suitable for office paces, educational institutes and also for private use in homes.

(3) *Benefits*: The main objective of this work is to provide motor rehabilitative aids for the stroke affected patients for different age groups by providing them necessary physical exercises.
(4) *Productivity*: It has already been shown that the current study is able to produce a high accuracy. Also, the work has been verified with respect to two statistical tests.
(5) *Convergence*: The algorithm is able to converge swiftly, thus within a small amount of time output is generated.
(6) *Flexibility*: As already told by robustness analysis, the work is applied for voluminous dataset, which makes the system flexible enough to implement for real-time applications.
(7) *Feasibility*: There is no question about the feasibility of this work as it finds its applications irrespective of ambience of the place where the patients are performing the rehabilitative exercises. Also, the attire of the patients does not affect the system.
(8) *Efficiency*: It is needless to say that the system having around 100% accuracy is efficient enough for implementation in real-time.
(9) *Computational time*: The computational burden for the classification task is in order of 5s on Intel Core 2 Duo processor @ 1.60 GHz and 8 GB RAM running MATLAB R2017.
(10) *Reliability*: Based on the points discussed in efficiency and computational time, it is evident that the proposed work is reliable.

The future work needs to be done in improving the accuracy level in the data acquisition technique to improve the efficacy of the results. Hence, electromyography is thought upon to be a choice to be added in this work in the future. Electromyography helps to fetch the electrical signals in the skeletal muscles thereby being a good choice in data acquisition in addition to the depth map fetched by Kinect so as to yield a more accurate understanding of the muscles of the limbs of stroke affected patients, thus becoming a more improved motor rehabilitation monitoring system for stroke affected patients.

References

[1] A.K. Roy, Y. Soni, S. Dubey, Enhancing effectiveness of motor rehabilitation using Kinect motion sensing technology, in: 2013 IEEE Global Humanitarian Technology Conference: South Asia Satellite (GHTC-SAS), IEEE, August 2013, pp. 298–304.
[2] D. Webster, O. Celik, Systematic review of Kinect applications in elderly care and stroke rehabilitation, J. NeuroEng. Rehabil. 11 (1) (2014) 108.
[3] P. Langhorne, J. Bernhardt, G. Kwakkel, Stroke rehabilitation, Lancet 377 (9778) (2011) 1693–1702.
[4] D. Piscitelli, Motor rehabilitation should be based on knowledge of motor control, Arch. Physiother. 6 (1) (2016) 1–3.
[5] H.R. Trankler, O. Kanoun, Recent advances in sensor technology, in: IMTC 2001. Proceedings of the 18th IEEE Instrumentation and Measurement Technology Conference. Rediscovering Measurement in the Age of Informatics (Cat. No. 01CH 37188), vol. 1, IEEE, 2001, May, pp. 309–316.
[6] B. Bonnechère, B. Jansen, L. Omelina, S. Van Sint Jan, The use of commercial video games in rehabilitation: a systematic review, Int. J. Rehabil. Res. 39 (4) (2016) 277–290.

[7] W.D. Russell, M. Newton, Short-term psychological effects of interactive video game technology exercise on mood and attention, J. Educ. Technol. Soc. 11 (2) (2008) 294–308.
[8] D.E. Warburton, S.S. Bredin, L.T. Horita, D. Zbogar, J.M. Scott, B.T. Esch, R.E. Rhodes, The health benefits of interactive video game exercise, Appl. Physiol. Nutr. Metabol. 32 (4) (2007) 655–663.
[9] Z. Zhang, Microsoft Kinect sensor and its effect, IEEE Multimed. 19 (2) (2012) 4–10.
[10] V.B. Semwal, M. Raj, G.C. Nandi, Biometric gait identification based on a multilayer perceptron, Robot. Autonom. Syst. 65 (2015) 65–75.
[11] D. Jack, et al., Virtual reality-enhanced stroke rehabilitation, Neural Sys. and Rehab. Engr., IEEE Trans. 9 (3) (2001) 308–318.
[12] B. Lange, S. Koenig, E. McConnell, C.Y. Chang, R. Juang, E. Suma, M. Bolas, A. Rizzo, Interactive game-based rehabilitation using the Microsoft Kinect, in: 2012 IEEE Virtual Reality Workshops (VRW), IEEE, March 2012, pp. 171–172.
[13] M. Pedraza-Hueso, S. Martín-Calzón, F.J. Díaz-Pernas, M. Martínez-Zarzuela, Rehabilitation using Kinect-based games and virtual reality, Procedia Comput. Sci. 75 (2015) 161–168.
[14] V. Nenonen, A. Lindblad, V. Häkkinen, T. Laitinen, M. Jouhtio, P. Hämäläinen, Using heart rate to control an interactive game, in: Proceedings of the SIGCHI Conference on Human Factors in Computing Systems, April 2007, pp. 853–856.
[15] C. Schuldt, I. Laptev, B. Caputo, Recognizing human actions: a local SVM approach, in: Proceedings of the 17th International Conference on Pattern Recognition, 2004. ICPR 2004, vol. 3, IEEE, August 2004, pp. 32–36.
[16] B. Lange, S. Flynn, R. Proffitt, C.Y. Chang, A. "Skip" Rizzo, Development of an interactive game-based rehabilitation tool for dynamic balance training, Top. Stroke Rehabil. 17 (5) (2010) 345–352.
[17] A. Askın, E. Atar, H. Koçyiğit, A. Tosun, Effects of Kinect-based virtual reality game training on upper extremity motor recovery in chronic stroke, Somatosens. Mot. Res. 35 (1) (2018) 25–32.
[18] R. Kayyali, S. Shirmohammadi, A. El Saddik, E. Lemaire, Daily-life exercises for haptic motor rehabilitation. 2007 IEEE International Workshop on Haptic, Audio and Visual Environments and Games, IEEE, 12 October 2007, pp. 118–123.
[19] D. González-Ortega, F.J. Díaz-Pernas, M. Martínez-Zarzuela, M. Antón-Rodríguez, A Kinect-based system for cognitive rehabilitation exercises monitoring, Comput. Methods Progr. Biomed. 113 (2) (2014) 620–631.
[20] M. Shaughnessy, B.M. Resnick, R.F. Macko, Testing a model of post-stroke exercise behavior, Rehabil. Nurs. 31 (1) (2006) 15–21.
[21] Y.J. Chang, S.F. Chen, J.D. Huang, A Kinect-based system for physical rehabilitation: a pilot study for young adults with motor disabilities, Res. Dev. Disabil. 32 (6) (2011) 2566–2570.
[22] P. Rakshit, S. Saha, A. Konar, S. Saha, A type-2 fuzzy classifier for gesture induced pathological disorder recognition, Fuzzy Set Syst. 305 (2016) 95–130.
[23] M. Riedmiller, Advanced supervised learning in multi-layer perceptrons—from backpropagation to adaptive learning algorithms, Comput. Stand. Interfaces 16 (3) (1994) 265–278.
[24] G. Lera, M. Pinzolas, Neighborhood based Levenberg-Marquardt algorithm for neural network training, IEEE Trans. Neural Netw. 13 (5) (2002) 1200–1203.
[25] W. Pedrycz, Conditional fuzzy clustering in the design of radial basis function neural networks, IEEE Trans. Neural Netw. 9 (4) (1999) 745–757.
[26] J. Ilonen, J.K. Kamarainen, J. Lampinen, Differential evolution training algorithm for feed-forward neural networks, Neural Process. Lett. 17 (1) (2003) 93–105.
[27] Q. McNemar, Psychological Statistics, 1955.
[28] T.G. Dietterich, Approximate statistical tests for comparing supervised classification learning algorithms, Neural Comput. 10 (7) (1998) 1895–1923.
[29] J.H. Zar, Biostatistical Analysis, Pearson Education India, 1999.

Appendix

Matlab code for feature extraction

```
clc
clear all
close all
dims=3;
joints=20;
fid = fopen('neha_re7_4.txt');
A = fscanf(fid, '%g', [1 inf]);
P=1;
total=length(A);
frames=total/(dims*joints);
F=zeros(frames,joints,dims);
for i=1:frames
  for j=1:joints
    for k=1:dims
      F(i,j,k)=A(P);
      P=P+1;
    end
  end
end
fclose(fid);
%%
figure;
hold on
axis equal
axis off
for j=1:frames
  if j==139%which no of frame to dispay
    iter=j;
    hold on
    for i=1:joints
      if i<13 && i~=8 && i~=12 && i~=4 && i~=1
        scatter3(F(iter,i,1),F(iter,i,2),F(iter,i,3),'sg','fill');
      else
        scatter3(F(iter,i,1),F(iter,i,2),F(iter,i,3),'+m');
      end
    end
    x=[3;    3;    5;    6;    9;    10;    3];
    y=[5;    9;    6;    7;    10;    11;    2];
    for i=1:length(x)
      plot3([F(iter,x(i),1)    F(iter,y(i),1)],[F(iter,x(i),2)    F(iter,y(i),2)],[F(iter,x(i),3)
F(iter,y(i),3)],'k','LineWidth',2);
```

```
        end
        x=[2; 3; 7; 11; 1; 1; 13; 14; 15; 17; 18; 19];
        y=[1; 4; 8; 12; 13; 17; 14; 15; 16; 18; 19; 20];
        for i=1:length(x)
          plot3([F(iter,x(i),1)            F(iter,y(i),1)],[F(iter,x(i),2)            F(iter,y(i),2)],[F(iter,x(i),3)
F(iter,y(i),3)],'m','LineWidth',1);
        end
    end
end
%%
temp=[];
%x=floor(frames/100);
temp=[];
s=size(temp);
for i= 1: frames
    if iter == 139 break;
    %if s(1)>1200 || s(1)==1200.
    % break
    else
    norm_dist=[sqrt((F(i,5,1)-F(i,9,1))^2)+((F(i,5,2)-F(i,9,2))^2)+((F(i,5,3)-F(i,9,3))^2)];
    feature1=[sqrt((F(i,5,1)-F(i,6,1))^2)+((F(i,5,2)-F(i,6,2))^2)+((F(i,5,3)-F(i,6,3))^2)]/norm_dist;
    feature2=[sqrt((F(i,9,1)-F(i,10,1))^2)+((F(i,9,2)-F(i,10,2))^2)+((F(i,9,3)-F(i,10,3))^2)]/norm_dist;
    feature3=[sqrt((F(i,6,1)-F(i,7,1))^2)+((F(i,6,2)-F(i,7,2))^2)+((F(i,6,3)-F(i,7,3))^2)]/norm_dist;
    feature4=[sqrt((F(i,10,1)-F(i,11,1))^2)+((F(i,10,2)-F(i,11,2))^2)+((F(i,10,3)-F(i,11,3))^2)]/
norm_dist;
    feature5=[sqrt((F(i,2,1)-F(i,6,1))^2)+((F(i,2,2)-F(i,6,2))^2)+((F(i,2,3)-F(i,6,3))^2)]/norm_dist;
    feature6=[sqrt((F(i,2,1)-F(i,10,1))^2)+((F(i,2,2)-F(i,10,2))^2)+((F(i,2,3)-F(i,10,3))^2)]/norm_dist;
    feature7=[sqrt((F(i,2,1)-F(i,7,1))^2)+((F(i,2,2)-F(i,7,2))^2)+((F(i,2,3)-F(i,7,3))^2)]/norm_dist;
    feature8=[sqrt((F(i,2,1)-F(i,11,1))^2)+((F(i,2,2)-F(i,11,2))^2)+((F(i,2,3)-F(i,11,3))^2)]/norm_dist;
    v1=[(F(i,5,1)-F(i,6,1)) (F(i,5,2)-F(i,6,2)) (F(i,5,3)-F(i,6,3))];
    v2=[(F(i,7,1)-F(i,6,1)) (F(i,7,2)-F(i,6,2)) (F(i,7,3)-F(i,6,3))];
    feature9=[((atan2(norm(cross(v1,v2)), dot(v1,v2)))*180/pi)];
    v3=[(F(i,9,1)-F(i,10,1)) (F(i,9,2)-F(i,10,2)) (F(i,9,3)-F(i,10,3))];
    v4=[(F(i,11,1)-F(i,10,1)) (F(i,11,2)-F(i,10,2)) (F(i,11,3)-F(i,10,3))];
    feature10=[((atan2(norm(cross(v3,v4)), dot(v3,v4)))*180/pi)];
    v5=[(F(i,3,1)-F(i,5,1)) (F(i,3,2)-F(i,5,2)) (F(i,3,3)-F(i,5,3))];
    v6=[(F(i,6,1)-F(i,5,1)) (F(i,6,2)-F(i,5,2)) (F(i,6,3)-F(i,5,3))];
    feature11=[((atan2(norm(cross(v5,v6)), dot(v5,v6)))*180/pi)];
    v7=[(F(i,3,1)-F(i,9,1)) (F(i,3,2)-F(i,9,2)) (F(i,3,3)-F(i,9,3))];
    v8=[(F(i,10,1)-F(i,9,1)) (F(i,10,2)-F(i,9,2)) (F(i,10,3)-F(i,9,3))];
    feature12=[((atan2(norm(cross(v7,v8)), dot(v7,v8)))*180/pi)];
      temp=[temp;feature1;feature2;
feature3;feature4;feature5;feature6;feature7;feature8;feature9;feature10;feature11;feature12];
    s=size(temp);
    end
end
```

Screenshot

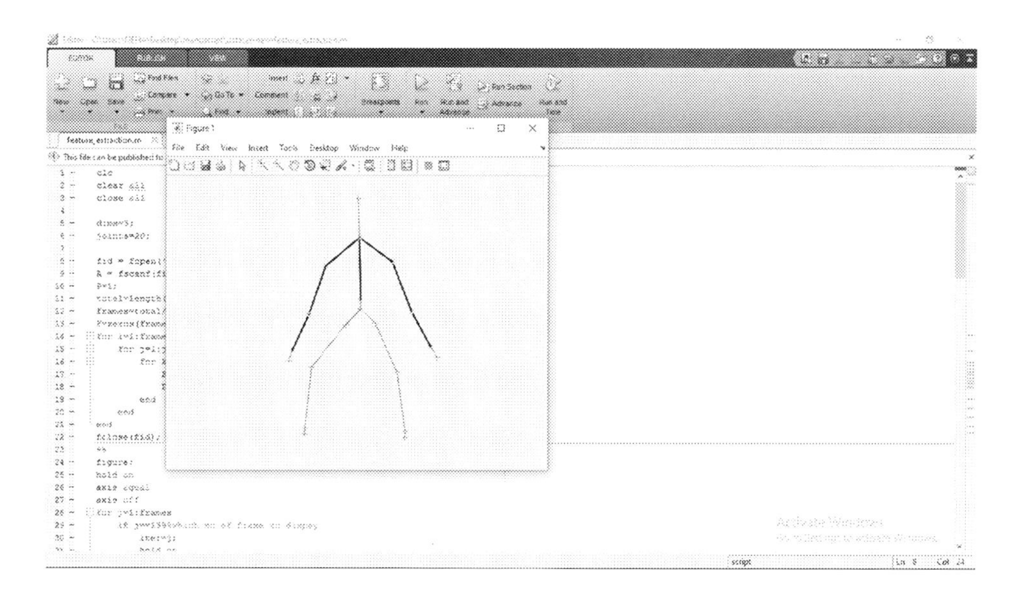

Matlab code for Neural Network Accuracy calculation

```
clear all;
close all;
clc;
input=xlsread('fv.xlsx');
target=xlsread('target.xlsx');
nntic=tic;
hiddenLayerSize = 10;
net = feedforwardnet(hiddenLayerSize,'traingd');
net.trainParam.lr = 0.05; %its not mandatory to give this value, automatic value will be taken
net.trainParam.epochs = 3000; %its not mandatory to give this value, automatic value will be taken
net.trainParam.goal = 1e-5; %its not mandatory to give this value, automatic value will be taken
net.divideParam.trainRatio = 70/100;
net.divideParam.valRatio = 15/100;
net.divideParam.testRatio = 15/100;
net=init(net);
[net,tr] = train(net,input,target); %training
output = sim(net,input); %simulation
figure, plotconfusion(target,output)
plotregression(target,output); %regresson plot
```

```
error = gsubtract(target,output);
performance = mse(error); %mean square error
figure, plotroc(target,output)
nntime=toc(nntic);
unknown=xlsread('unknown.xlsx');%let it is the unknown feature value
y = net(unknown);%results obtained for all classes
```

Screenshot

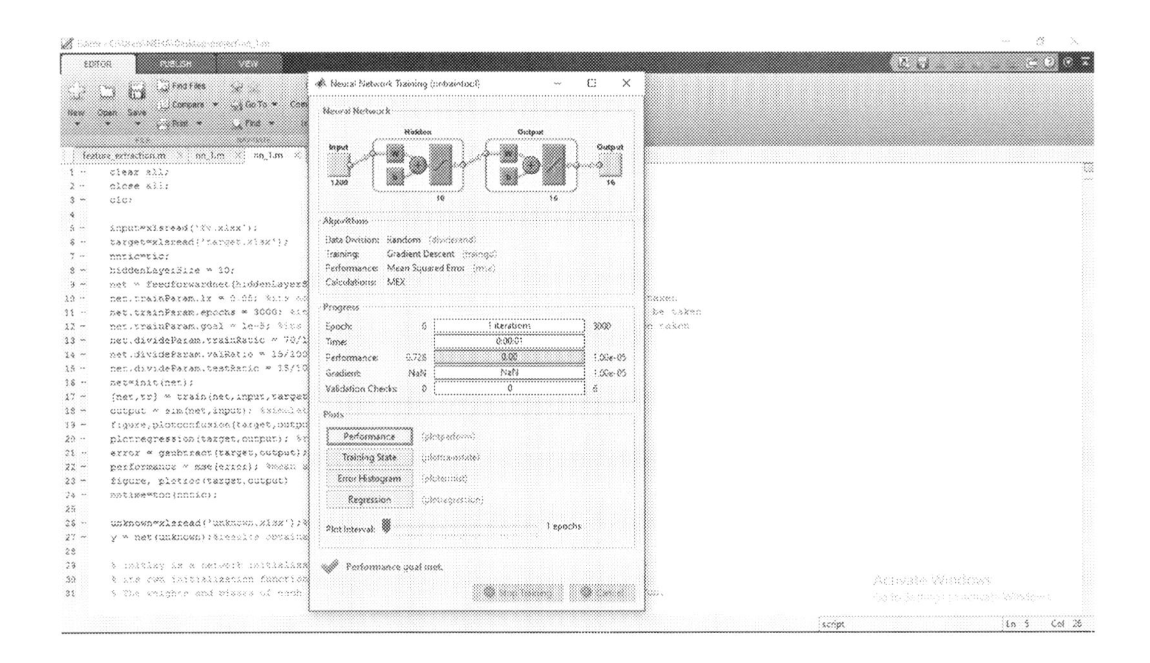

CHAPTER

12

Empirical study on Uddanam chronic kidney diseases (UCKD) with statistical and machine learning analysis including probabilistic neural networks

T. PanduRanga Vital

Department of Computer Science and Engineering, Aditya Institute of Technology and Management, Srikakulam, Andhra Pradesh, India

1. Introduction

Chronic kidney disease (CKD) covers a collection of various pathophysiological forms related to unusual kidney utility, and a dynamic decline in Glomerular Filtration Rate (GFR). It gives a generally acknowledged classification, in light of ongoing rules of the National Kidney Foundation, in which phases of CKD are characterized by the evaluated GFR [1]. The term Chronic Renal Failure applies to the procedure of proceeding with noteworthy irreversible decrease in number of nephrons, and ordinarily compares to CKD stages 3–5. The pathophysiological procedures and adjustments related with ceaseless renal disappointment will be of core interest. It is critical to recognize factors that precipitate risk for CKD, even in people with typical GFR [2]. Risk possibility factors incorporate hypertension, immune system infections, very old age, and family ancestry of renal malady, diabetes mellitus, a past scene of intense renal disappointment, and the nearness of proteinuria, strange urinary residue, or auxiliary variations from the norm of the urinary tract. CKD does not just expand the mortality and dreariness because of its vascular difficulties bringing about cardiovascular and cerebrovascular occasions and CKD movement to end-arrange kidney disappointment; yet in addition due to its antagonistic effect on the economy of the nation. Taking into account that predominance of CKD in India is noted to be 13.8% which

Handbook of Computational Intelligence in Biomedical Engineering and Healthcare
https://doi.org/10.1016/B978-0-12-822260-7.00013-3

itself is extremely high; early location, assessment and preventive administration will be the way to postpone movement and to counteract unfavorable results. In India, nearly 90% patients can't bear the cost of the expense.

CKD is an overall medical issue, influencing a huge number of individuals. The importance of the issue is ailing depicted by the quantity of individuals that will start renal substitution treatment (hemodialysis, peritoneal dialysis, and renal transplantation), as the occurrence of 1 to 3 per 10,000 every year in the all-inclusive community [3]. The stages of CKD depends on the intensity of kidney work estimated by GFR, where stage 1 or step 1 constitutes to harm kidney with elevated or regular GFR and stage 5 or step 5 speaks to a GFR of under 15 mL/min and require treatment with dialysis. The CKD-Epidemiological Collaboration has been presented that the CKD-EPI condition has lower inclination, particularly at an expected GFR more prominent than 60 mL/min per 1.73 m^2; nonetheless, accuracy stays constrained. The improved exactness of the CKD-EPI condition conquers a portion of the confinements of the modification of diet in renal disease (MDRD) study condition and has significant ramifications for general wellbeing and clinical practice. Table 12.1 represents the kidney functionality related to GFR value. It shows stages of the kidney disease with GFR value. If the value is less than 15 then failure of kidneys occurs. If GFR value is above 90 then the kidneys are safe. I will conduct other tests to identify the kidney diseases.

In India, given its populace greater than 1 billion, the rising frequency of CKD is probably going to present serious issues for both social insurance and economical conditions. To be sure, it has been as of late assessed that the age-balanced occurrence pace of end-stage kidney disease (ESRD) in India to be 229 for every million populace and >100,000 [4] new patients enter renal substitution programs yearly in India. The most well-known reason for CKD in populace based investigations is diabetic nephropathy [5,6].

In organization of the paper, the second section covers the literature survey. In this section, I review the so many papers from reputed journals and discuss their research work and results related to this chronic kidney disease. In the third section, I discuss about proposal model and materials in detail. The fourth section provides results and discussions. In this, the Uddanam clinical CKD and non-CKD statistical analysis describes the essential attributes that are identified for causing CKD in Uddanam as well as compare all applied ML algorithms include PNN set-up with incremental neurons hidden layer. Lastly, I provide conclusion section with limitations and scope of work.

TABLE 12.1 Kidney functionality related to GFR values.

Steps or stages	Affect description	GFR ml/min/1.73 m^2
1	Normal or expanded	GFR >90
2	Mildly decreased	GFR 60–89
3	Kidney disease in moderate	30–59
4	Severe kidney disease	15–29
5	Kidney failure	<15 or dialysis

2. Literature survey

In the literature survey section, I collected and referred so many high-quality reputed journal papers concerned with the nephrology problems. Many researchers research in divergent ways present their work for predicting the CKD with using MLs, NNs, and other hybrid algorithms. The medicinal services industry is delivering huge measures of information which should be mine to find concealed data for compelling prediction, investigation, conclusion, and decision making [7]. ML procedures can help and gives some remedy to deal with these conditions. Additionally, CKD problem forecast is a standout among the most focal issues in medical area, since it is a powerful reason for death. So, artificial mechanism for early anticipation of this illness will be helpful to fix. In this paper, the trials were directed for the task of expectation of CKD data acquired from University of California, Irvine Academic & Science (UCI) utilizing the six algorithms in ML (machine learning), to be specific: Random Forest classifiers, SVM, NBs, RBF, and FFN algorithms like Multilayer Perceptron (MLP) and Simple Logistic Regression classifier. The attribute chosen is utilized for training and trying of every classifier independently with 10 times cross-approval. The outcomes got a demonstration that the RF classifier outflanks different classifiers as far as Area under the ROC (AUC), precision, and MCC with qualities differently like 1.0, 1.0 and 1.0.

Sharma et al. [8], used ML algorithms to an issue in the domain of medical diagnosis. The investigation is the determination of the CKD. The dataset utilized for the examination comprises of 400 instances and 24 attributes. The different measurements utilized for performance calculation are analytical accuracy, exactness, affectability, and explicitness. The outcomes show that the decision tree performed best with almost the exactness of 98.6%, affectability of 0.9720, and accuracy of 1 and explicitness of 1. CKD is universal curative problem that approximately 10% of the people were suffered with this in the universe. Chaitanya et al. [4], distinguished hybrid algorithms on CKD using neural networks (ANN) concepts. In this, the algorithms like ANN with GSA, ANN with GA, and kNN are utilized. The aim of this paper is to think about the execution of these two algorithms based on precision and specificity.

CKD is growing gradually, so that the remedies are only with good treatment and early detection of the disease. ML procedures are very important in medicinal analysis on account of their categorization capacity with high precision rates. The precision of organization algorithm rely upon the utilization of exact component determination algorithms to diminish the element of datasets. Polat et al. [7] analyzed CKD using SVM algorithm. The outcomes demonstrated that the SVM classifier by utilizing clarified subset evaluator with the Best First web index choice technique has greater precision rate (98.5%) in the analysis of CKD.

Table 12.2 shows the contribution of author's research work on CKD in detailed that it describes application area, contribution and year of the paper.

CKD is a threatening state of living, can be induced due to malfunctioning of kidney or its pathology. Aljaaf et al. [26], described a multistaged disease, early detection, and the measures of precautions can guarantee the weakening of its progression and can even lead the disease to its demise before reaching to its final stage, where dialysis and transplantation are the only ways to safeguard the patient's life. Their broad study and its associative

TABLE 12.2 Authors reference and their contributions.

Ref. No.	Reference	Contribution	Area of application	year
[7]	Polat et al.	SVM (98.5%)	Diagnosis of CKD	2017
[9]	Abdelaziz et al.	Linear regression (97.8%.)	Predicting of CKD with MLs, IOT cloud computing	2019
[4]	Chaitanya et al.	ANN (56%) and KNN (67%)	Detection of CKD by using ANN and GA	2019
[10]	Kriplani et al.	SVM (97%)	Prediction of CKD using deep ANN technique	2019
[11]	Mani et al.	Random forest (94%)	Chronic renal failure predictions by random subspace	2016
[12]	Hore et al.	Random Forest (92%) MLP (98.33%)	Detection of CKD: An NN-GA-based approach	2017
[13]	Kunwar et al.	ANN (72.73%)	CKD analysis using DM classification techniques	2016
[14]	Salekin et al.	SVM (85.3%)	Detection of CKD and selecting important predictive attributes	2016
[15]	Subas et al.	ANN (98.5%) RF (100%)	Diagnosis of CKD by using random forest	2017
[16]	Vijayarani et al.	ANN (87.70%) SVM (76.32%)	Prediction of KD by SVM and ANN algorithms	2015
[17]	Sinha et al.	SVM (73.75%) KNN (78.75%)	Comparative study of CKD using KNN and SVM	2015
[18]	Dhayanand et al.	NBs (70.96%) SVM (76.32%)	Prediction and classification for KD	2015
[19]	Anantha Padmanaban et al.	Decision tree (91%)	Predicting CKD applying ML techniques	2016
[20]	Borisagar et al.	BPNN	Prediction using BPNN algorithm for CKD	2017
[21]	Hussein et al.	Fuzzy (93%)	Diagnosing KD applying NNs and wavelet analysis	2013
[22]	Chatterjee et al.	NN-MCS(99.6)	Hybrid modified Cuckoo Search-NN in CKD classification	2017
[23]	Kumar et al.	RF (Random forest) (93%)	Prediction of CKD using RF-ML algorithm	2016
[24]	Al-Hyari et al.	ANN (92.2%)	Diagnosis of CKD enforcing NNs	2013
[8]	Sharma et al.	DTs (98.6%)	Classification techniques for CKD diagnosis	2016
[25]	Gunarathne et al.	Multiclass decision forest (99.1%)	Performance evaluation and forecasting through data analytics for CKD	2017

examination of machine learning algorithms have suggested some methods for its early prediction. As this is the motivation for many of the studies, authors support their methodology with the help of predictive analytics that examine the data parameters and their relationship with the target class attribute. The authors finally arrived at the optimum set of attributes using the predictive analysis which enables us to use them for building a set of machine learning predictive models.

In the data-centric world, the data is considered to be the best resource for betterment of living standards. Every industry has been collecting large amounts of data that can be mined; health care industry is not an exception. The data that is being collected can be mined for the diagnosis, exploration, decision making, and effective prediction. It is obvious that machine learning can accomplish this task for them. CKD, a primary reason of demise in the recent past and its prediction being the centered problem in medical decision making, needs the best attention, and any automated tool that can do its early protection will be helpful for the cure. In this observatory, Kumar et al. have examined the performance of different MLs namely random forest, Naïve Bayes, RBF and MLP classifier, and simple logistic on the UCI machine respiratory [23].

Technological evidences show that Internet of Things (IOT) and cloud computing are playing a vital role in health care services and have a decent contribution in predicting the majority of the disease in urban areas. Cloud computing accepts and stores the big data collected by the IOT devices onto CKD. Such that the data can be used in the future and can improve the accuracy in predicting the disease on the environment. It has been a challenging task for the people who interact with health care services to use the prediction of dangerous diseases like that of CKD-based cloud-IOT. Patients are being greatly helped by the cloud computing for the prediction of CKD in the smart cities. They make use of linear regression and neural network techniques to propose a cloud-IOT based hybrid intelligent model for predicting CKD. Abdelaziz et al. [9] use linear regression and neural network in assessing the critical factors that is influencing CKD and for the prediction of CKD, respectively. Their studies have been ended up with a hybrid model with 97.8% accuracy.

As per some surveys, CKD is a global health issue where 10% of the world population is suffered from CKD. So, it is essential to diagnose CKD all over the world in a systematic way and automated diagnosis. To contribute through this paper, the above said automation and the application of systematic analysis uses different types of machine learning. It was obvious that machine learning algorithms are the driving force for many of the optimized analysis and even in search of the abnormalities in the large chunks of psychological data that is the reason behind their application in many of the classification problems with an adequate success rate. In this focused study on detection of the chronic kidney disease, they have tested different machine learning models against the real data set taken from the UCI machine learning repository and their findings has been compared with those of the recent literature. According to their study the random forest classifier has achieved comparatively high success rate in detection of CKD and can even be used for the detection of similar diseases [10].

Machine learning provides a great number of classifiers, which can be used in analyzing the abnormalities of different diseases. The variations in the people, taking kidney transplantation is so wide that their graft outcomes analyzation became very hectic task. Literature has a very good variety of predictive machine learning algorithms for analyzing the graft outcomes. But, to date their systematic analysis hasn't been successfully applied over the kidney transplant. Their study has been focusing on this point and analyzing the best, most suitable ones out of the great variety of machine learning algorithms and asses the best available to make the better predictions such that they can be used as an aid for the decision making in the future [27]. CKD has the most alarming statistics among all other diseases. It is a state of kidney which has either been malfunctioned or subjected to pathology. The phenomenon was uncertain since it can be hidden or go undetected over a few years after being attacked; it is purely depending on the patient's living conditions. Per the current statistical reports there are around 10 percent of the world's population who are suffering from this deadly disease [28]. CKD is a life-threatening state of an individual which may be caused by weakening of kidney functioning or its pathology whose early detection and effective treatment can reduce the chances of reaching the final stage where transplantation or dialysis is the only way to save patient's life. The performance of an algorithm depends on its ability in reducing the dimensions of the data set. SVM classification, a machine learning algorithm, has been used in this study for the early detection of CKD. Wrapper and filter, two types of feature selection methods are used in this study to reduce the dimensions of the CKD data set [3]. The primary concern has been turned to be the early detection and early treatment with preventive medicine at earliest stage that can be possibly detected as the health care costs are being rising to new heights. In their focused study, they have gathered up with an innovative application of deep learning which can be used for the early detection of future diseases. Their model has outperformed all the other existing models with an impeccable performance results in detecting the deadliest diseases, namely, liver cancer, diabetes, and heart failure [29].

Classification and DM concepts, which assume the preclassified data as referenced one, eventually develop a data model which can further be used to classify the large sets of data [30]. The accurate prediction of the target class for every case present in the data is the major objective of this classification model. They have gone through a detailed study in the field of bioinformatics and medical sciences where actually they felt the need of its usage and application, fortunately Weka have provided the basic data mining tool for their study on the detection of CKD. Weka is a great mining tool which initially classifies the data and also makes a great search among the available algorithms which can assure preliminary results for the dialysis of CKD. Their model starts learning from the edge of symptoms of the patients, then tries to identify patients in a large set of patients and healthy beings. So their study has been focused on the accurate prediction of target class using the classification technique through the analysis of the chronic kidney disease data set in each case. The Weka tool is the most widely used DM tool for the analysis of DM models, classifiers applied on different datasets in the view of their performance. Unfortunately, none have accommodated a CKD data set in to these data models using Weka. That is the challenge they have accepted, and they considered six numbers of classifiers applying them in the data set based on various parameters [19].

The life time of the people are decreased due to CKD. Medical treatment considers this an alarming situation to look into and to be solved at the earliest. Through this paper they want to bring up an integrated system for kidney prognosis and diagnosis including case-based reasoning. Their system comprises four major processes to be performed, namely drawing meaningful implicit rules from the health examination data using different data mining tools, performing prognoses based on the rules produced in the early process, diagnosis, and treatment of chronic diseases using case-based reasoning (CBR), expansion of which can conveniently accommodated into an existing system for creating, refining, organizing and shearing of this knowledge. MJ health screening center has supported us with the experimental data, which gone through a systematic analysis. Their detailed reasoning on these aspects may help the people involving in the premises of chronic disease treatment [31].

3. Proposal model and materials

Fig. 12.1 shows the CKD prediction proposal model. In this, I collected the CKD and non-CKD patient's personal and clinical data and environmental records from Uddanam region (area of Srikakulam district, A.P., India) using good questionnaires. The personal information about patients is food habits, drinking and smoking habits, BMI reports, occupation, and so on. The clinical report attributes are like blood reports, urinal, scanning, and so on. For the

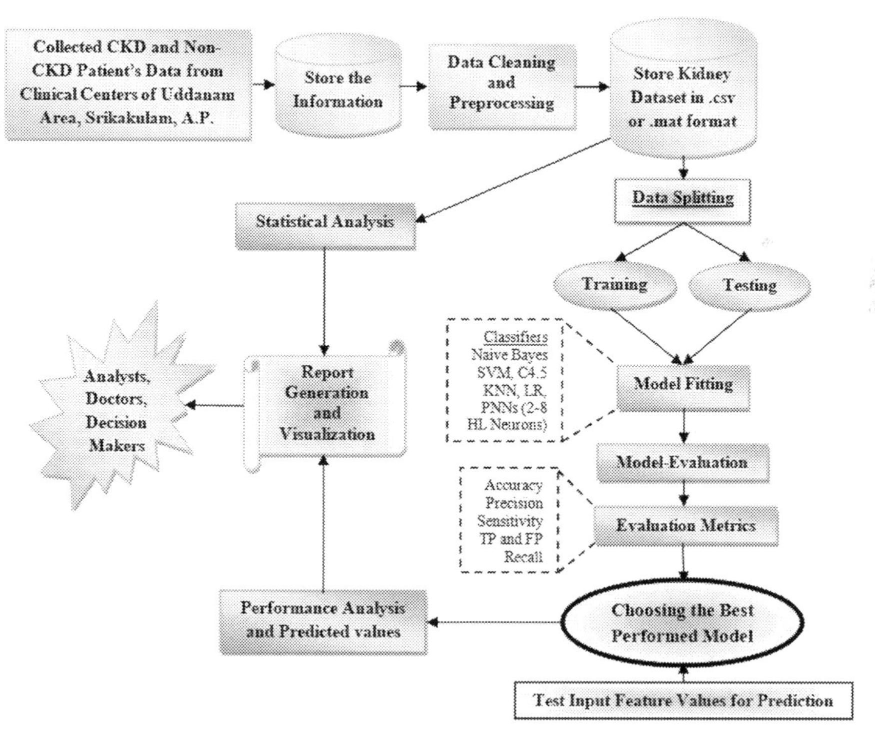

FIGURE 12.1 Chronic kidney disease prediction proposal model.

environmental reports, I gathered water sample reports, soil tests, weather statistics, and so on. The whole information stored into the secondary storage section. In this research, I focused on clinical data for statistical analysis and classification. After data cleaning and preprocessing, I construct the dataset in *.csv format. The total data set is split into two parts that are training data set and testing data set for model fitting. I used ML algorithms and PNNs (2–8 hidden layer neuron PNNs) for classification. Evaluate the model with performance parameters like AC, precision, sensitivity, TP, FP rates, and so on. As per classification analysis, I choose the best performed algorithm for predicting CKD. Lastly, generate the statistical and classification performance reports and predicting values have been send to the analysts, doctors and government for further actions on CKD.

3.1 Sampling the data

The Uddanam area collectedly is around 100 villages. As per statistical CKD severity ratio and rules, collect the CKD and non-CKD data with all age groups, gender, and social status from all villages of Uddanam area people. I also analyzed some of statistically parameters causing CKD in Uddanam like age, sex, height, weight, occupation, sweat formation, food-dirking-smoking habits, tap-ground-mineral-water used, water pH values, soil test analysis reports, and so on. As well, I also collected clinical data with essential 37 attributes. Mainly the statistical values are measured in maximum, minimum, mean, mean deviation, standard deviation, correlations, and so on.

3.2 Algorithm principal component analysis (PCA) algorithm

PCA is an orthogonal linear transformation that transfers the data to a new coordinate system such that the greatest variance by any projection of the data comes to lie on the first coordinate (first principal component), the second greatest variance lies on the second coordinate (second principal component), and so on. PCA used to reduce dimensions of data without much loss of information. It is used in machine learning in very much.

1. Get the mean vector.

Numerically, this should be possible by subtracting the mean and dividing standard deviation (SD) for every value of each factor or variable.

$$Z = \frac{\text{Value} - \text{Mean}}{\text{SD}}$$

Where SD is the standard deviation. In this, every one of the factors or variables will be changed into same scale value.

2. Assemble the every data point or sample in a mean balanced matrix and make the covariance matrix.

This progression is to see how the factors of the data set are changing from the mean as for one another. Sometimes, the factors are very associated or correlated so that they contain excess or redundant data. In this way, so as to distinguish these correlations, I register the covariance matrix. The covariance matrix is an $n \times n$ symmetric matrix with n dimensions.

The covariance $n \times n$ matrix is

$$\begin{bmatrix} Cov(x_0, x_0) & Cov(x_0, x_1) & \ldots & Cov(x_0, x_n) \\ Cov(x_1, x_0) & Cov(x_1, x_1) & \ldots & Cov(x_1, x_n) \\ \ldots & \ldots & \ldots & \ldots \\ Cov(x_n, x_0) & Cov(x_n, x_1) & \ldots & Cov(x_n, x_n) \end{bmatrix}_{n \times n}$$

3. Register the Eigenvalues and Eigenvectors.
4. Register the premise vectors.
5. Each data point is a linear combination (LC) of premise vectors.

3.3 Probabilistic neural networks (PNN) algorithm

The PNN is said to Bayesian theory and also is an estimation of probability density functions (PDFs). It classifies the input vectors X_1, X_2 ..., Xn into two classes that are Class 1 and Class 2 (in the present study) like optimal Bayesian theory. This theory represents the fact with a cost function. As per Bayes rule, the input vector class 1 is classed as in Eq. (12.1).

$$P_a C_a f_a(x) > P_b C_b f_b(x), \quad \forall a \neq b \tag{12.1}$$

Where P_a is the prior probability of occurring samples in class 1, C_a is the cost value of classified input vectors or options, and f_a is PDF of class 1 (C1). Also, as well as P_b, is prior probability of occurring samples in class 2, C_b is cost value of classified input vectors or attributes and f_b is PDF of class 2 (C2). Bayes decision rule is employed to estimate the PDF. The PDF is positive (+ve) for all and it is right to be integral overall input that x is one. The class 1 contains P-sampler feature vectors xi, where $i = \{1, 2, \ldots P\}$ and class 2 contains Q-sampler feature vectors X_j where $j = \{1,2, \ldots Q\}$. So the pattern (second) layer contains P nodes that $P_{1,1}$, $P_{1,2}$, ..., $P_{1,i}$ .... $P_{1,P}$ are grouped under class 1 and Q nodes that are $P_{2,1}$, $P_{2,2}$, ... $P_{2,j}$, ... $P_{2,Q}$ grouped under class 2. If the class 1 point is x_i ($P_{1,i}$) and class 2 point is X_j ($P_{2,j}$) then Gaussian focused equations $g_a(x)$ and $g_b(x)$ is computed as Eqs. (12.2) and (12.3) with N dimension vectors for class 1 and class 2 groups.

$$g_a(x) = \frac{1}{\sqrt{(2\pi\sigma^2)^N}} e^{\left[-\frac{(x-x_i)^T (x-x_i)}{2\sigma^2}\right]} \tag{12.2}$$

$$g_b(x) = \frac{1}{\sqrt{(2\pi\sigma^2)^N}} e^{\left[-\frac{(x-x_j)^T (x-x_j)}{2\sigma^2}\right]} \tag{12.3}$$

Where x is unknown input, x_i is ith sample or parameter from class1, X_j is jth parameter or sample from class 2, σ is smoothing parameter and N specifies the vector length or dimensions.

The average of the PDFs for the N samples in the kth population is measured as follows. The function of $f_a(x)$ value is calculated with sample vectors average distance at each sample in the class 1 group. The $f_b(x)$ value is also calculated like $f_a(x)$, in class 2 group. Like this, any kth node in the summation layer is sums of the values from the class kth group nodes. This process is called Parzen windows or mixed Gaussians. The sums are calculated with Eqs. (12.4) and (12.5).

$$f_a(x) = \left(\frac{1}{\sqrt{(2\pi\sigma^2)^N}}\right)\left(\frac{1}{P}\right)\sum_{i=1}^{P} e^{\left[-\frac{(x-x_i)^T(x-x_i)}{2\sigma^2}\right]} \quad (12.4)$$

$$f_b(x) = \left(\frac{1}{\sqrt{(2\pi\sigma^2)^N}}\right)\left(\frac{1}{Q}\right)\sum_{j=1}^{Q} e^{\left[-\frac{(x-x_j)^T(x-x_j)}{2\sigma^2}\right]} \quad (12.5)$$

Where σ is the smoothing parameter and N intends the vector length or dimensions. P is the number of feature vectors in the class 1 group and Q indicates the number of feature vectors under class 2 group. The weights between the summation layer and output layer are Y_a and Y_b (Eqs. 12.6 and 12.7).

$$Y_a = 1 \quad (12.6)$$

$$Y_b = -\frac{P_b C_b Q}{P_a C_b P} \quad (12.7)$$

3.3.1 Steps for PNN training net as follows

Step 1: For every training input vector X_i, for $i = 1, 2, 3 \ldots, N$ execute Steps 2 to 3.

Step 2: Create Pattern net P_i. P_i weight vector: $w_i = x_i$, where the P_i is in class 1 or in class 2 group.

Step 3: Pattern layer added to the Summation layer.

3.3.2 Steps for PNN application algorithm as follows

Step 1: Initialize weights by utilizing the PNN training algorithm.

Step 2: For classification of input vectors, perform three to five steps.

Step 3: Pattern layer: Compute net input with weights W_j using Eq. (12.8).

$$p_{inj} = x \cdot w_j = x^T \cdot w_j \quad (12.8)$$

Compute net output using Eq. (12.9)

$$P = e^{\left[\frac{p_{inj}-1}{\sigma^2}\right]} \tag{12.9}$$

Step 4: Summation layer: The summation layer for class 1 and class 2 with using weights as Eqs. (12.10) and (12.11)

$$Y_a = 1 \tag{12.10}$$

$$Y_b = -\frac{P_b C_b Q}{P_a C_b P} \tag{12.11}$$

Step 5: Output layer: The output is the sum of the signal from f_a (C1) and f_b (C2).

4. Results and discussions

In this section, I will discuss about statistical analysis, machine learning, and PNN results for prevention, prediction, and protection of the Uddanam CKD. I have analyzed some of parametric statistics for causing CKD in Uddanam with personal information as well as clinical information. In the collected data has personal information like age, sex, height, weight, occupation, sweat formation, food-dirking-smoking habits, tap-ground-mineral-water used and so on. Most of the Uddanam CKD individuals are field workers. CKD people in Uddanam have drinking alcohol and smoking habits than NON-CKD people in there. As per analysis, most of the people used ground water in previous years. In these days they are using mineral water.

4.1 Statistical analysis Uddanam CKD dataset

The clinical data information attributes are 38 include class attribute (CKD and non-CKD) such as age, sex, blood pressure, specific gravity, and sugar consumption; blood hemoglobin, PCV, RBC count, WBC count, platelet count, pus cells, pus cell clamps, differential leukocytes count, glucose, urea, creatinine, uric acid, and bilirubin; serum albumin, sodium, potassium, and chloride; bacteria, etc. Mainly the statistical values are measured in maximum, minimum, mean, mean deviation, standard deviation, correlations, and so on. As per statistics of clinical data, mean age of the people with CKD in Uddanam is 52.15 ± 16.08 and more CKD people are men (74.7%) than women. The main indication of CKD is high B.P. or low B.P. that 543 (82.8%) CKD individuals are suffered with high B.P. (272 (41.5%)) and low B.P. (271 (41.3%)) compared to non-CKD individuals.

Table.12.3 discloses the statistical analysis of every attribute. Most of the attributes like albumin, bacteria, neutrophils, and puss-cells disclosed CKD identification. Another factor

TABLE 12.3 Statistical analysis of Uddanam data set non-CKD and CKD.

Data feature	Category	Ckd measures	Non_Ckd measures	Data feature	Category	Ckd measures	Non_Ckd measures
Age	Mins	2.00	4.00	**Neutrophils**	Mins	1.40	1.40
	Means	52.15 ± 16.08	46.94 ± 15.95		Means	59.17 ± 13.15	59.42 ± 13.30
	Medians	55.00	47.00		Medians	61.20	61.40
	Maxs	90.00	80.00		Maxs	84.90	84.90
Sex	Male	490(74.7%)	301(75.4%)	**Red blood cells**	Normal	369(56.2%)	370(92.7%)
	Female	166(25.3%)	98(24.6%)		Abnormal	287(43.8%)	29(7.3%)
Blood pressure	Normal	113(17.2%)	260(65.2%)	**Bacteria**	Present	222(33.8%)	12(3.0%)
	High	272(41.5%)	8(2.0%)		Not present	434(66.2%)	387(97.0%)
	low	271(41.3%)	131(32.8%)				
Specific gravity	Min	1.00	1.00	**Potassium**	Min	3.00	3.30
	Mean	1.01 ± 0.00	1.02 ± 0.00		Mean	6.03 ± 8.35	4.37 ± 0.59
	Median	1.01	1.02		Median	4.90	4.50
	Max	1.02	1.02		Max	116.30	5.50
Albumin	Min	1.44	0.00	**Chloride**	Min	10.20	93.60
	Mean	4.57 ± 10.87	0.18 ± 0.54		Mean	106.83 ± 8.82	103.31 ± 4.81
	Median	3.47	0.00		Median	105.80	102.30
	Max	14.00	5.00		Max	144.70	121.00
Sugar	Min	0.00	0.00	**Hemoglobin**	Min	4.00	10.30
	Mean	1.57 ± 1.777	0.09 ± 0.29		Mean	9.37 ± 2.88	14.98 ± 1.53
	Median	1.00	0.00		Median	8.70	15.00
	Max	5.00	1.00		Max	26.20	18.50
Mcv	Min	7.40	49.00	**Packed cell**	Min	10.70	29.00
	Mean	81.21 ± 10.63	80.84 ± 8.55		Mean	26.73 ± 7.40	45.31 + 4.89
	Median	82.00	80.00		Median	25.00	45.00
	Max	110.00	100.00		Max	50.30	54.00
Platelet count	Min	0.10	1.00	**White blood cells**	Min	1290.0	3200.00
	Mean	2.31 ± 0.92	2.42 ± 0.84		Mean	7050.05 ± 375.8	7551.88 ± 1908.55
	Median	2.20	2.30		Median	6560.00	7300.00
	Max	7.90	7.90		Max	58,802.00	13,200.00
Pus cell clumps	Present	435(66.3%)	20(5.0%)	**Pus cell**	Normal	334(50.9%)	371(93.0%)
	Not present	221(33.7%)	379(95.0%)		Abnormal	322(49.1%)	24(7.0%)

Blood glucose	Min	22.00	70.00	**Lymphocytes**	Min	10.50	10.50
	Mean	180.29 ± 88.21	114.72 ± 33.16		Mean	30.79 ± 10.83	30.75 ± 10.93
	Median	156.00	111.00		Median	29.20	29.20
	Max	490.00	312.00		Max	68.90	68.90
Blood urea	Min	1.50	10.00	**Eosinophils**	Min	0.00	0.00
	Mean	72.30 ± 57.87	35.37 ± 14.94		Mean	3.78 ± 3.97	3.64 ± 3.80
	Median	53.00	34.00		Median	2.60	2.60
	Max	391.00	95.00		Max	22.50	22.50
Serum	Min	0.50	0.40	**Monocytes**	Min	0.60	0.60
	Mean	3.29 ± 4.34	1.01 ± 0.60		Mean	5.55 ± 3.39	5.45 ± 3.22
	Median	2.40	0.90		Median	4.80	4.80
	Max	48.10	5.90		Max	44.00	44.00
Urea	Min	10.96	10.96	**Basophils**	Min	0.00	0.00
	Mean	204.74 ± 1812.12	217.52 ± 1906.06		Mean	0.13 ± 0.20	0.13 ± 0.20
	Median	77.94	77.94		Median	0.10	0.10
	Max	27,066	27,066.00		Max	1.80	1.80
Creatinine	Min	0.26	0.26	**Bilirubin**	Min	0.02	0.02
	Mean	26.09 ± 335.30	22.78 ± 304.09		Mean	0.73 ± 1.26	0.71 ± 1.21
	Median	7.93	7.83		Median	0.49	0.48
	Max	6089.00	6089.00		Max	18.22	18.22
Uric_Acid	Min	0.22	0.22	**Red blood cells count**	Min	1.27	2.10
	Mean	6.73 ± 2.08	6.68 ± 2.11		Mean	3.42 ± 0.89	5.19 ± 0.77
	Median	6.76	6.75		Median	3.29	5.20
	Max	16.54	16.54		Max	6.29	6.50
Diabetes	Yes no	396(60.4%)	9(2.3%)	**Anemia**	Yes	302(46.0%)	13(3.3%)
		260(39.6%)	390(97.7%)		No	354(54.0%)	386(96.7%)
Hypertension	Yes no	396 (60.4%)	15 (3.8%)	**Coronary Artery**	Yes	191 (29.1%)	17 (4.3%)
		260 (39.6%)	384 (96.2%)		No	465 (70.9%)	382 (95.7%)

for the identification of the CKD is the serum test, the mean value of the CKD serum is more (3.29 ± 4.34) than that of non-CKD (1.01 ± 0.60). As well, hypertension is the cause of CKD that the statistics are 60.4% people with CKD and only 3.6% of cases in non-CKD.

4.2 Classification algorithms

In this research, I compared the five mostly high performed algorithms for the Uddanam Nephrology Dataset like SVM, C4.5, naïve Bayes, kNN and logistic regression measured with performance measures like Classification accuracy, Sensitivity, Specificity, Area under ROC curve, Information score F-measure Precision Recall, and Brier score Mathew correlation coefficient. Every ML algorithm gives the very high accuracy for predicting the CKD with this training data set.

4.2.1 Naivey Bayes classifier

As per analysis of this model on Uddanam CKD data set, the correctly Classified Instances are 1033(97.91%) and Incorrectly Classified Instances are 22 (2.09%). The Kappa statistic values is 0.9655 and Mean absolute error value is 0.0164. The Root mean squared error is 0.1245, Relative absolute error is 3.4958%, and Root relative squared error is 25.6649%. As per analysis of confusion matrix TP values is 647, FP value is 9, the TN value is 13 the true negative value is 386. I measured the characteristics of accuracy measured TP Rate, FP rate, Precision, Recall and F-measure values of CKD and NON-CKD. Table 12.4 shows the computed values of accuracy measures. Table 12.4 shows the Confusion Matrix of the model Naïve Bayes. The total data points are 1055 split into classified instances (647 + 386) and error classified instances (9 + 13).

Table 12.5 depicts the performance parameters like Classification accuracy (CA), Sensitivity (Sens), Specificity (Spec.), Area under ROC curve (AUC), Information score (IS), F-measure (F1), Precision (Prec.), Recall, Brier score (BC), Mathew correlation coefficient (MCC) of the model Naïve Bayes on Uddanam CKD dataset. The total data points are 1055 split into classified instances (647 + 386) and error classified instances (9 + 13), so the avg CA value is 0.9791 and AUC values is 0.9986.

TABLE 12.4 Confusion matrix Naïve Bayes.

Class	CKD(P)	non_CKD(N)	Total
CKD(T)	647	9	**656**
NON-CKD(F)	13	386	**399**
Total	**660**	**395**	**1055**

TABLE 12.5 Naïve Bayes classification CKD and non-CKD performance values.

Class	CA	Sens	Spec	AUC	IS	F1	Prec	Recall	Brier	MCC
CKD	0.9791	0.9695	0.9599	0.9986	0.9141	0.9725	0.9755	0.9695	0.0362	0.9276
Non-CKD	0.9791	0.9674	0.9863	0.9986	0.9141	0.9723	0.9772	0.9674	0.0362	0.9556
AVG	**0.9791**	**0.9685**	**0.9731**	**0.9986**	**0.9141**	**0.9724**	**0.9764**	**0.9685**	**0.0362**	**0.9416**

4.2.2 Support vector machine (SVM) classifier

As per analysis of SVM model that the correctly classified Instances are 1026 (97.25% of accuracy) out of 1055 instances and Incorrectly Classified Instances are 29 (2.75% of the error value). The Kappa statistic values is 0.9536 and mean absolute error value is 0.0218. The root mean squared error values is 0.1477 and the relative absolute error value is 4.6347%. The last measuring value is 30.4475% for root relative squared error. As per these measurements, the SVM model with RBF kernel is somewhat efficient. SVM classifier classifies 643 instances are classified as true positive, 13 instances are as in false positive. 383 data points are under true negative and 16 point instances are in false negative. Table 12.6 shows the SVM classifier accuracy measuring with TP Rate, FP rate, Precision, Recall and F-measure values of CKD and non-CKD. Table 12.6 describes the confusion matrix SVM.

Table 12.7 depicts the performance parameters like Classification accuracy (CA), Sensitivity (Sens), Specificity (Spec.), Area under ROC curve (AUC), Information score (IS), F-measure (F1), Precision (Prec.), Recall, Brier score (BC), Mathew correlation coefficient

TABLE 12.6 Confusion matrix SVM.

Class	CKD (P)	NON-CKD(N)	Total
CKD (T)	643	13	**656**
NON-CKD (F)	16	383	**399**
Total	**659**	**396**	**1055**

TABLE 12.7 SVM classification accuracy values of CKD and non-CKD of data set.

Class	Acc.	Sensi	Speci	AUC	IS	F1	Preci	Recall	Brier	MCC
CKD	0.9725	0.9878	0.9850	0.9951	0.8692	0.9893	0.9908	0.9878	0.0445	0.9718
Non-CKD	0.9725	0.9599	0.9802	0.9951	0.8692	0.9635	0.9672	0.9599	0.0445	0.9415
AVG	**0.9725**	**0.9739**	**0.9826**	**0.9951**	**0.8692**	**0.9764**	**0.9790**	**0.9739**	**0.0445**	**0.9567**

(MCC) of the model SVM. The total data points are 1055 split into classified instances correctly (643 + 383) and error classified instances (16 + 13), so the avg CA value is 0.9725 and AUC values is 0.9951. The remaining parameters are shown in Table 12.7.

4.2.3 C4.5 decision tree classifier

As per analysis of this model on Uddanam CKD data set, the correctly Classified Instances are 1050 (99.53%) and Incorrectly Classified Instances 5 (0.47%). The Kappa statistic values is 0.998 and Mean absolute error value is 0.0016. The root mean squared error is 0.0281, Relative absolute error is 0.3358% and Root relative squared error is 5.7956%. As per analysis of confusion matrix TP values is 654, FP value is 3, the TN value is two the false negative value is 396. I measured the characteristics of accuracy measured TP Rate, FP rate, Precision, Recall and F-measure values of CKD and non-CKD. Table 12.8 shows the confusion matrix of the model C4.5. The total data points are 1055 split into correctly classified instances (654 + 396) and error classified instances (2 + 3). Table 12.8 shows the Confusion matrix of C4.5.

Table 12.9 depicts the performance parameters like Classification accuracy (CA), Sensitivity (Sens), Specificity (Spec.), Area under ROC curve (AUC), Information score (IS), F-measure (F1), Precision (Prec.), Recall, Brier score (BC), Mathew correlation coefficient (MCC). The total data points are 1055 split into classified instances correctly (654 + 396) and error classified instances (2 + 3), so the average CA value is 0.9953 and AUC values is 0.9975. The remaining parameters are shown in Table 12.9.

As per analysis of C4.5 visualization tree (Fig. 12.2), the main factors of CKD are albumin and packed cells. If the albumin is less than or equal to one then it derives as NON-CKD other than it depends the packed cell. If the values of packed cell less than or equal to 0.9 values

TABLE 12.8 Confusion matrix C4.5.

Class	CKD	Non-CKD	Total
Ckd	654	2	**656**
Non-CKD	3	396	**399**
Total	**657**	**398**	**1055**

TABLE 12.9 Accuracy measures values of C 4.5 decision tree algorithm.

Class	CA	Sens	Spec	AUC	IS	F1	Prec	Recall	Brier	MCC
CKD	0.9953	0.9970	0.9925	0.9975	0.9470	0.9962	0.9954	0.9970	0.0081	0.9899
Non-CKD	0.9953	0.9925	0.9970	0.9975	0.9470	0.9937	0.9950	0.9925	0.0081	0.9899
AVG	**0.9953**	**0.9948**	**0.9948**	**0.9975**	**0.9470**	**0.9949**	**0.9952**	**0.9948**	**0.0081**	**0.9899**

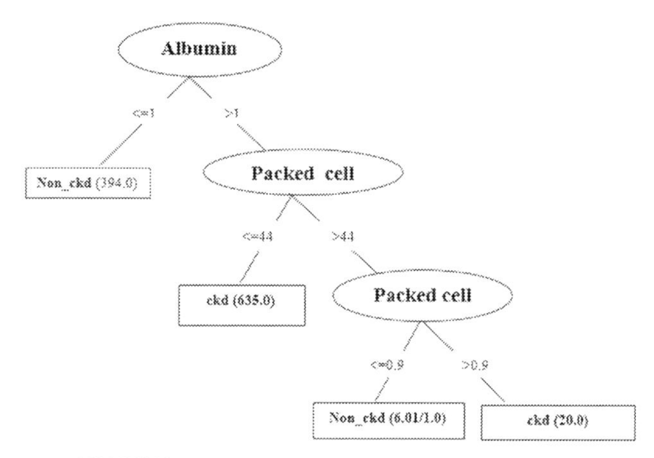

FIGURE 12.2 Visualization tree of C4.5 algorithm.

then it declares the NON-CKD. Fig. 12.2 shows the description about C 4.5 algorithm visualization tree classifies the CKD and NON-CKD. With this algorithm I predict the disease 100% testing accuracy.

4.2.4 K- nearest neighbors (kNN) classifier

As per analysis of kNN model that the correctly classified Instances are 1019 (96.59% of accuracy) out of 1055 instances and Incorrectly Classified Instances are 29 (2.75% of the error value). The kNN classifier classifies that 636 instances are classified as true positive, 20 instances are as in false positive. 383 data points are under true negative and 16 point instances are in false negative. Table 12.10 shows the KNN classifier accuracy measuring with TP Rate, FP rate, Precision, Recall, and F-measure values of CKD and non-CKD utilizing confusion matrix. Table 12.10 shows the Confusion matrix of k-NN.

Table 12.11 depicts the performance parameters like Classification accuracy (CA), Specificity (Spec.), Sensitivity (Sens), Area under ROC curve (AUC), F-measure (F1), Precision (Prec.), Recall, Brier score (BC), Information score (IS), Mathew correlation coefficient (MCC) of the model k-NN. The total data points are 1055 split into classified instances correctly (636 + 383) and error classified instances (16 + 20), so the average CA value is 0.9659 and AUC values is 0.9954. The remaining parameters are shown in Table 12.11.

TABLE 12.10 Confusion matrix kNN.

	Ckd (P)	non_CKD (N)	Total
Ckd(T)	636	20	**656**
Non-CKD(F)	16	383	**399**
Total	**652**	**403**	**1055**

TABLE 12.11 kNN classification accuracy values of CKD and non-CKD of data set.

Class	CA	Sens	Spec	AUC	IS	F1	Prec	Recall	Brier	MCC
CKD	0.9659	0.9802	0.9599	0.9954	0.8800	0.9779	0.9757	0.9802	0.0468	0.9415
Non-CKD	0.9659	0.9599	0.9695	0.9954	0.8800	0.9551	0.9504	0.9599	0.0468	0.9276
AVG	**0.9659**	**0.9701**	**0.9647**	**0.9954**	**0.8800**	**0.9665**	**0.9631**	**0.9701**	**0.0468**	**0.9346**

4.2.5 Logistic Regression (LR) classifier

As per analysis of LR model that the correctly classified Instances are 1041 (98.67% of accuracy) out of 1055 instances and Incorrectly Classified Instances are 14 (1.33% of the error value). LR classifier classifies that 648 instances are classified as True Positive, 8 instances are as in false positive. 393 data points are under true negative and 6 point instances are in false negative. Table 12.12 shows the LR classifier accuracy measuring with TP Rate, FP rate, Precision, Recall, and F-measure values of CKD and NON-CKD with confusion matrix.

Table 12.12 depicts the performance parameters like Classification accuracy (CA), Sensitivity (Sens), Specificity (Spec.), Area under ROC curve (AUC), Information score (IS), F-measure (F1), Precision (Prec.), Recall, Brier score (BC), Mathew correlation coefficient (MCC) of the model Logistic Regression. The total data points are 1055 split into classified instances correctly (648 + 393) and error classified instances (8 + 6), so the avg CA value is 0.9867 and AUC values is 0.9990. The remaining parameters are shown in Table 12.13.

TABLE 12.12 Confusion matrix Logistic Regression.

Class	Ckd (P)	non_CKD (N)	Total
Ckd(T)	648	8	**656**
Non-CKD(F)	6	393	**399**
Total	**654**	**401**	**1055**

TABLE 12.13 Logistic Regression classification accuracy values of CKD and NON-CKD of data set.

Class	CA	Sens	Spec	AUC	IS	F1	Prec	Recall	Brier	MCC
CKD	0.9867	0.9863	0.9674	0.9990	0.8747	0.9833	0.9803	0.9863	0.0255	0.9556
Non-CKD	0.9867	0.9850	0.9878	0.9990	0.8747	0.9825	0.9800	0.9850	0.0255	0.9718
AVG	**0.9867**	**0.9857**	**0.9776**	**0.9990**	**0.8747**	**0.9829**	**0.9802**	**0.9857**	**0.0255**	**0.9637**

4.3 Comparative MLs analysis

Table 12.14 shows the Comparison Accuracy Analysis for all MLs Accuracy Measures including time that are Classification accuracy (CA), Sensitivity (Sens), Specificity (Spec.), Area under ROC curve (AUC), Information score (IS), F-measure (F1), Precision (Prec.), Recall, Brier score (BS), Mathew correlation coefficient (MCC) on Uddanam CKD dataset. In Table 12.14, I describe every classification accuracy values in detail. The ROC value is the crucial for the performance of the algorithm. If the ROC value is nearer to 1 then that algorithm accuracy is very high. All algorithms performances are good where all algorithms CA and AUC values are above 0.97. As per comparative analysis, C4.5 is very accurate than other specified algorithms that CA and AUC values are 0.9953 and 0.9948 respectively. The Naïve Bayes takes 0.09 s only that it is very less time than other experimental ML algorithms.

Fig. 12.3 shows the time taken for building of each model. Naïve Bayes takes least time (0.09 s) for building a model as well C4.5 takes the more time (0.54 s) than other ML algorithms. LR model takes the 0.24 s that the second least times. The SVM takes the 0.26 s and k-NN takes the 0.42 s. As per time analysis, the least time 0.09 s taken by Naïve Bayes with 0.9791 accuracy.

Fig. 12.4 shows the Area Under ROC curves of the experimental models SVM, C4.5 Naïve Bayes and LR on Uddanam CKD data set. The AUC values of SVM, C4.5 Naïve Bayes and LR are 0.9986, 0.9951, 0.9975, 0.9954, and 0.9990, respectively. The best ROC values are C4.5 and LR models. All ROC curves represents in Fig. 12.4 with different color modes. Most accurate model C4.5 curve represents the yellow (dark gray in printed version) color and the LR mode specified with blue (light gray in printed version) color.

Fig. 12.5 shows the Accuracy Analysis with CA and ROC values of the ML algorithms. C4.5 model is the best compare to other experimental ML algorithms where AUC value is 0.9986 and CA value is 9953. The second highest algorithm is LR that the AUC value is 0.999 and accuracy is 0.9867 and takes 0.24 s of the average time for predicting CKD correctly. The KNN algorithm is least performance according to CA (0.9659) and AUC values (0.9954).

TABLE 12.14 Machine classification accuracy comparative analysis.

ML algorithm	Time	CA	Sens	Spec	AUC	IS	F1	Prec	Recall	Brier	MCC
Naïve Bayes	**0.09**	0.9791	0.9685	0.9731	0.9986	0.9141	0.9724	0.9764	0.9685	0.0362	0.9416
SVM	0.26	0.9725	0.9739	0.9826	0.9951	0.8692	0.9764	0.9790	0.9739	0.0445	0.9567
C4.5	0.54	**0.9953**	**0.9948**	**0.9948**	**0.9975**	**0.9470**	**0.9949**	**0.9952**	**0.9948**	**0.0081**	**0.9899**
KNN	0.42	0.9659	0.9701	0.9647	0.9954	0.8800	0.9665	0.9631	0.9701	0.0468	0.9346
Logistic regression	**0.24**	**0.9867**	**0.9857**	**0.9776**	**0.9990**	**0.8747**	**0.9829**	**0.9802**	**0.9857**	**0.0255**	**0.9637**

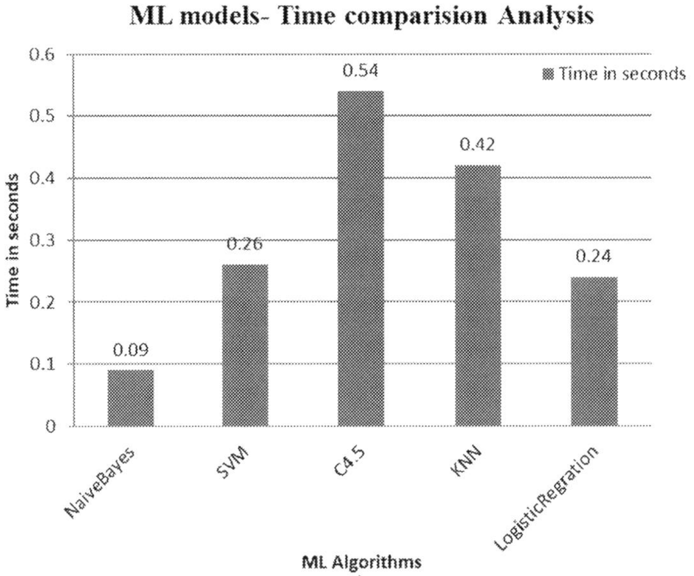

FIGURE 12.3 Time comparison analysis of ML algorithm for build the models.

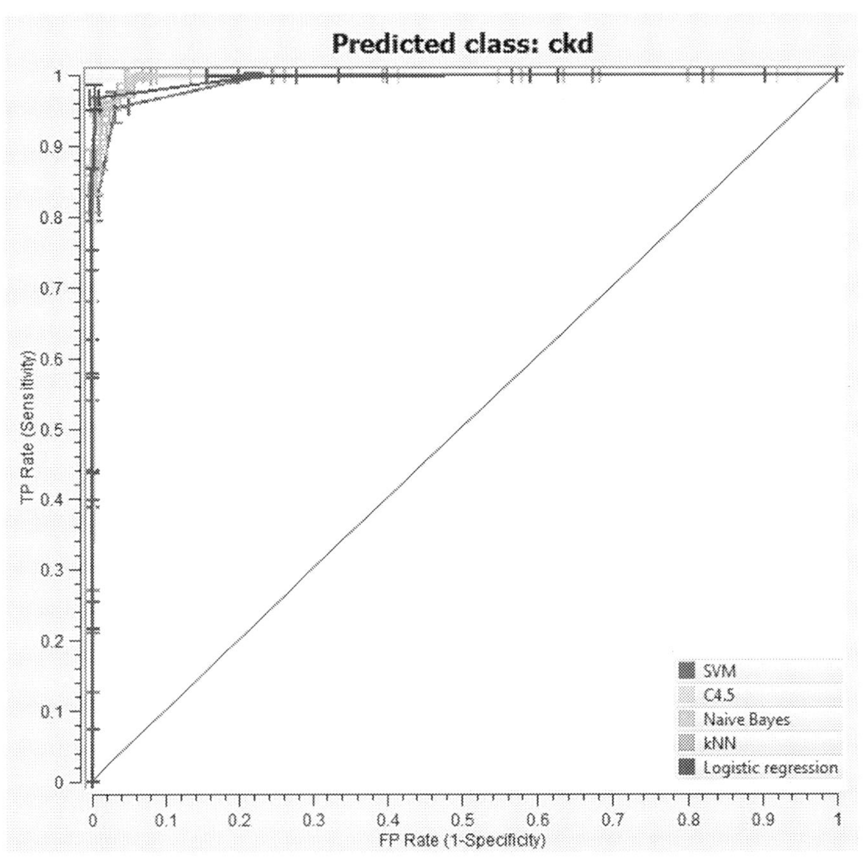

FIGURE 12.4 Comparative analysis of ML models with respect to AUC.

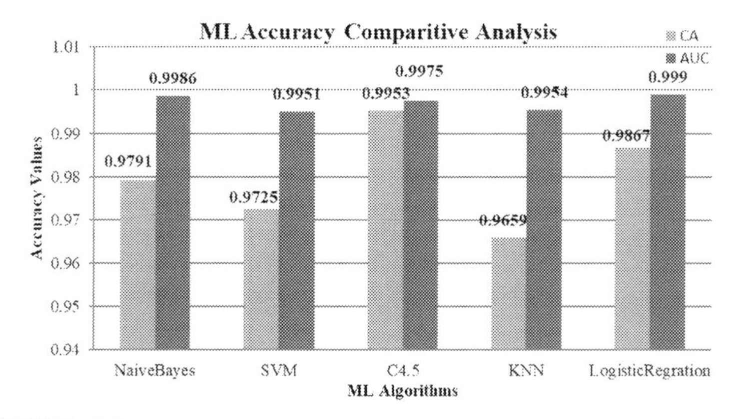

FIGURE 12.5 Accuracy analysis with CA and AUC values of the ML algorithm.

4.4 Probabilistic neural networks analysis with incremental hidden neurons

The data set of Uddanam CKD is analyzed by the PNN. In this experiment, the hidden layer of the PNN neurons are increased until to get the goal that the accuracy value in peak and error rate is in minimum. For this, I set up the PNN model that the data set is splited in to two parts, training and testing in 70:30 ratio and compute the performance parameters like accuracy, performance with error ratio and gradient value in each stage. The input training data set with 37 features of the Uddanam CKD is given to the PNN model and traced with the binary target class (CKD and non-CKD) for the process. I started the experiment with two neurons of the hidden layer of the PNN. I get the peak accuracy value at eight hidden neurons of the PNN.

4.4.1 Confusion matrix comparative analysis

Fig. 12.6 shows the collective confusion matrices after processing PNN model with increment hidden neurons 2 to 8. I can compute performance parameters like accuracy, specificity; recall and so on with help of confusion matrix. As per analysis, the two neurons hidden layer PNN gives the accuracy 0.9887(98.9%). The three neurons hidden layer PNN gives the accuracy 0.9895(99.0%). PNN four neurons hidden layer process some better accuracy that is 0.9915(99.1%). The five, six, and seven neurons of hidden layer PNN performs 99.4%, 99.8%, and 99.9%, relatively. Finally, the eight neurons hidden layer PNN performs peak accuracy 100%. In this, all 656 CKD instances and 399 non-CKD instance are classified correctly.

4.4.2 Best training performance analysis

Fig. 12.7 shows the collective best performances after processing PNN model with increment hidden neurons 2 to 8. In this, I compute the performance using cross-entropy. The X-axis specifies number of epochs and Y-axis represents the cross-entropy value. The splitting of data set is performed into training and testing dataset. The blue (dark gray in printed version) line specifies the training performance and the red (gray in printed version) line chooses the testing performance line. The best training performance value is calculated

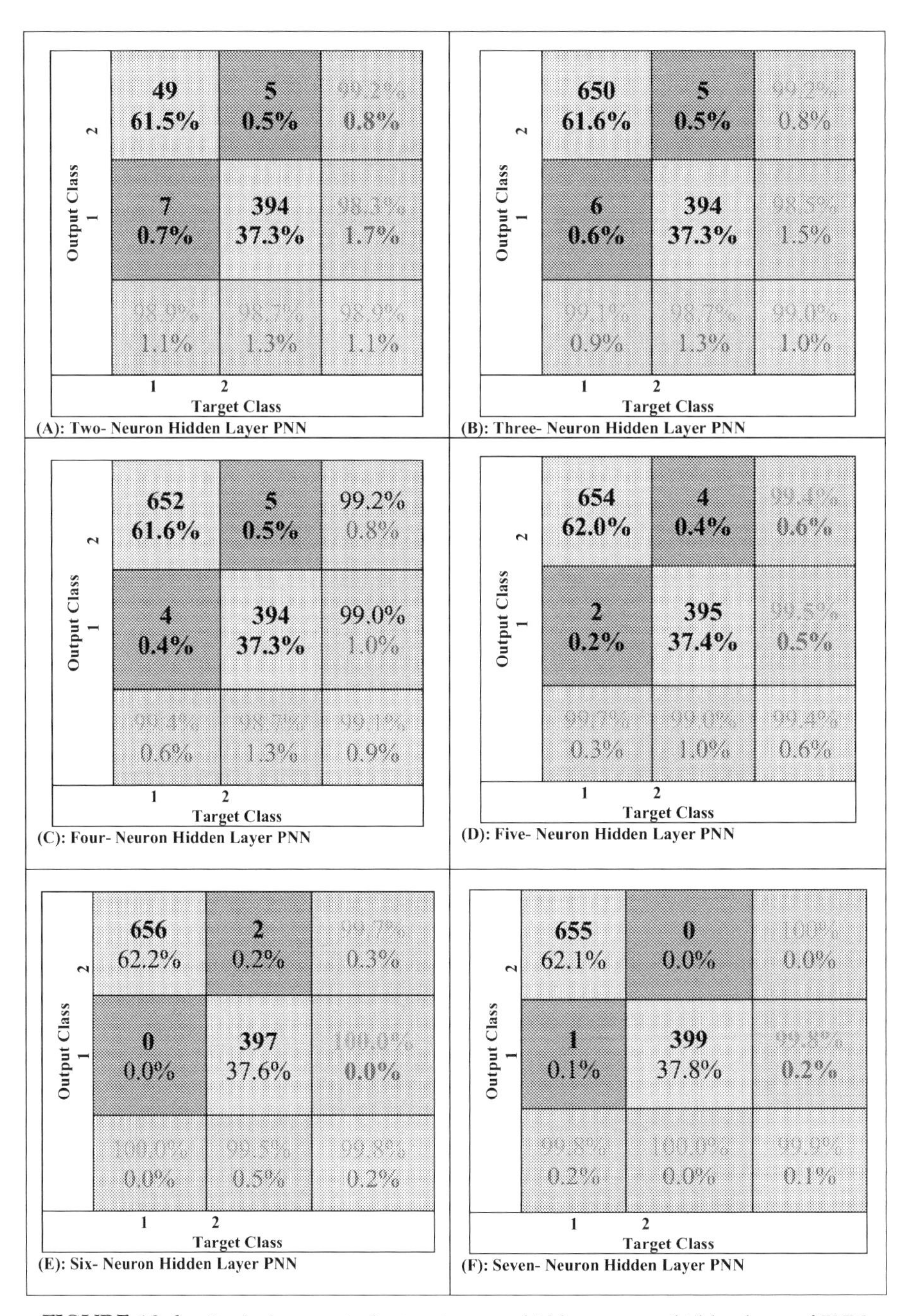

FIGURE 12.6 Confusion matrix for two to seven hidden neurons hidden layer of PNN.

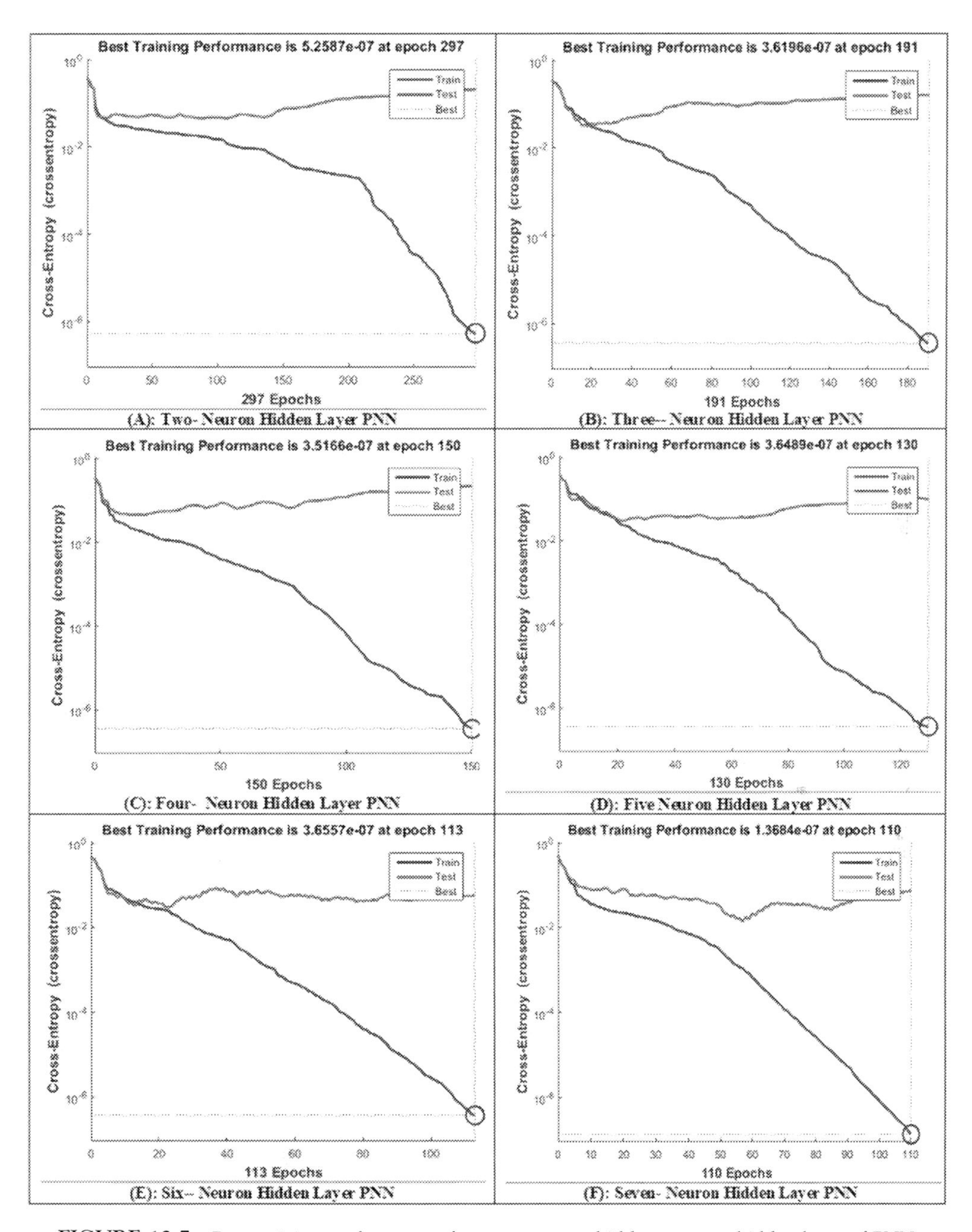

FIGURE 12.7 Best training performances for two to seven hidden neurons hidden layer of PNN.

by specific point of the parameters that are cross-entropy and epochs. As per analysis, the two neurons hidden layer PNN gives the best training performance value ($5.2587e^{-07}$) at epoch 297 within 0.029 s. The three neurons hidden layer PNN gives the best training performance value ($3.6196e^{-07}$) at epoch 191 within 0.023 s. The 4, 5, 6, and 7 HL neurons PNNs, best performed values are $3.5166e^{-07}$ at epoch 150, $3.6489e^{-07}$ at epoch 130, $3.6557e^{-07}$ at epoch 113, and $1.3684e^{-07}$ at epoch 110, respectively. The time taken for each process of four, five, six, and seven hidden neurons PNN is 0.029, 0.023, 0.021, 0.019, 0.015, 0.013, and 0.01.

4.4.3 *Gradients analysis*

Fig. 12.8 shows the collective Best Computation of Gradient Values for two to seven hidden neurons of PNN model. In this, I compute the performance of the model with using gradient value. The X-axis specifies number of epochs and Y-axis represents the gradient value. As per analysis, the two neurons hidden layer PNN gives the best gradient value that is $7.3406e^{-07}$ at epoch 297 within 0.028 s. The three neurons hidden layer PNN gradient value ($8.0445e^{-07}$) at epoch 191 within 0.021 s. The 4, 5, 6 and 7 HL neurons PNNs, best gradient values are $3.5166e^{-07}$ at epoch 150, $3.6489e^{-07}$ at epoch 130, $3.6557e^{-07}$ at epoch 113, and $1.3684e^{-07}$ at epoch 110, respectively.

4.4.4 *Error instances analysis with histograms*

Fig. 12.9 shows the collective error histograms with 20 bins for two to seven hidden neurons hidden layer of PNN model; I compute the instance of error values and plot the histograms. In this analysis, The X-axis specifies Errors and Y-axis represents the number of instances. The blue (black in printed version) bars represents the error value is occurred to the instances. The orange (light gray in printed version) line depicts to the zero error rate. Most of the blue (black in printed version) bars are near to zero error that the error rate is very less. Per the analysis, the two neurons hidden layer PNN error histogram shows that the most of instances error value is +0.05 nearly 1200. Some of instance error values are nearer to −0.05 nearly 900. Minor instances error values are +0.95 and −0.95. In the three neurons, hidden layer PNN error histogram represents minimize the errors values of −0.95 and +0.95 and most of error valued instances are −0.05. The 4, 5, 6, and 7 neurons of hidden layer PNN error histograms are shown in Fig. 12.9.

4.4.5 *ROC analysis*

Fig. 12.10 shows the collective ROC curves for two to seven hidden neurons hidden layer of PNN after processing PNN model with increment hidden neurons two to eight. In this, I compute the performance values. The X-axis true positive ratio value and Y-axis represents the false positive values. The blue (dark gray in printed version) line specifies the CKD class-1 and the yellow (gray in printed version) line represents the NON-CKD class. The ROC values of two to seven hidden neurons PNNs are 0987, 0.989, 0.991, 0.995, 0.998, 1.000, and 1.000, respectively.

4.4.6 *8-Hidden neurons PNN analysis*

Fig. 12.11 shows the eight hidden neurons hidden layer PNN performance analysis. In this, the performance of the PNN is in peak that the accuracy and ROC values are 1.0. The best training performance is $1.61e^{-07}$ at epoch 113 within the 0.014 s. All the instances error values are near to zero (shown in Fig. 12.11D). So, 8-hidden neurons PNN are the best model to predict the Uddanam CKD with 100% of accuracy.

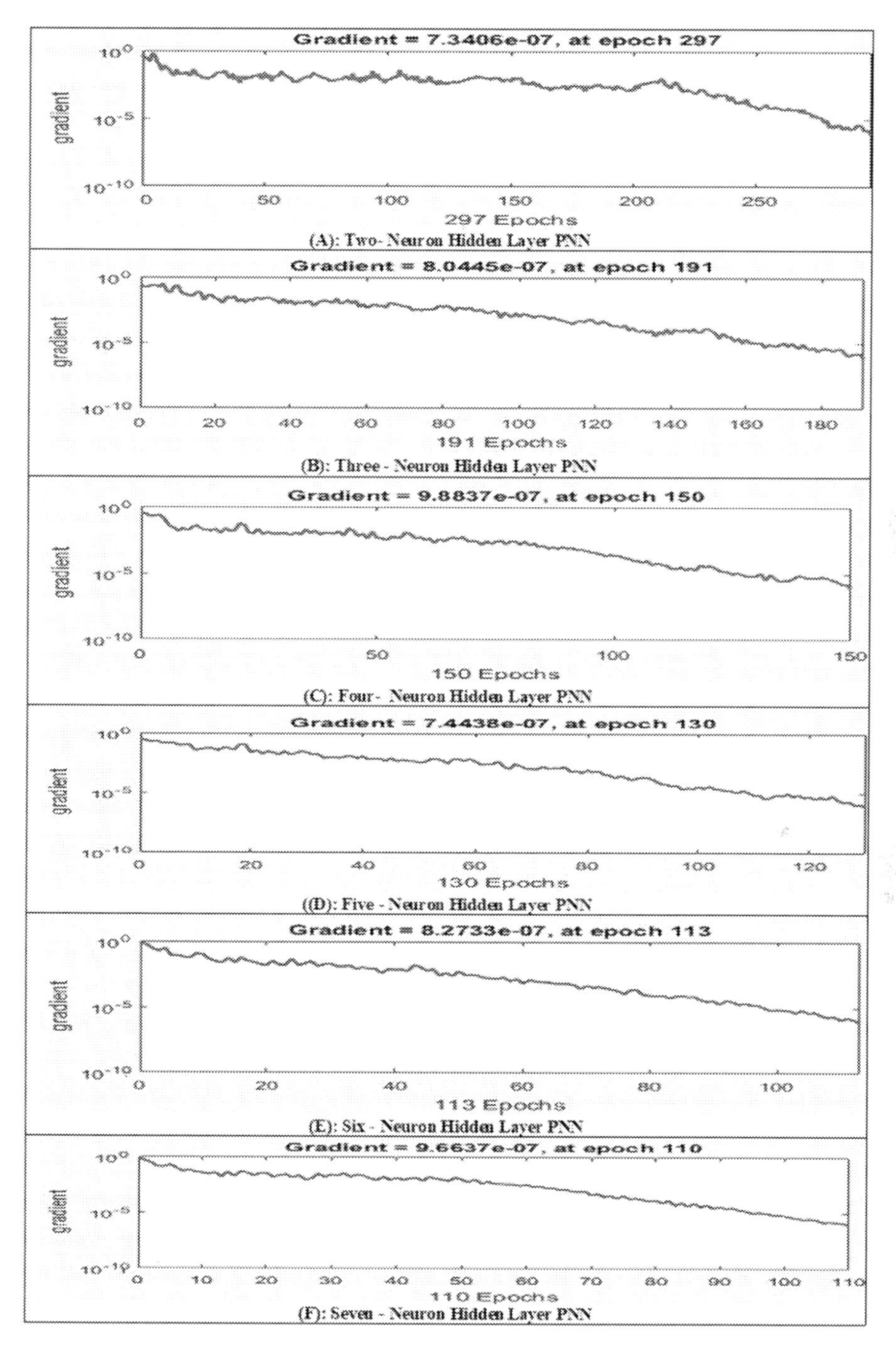

(A): Two- Neuron Hidden Layer PNN

(B): Three - Neuron Hidden Layer PNN

(C): Four- Neuron Hidden Layer PNN

(D): Five - Neuron Hidden Layer PNN

(E): Six - Neuron Hidden Layer PNN

(F): Seven - Neuron Hidden Layer PNN

FIGURE 12.8 Computation of gradient values for two to seven hidden neurons hidden layer of PNN.

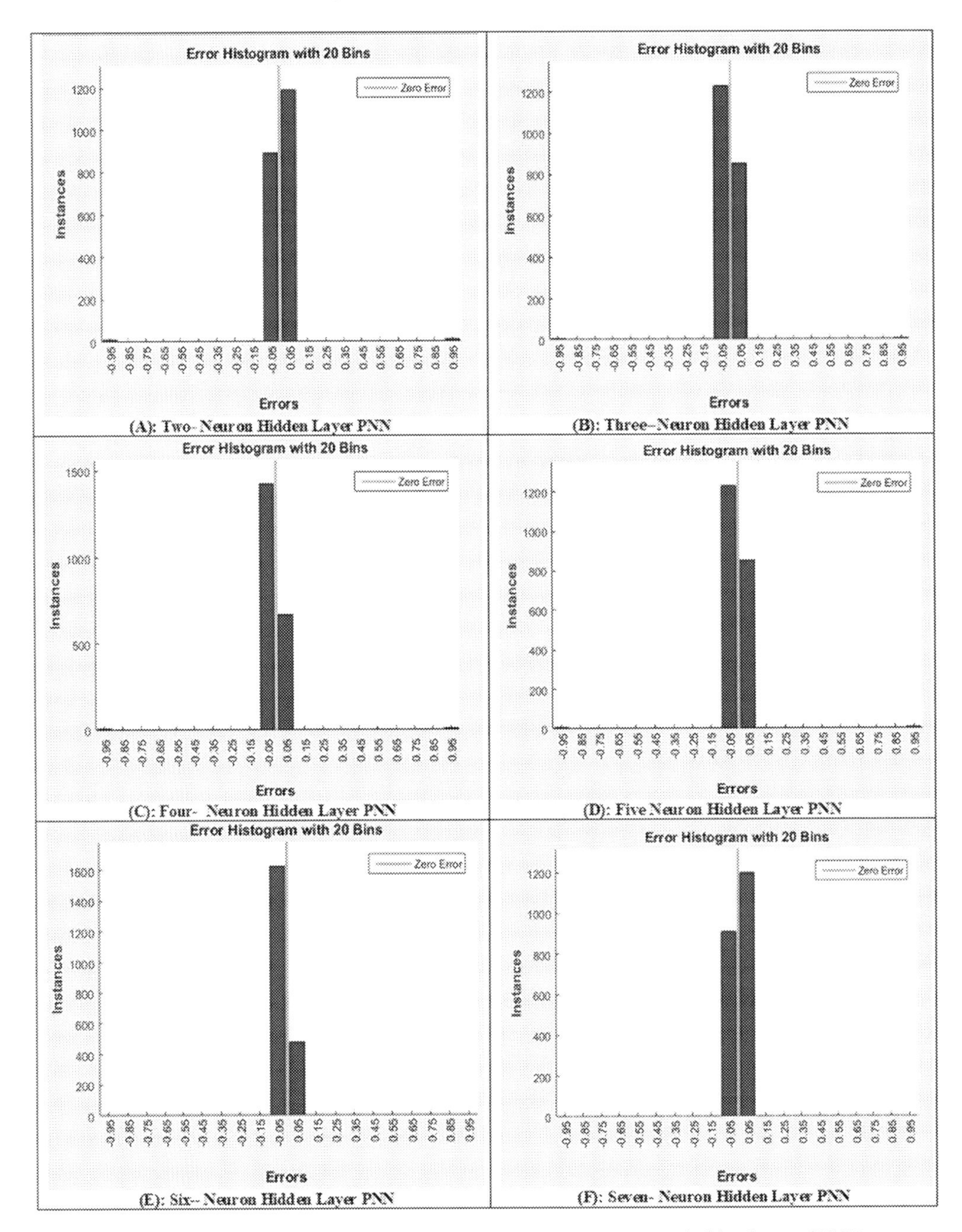

FIGURE 12.9 Error histograms for two to seven hidden neurons hidden layer of PNN.

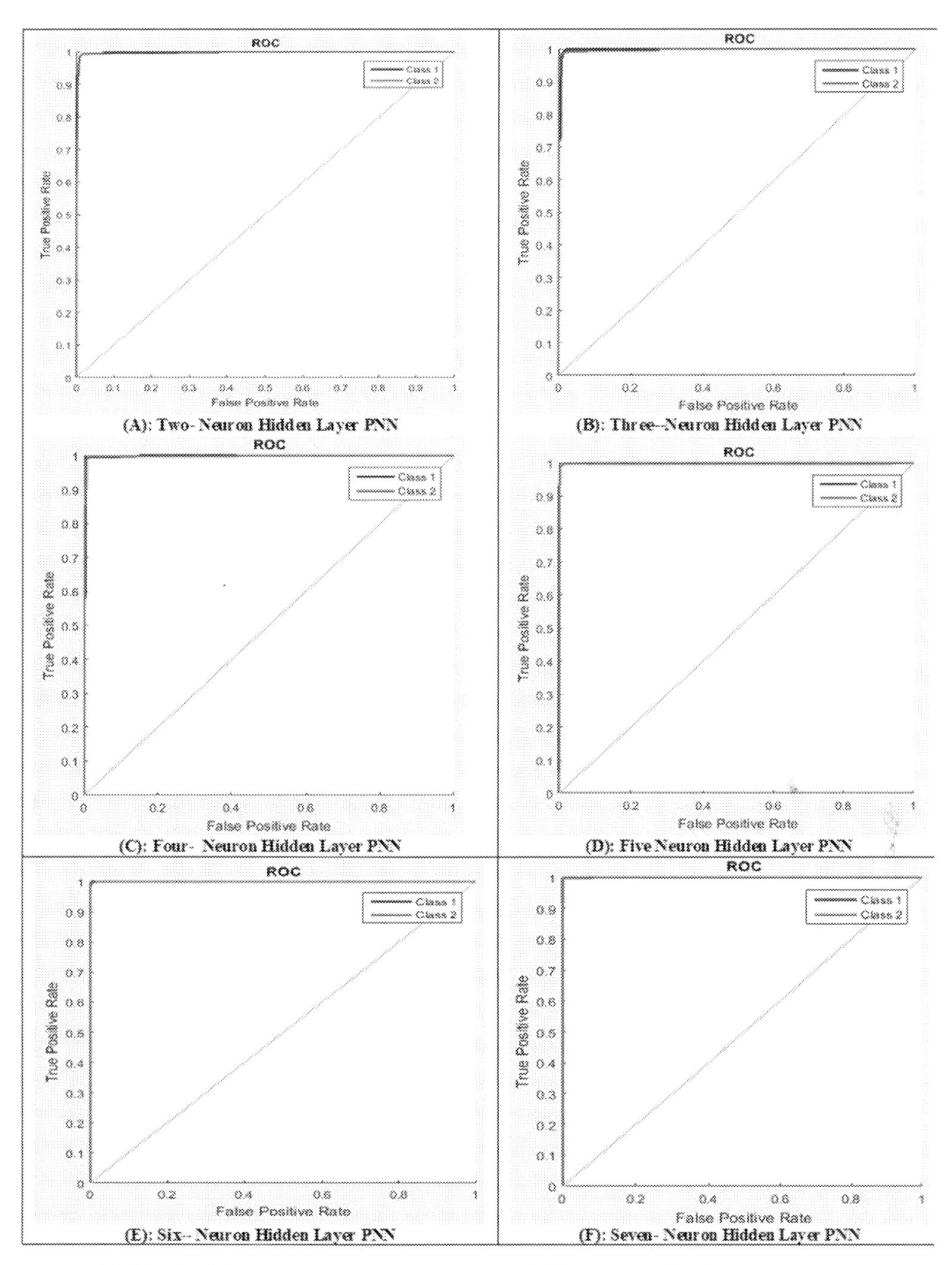

FIGURE 12.10 ROC curves for two to seven hidden neurons hidden layer of PNN.

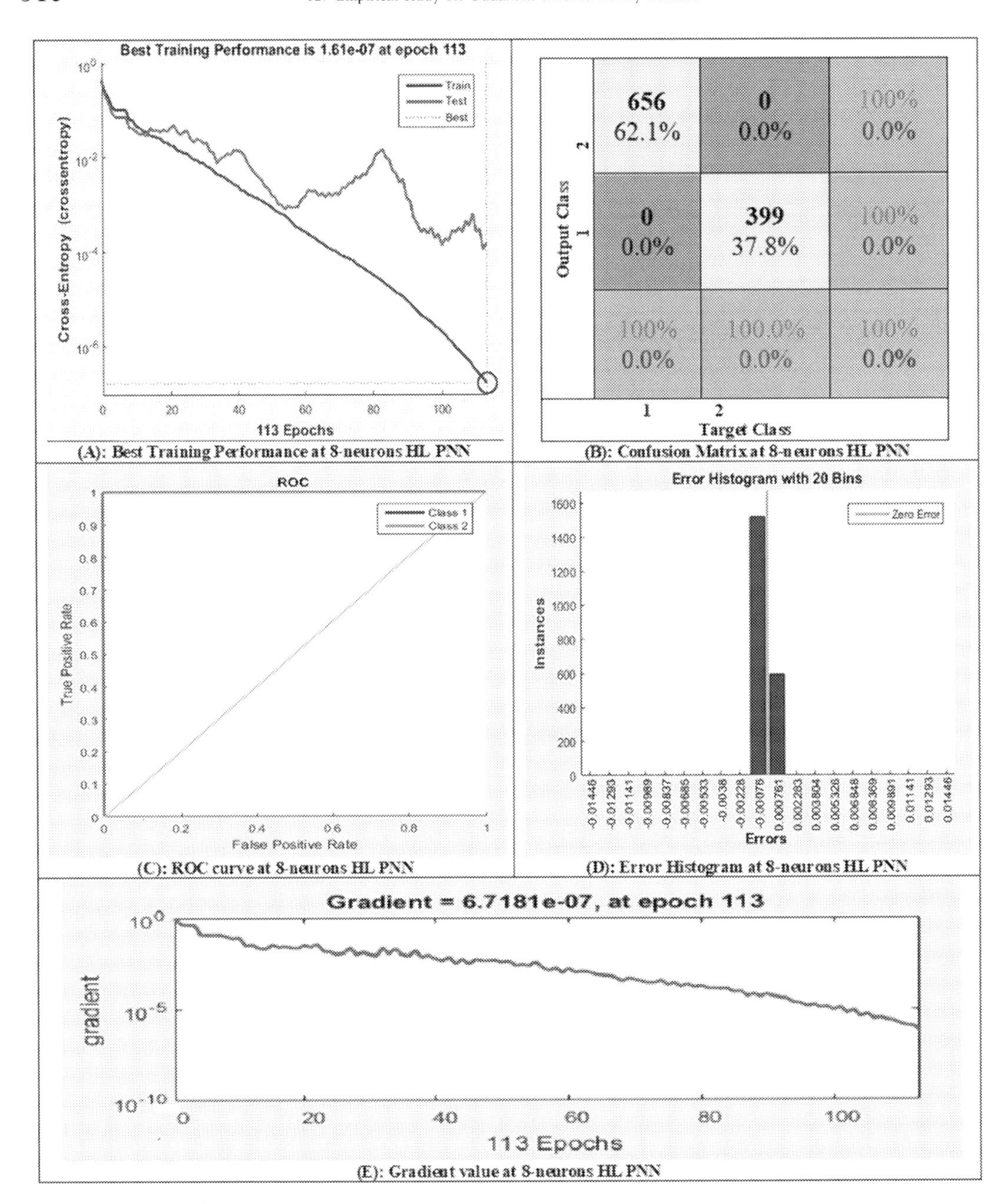

FIGURE 12.11 8 hidden neurons hidden layer PNN performance analysis.

4.4.7 2 to 8 hidden neurons PNN comparative analysis

Table 12.15 shows the PNNs with two to eight hidden neurons classification accuracy comparative analysis. As per analysis the eight hidden neuron PNN model is the best to other NN and ML algorithms where it discovers the peak accuracy value within 0.014 s.

Fig. 12.12 shows the Processing Time analysis with respect to PNN hidden neurons. In this, the X-axis represents the No. of hidden neurons in PNN and Y-axis represents the

TABLE 12.15 Two to eight hidden neuron PNNs classification accuracy comparative analysis.

Hidden neurons in PNN	Processing time in seconds	Epochs	Best training performance	Accuracy	Roc value	Gradient values
2	0.029	297	$5.2587e^{-07}$	0.9886	0.987	$7.3406e^{-07}$
3	0.023	191	$3.6196e^{-07}$	0.9895	0.989	$8.0445e^{-07}$
4	0.021	150	$3.5166e^{-07}$	0.9914	0.991	$9.8837e^{-07}$
5	0.019	130	$3.6489e^{-07}$	0.9943	0.995	$7.4438e^{-07}$
6	0.015	113	$3.6557e^{-07}$	0.9981	0.999	$8.2733e^{-07}$
7	0.013	110	$1.3684e^{-07}$	0.9990	1.000	$9.6637e^{-07}$
8	0.014	113	$1.6100e^{-07}$	1.0000	1.000	$8.2733e^{-07}$

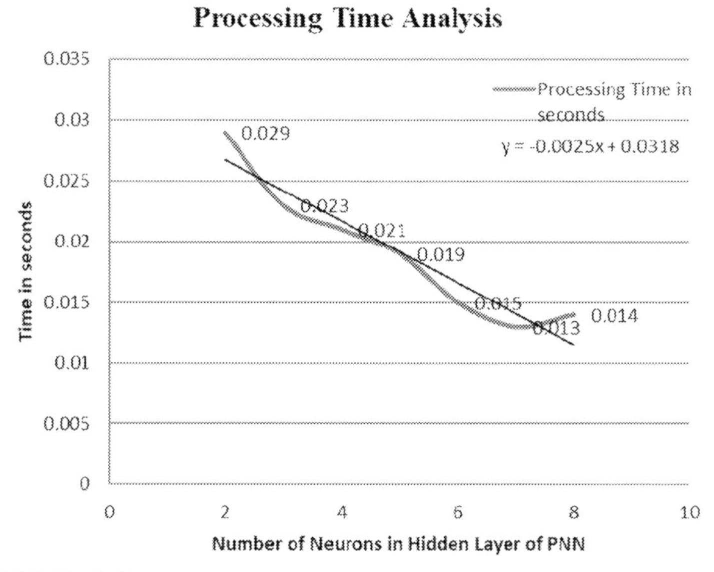

FIGURE 12.12 Processing time analysis with respect to PNN hidden neurons.

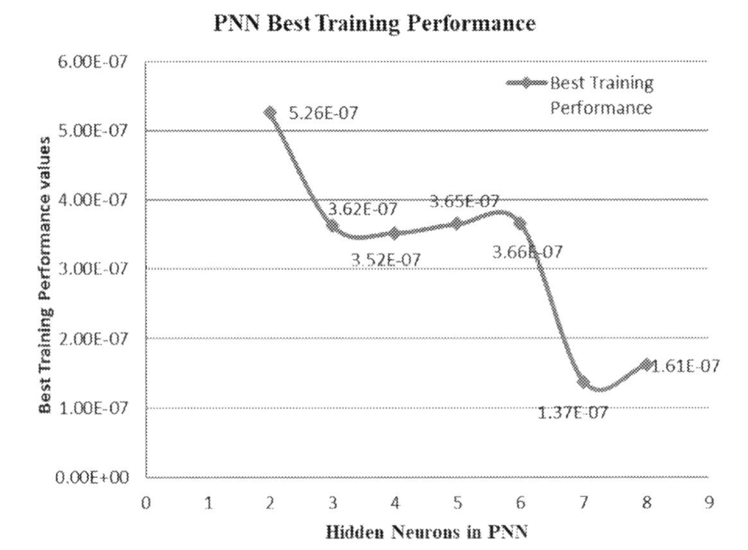

FIGURE 12.13 Best training performance values respect to PNN hidden neurons.

Time in seconds. The blue (dark gray in printed version) color curve represents the processing time according to the no. of hidden PNN neurons. The process time is decreased according to incremental neurons. The line is very much adequate to the linear line equations $y = -0.002 * X + 0.031$.

Fig. 12.13 shows best training performance values respect to PNN hidden neurons. In this, the X-axis represents the No. of hidden neurons in PNN and Y-axis represents the Best training performance values. The blue (dark gray in printed version) color curve represents the Best training performance according to the no. of hidden PNN neurons. It shows the best training.

5. Conclusion and social benefits

Per the ML analysis, C4.5 decision tree algorithm is very accurate to predict CKD in Uddanam area. The neural networks are very efficient and more accurate than all choosing algorithms. In this, eight-hidden PNN is effective algorithm where it works with 100% accuracy for the Uddanam CKD data set. This automatic diagnosis CKD tool will be able to help the doctor to identify the disease in the early stage of CKD with less time and low cost, and ease the patients' survival. In this empirical study, I present the prevalence of CKD in area of Uddanam. This study will be very useful to analysts and governments to make decisions for further steps about Uddanam CKD. Furthermore, I will analyze the Uddanam CKD on images of kidney using deep convolution neural networks for more effectiveness and efficiency.

References

[1] S.I. Hallan, J. Coresh, B.C. Astor, A. Åsberg, N.R. Powe, S.,. Romundstad, J. Holmen, International comparison of the relationship of chronic kidney disease prevalence and ESRD risk, J. Am. Soc. Nephrol. 17 (8) (2006) 2275–2284.

[2] R. Ani, G. Sasi, U.R. Sankar, O.S. Deepa, Decision support system for diagnosis and prediction of chronic renal failure using random subspace classification, in: 2016 International Conference on Advances in Computing, Communications and Informatics (ICACCI), IEEE, September 2016, pp. 1287–1292.

[3] E. Di Angelantonio, J. Danesh, G. Eiriksdottir, V. Gudnason, Renal function and risk of coronary heart disease in general populations: new prospective study and systematic review, PLoS Med. 4 (9) (2007) e270.

[4] S.M.K. Chaitanya, P.R. Kumar, Detection of chronic kidney disease by using artificial neural networks and gravitational search algorithm, in: Innovations in Electronics and Communication Engineering, Springer, Singapore, 2019, pp. 441–448.

[5] B. Wikström, M. Fored, M.A. Eichleay, S.H. Jacobson, The financing and organization of medical care for patients with end-stage renal disease in Sweden, Int. J. Health Care Finance Econ. 7 (4) (2007) 269–281.

[6] S.K. Agarwal, R.K. Srivastava, Chronic kidney disease in India: challenges and solutions, Nephron Clin. Pract. 111 (3) (2009) c197–c203.

[7] H. Polat, H.D. Mehr, A. Cetin, Diagnosis of chronic kidney disease based on support vector machine by feature selection methods, J. Med. Syst. 41 (4) (2017) 55.

[8] S. Sharma, V. Sharma, A. Sharma, Performance Based Evaluation of Various Machine Learning Classification Techniques for Chronic Kidney Disease Diagnosis, 2016 arXiv preprint arXiv:1606.09581.

[9] A. Abdelaziz, A.S. Salama, A.M. Riad, A.N. Mahmoud, A machine learning model for predicting of chronic kidney disease based internet of things and cloud computing in smart cities, in: Security in Smart Cities: Models, Applications, and Challenges, Springer, Cham, 2019, pp. 93–114.

[10] H. Kriplani, B. Patel, S. Roy, Prediction of chronic kidney diseases using deep artificial neural network technique, in: Computer Aided Intervention and Diagnostics in Clinical and Medical Images, Springer, Cham, 2019, pp. 179–187.

[11] M.K. Mani, Prevention of chronic renal failure at the community level, Kidney Int. 63 (2003) S86–S89.

[12] S. Hore, S. Chatterjee, R.K. Shaw, N. Dey, J. Virmani, Detection of chronic kidney disease: a NN-GA-based approach, in: Nature Inspired Computing, Springer, Singapore, 2018, pp. 109–115.

[13] V. Kunwar, K. Chandel, A.S. Sabitha, A. Bansal, Chronic kidney disease analysis using data mining classification techniques, in: 2016 6th International Conference-Cloud System and Big Data Engineering (Confluence), IEEE, January 2016, pp. 300–305.

[14] A. Salekin, J. Stankovic, Detection of chronic kidney disease and selecting important predictive attributes, in: 2016 IEEE International Conference on Healthcare Informatics (ICHI), IEEE, October 2016, pp. 262–270.

[15] A. Subasi, E. Alickovic, J. Kevric, Diagnosis of chronic kidney disease by using random forest, in: CMBEBIH 2017, Springer, Singapore, 2017, pp. 589–594.

[16] S. Vijayarani, S. Dhayanand, M. Phil, Kidney disease prediction using SVM and ANN algorithms, Int. J. Comput. Bus. Res. (IJCBR) 6 (2) (2015).

[17] P. Sinha, P. Sinha, Comparative study of chronic kidney disease prediction using KNN and SVM, Int. J. Eng. Res. Technol. 4 (12) (2015) 608–612.

[18] Dhayanand, S.,S. Vijayarani, Data mining classification algorithms for kidney disease prediction, Int. J. Cybern. Inform. (IJCI) 4 (4) (2015) 13–25.

[19] K.A. Padmanaban, G. Parthiban, Applying machine learning techniques for predicting the risk of chronic kidney disease, Indian J. Sci. Technol. 9 (29) (2016) 1–6.

[20] N. Borisagar, D. Barad, P. Raval, Chronic kidney disease prediction using back propagation neural network algorithm, in: Proceedings of International Conference on Communication and Networks, Springer, Singapore, 2017, pp. 295–303.

[21] S.E. Hussein, O.A. Hassan, M.H. Granat, Assessment of the potential iridology for diagnosing kidney disease using wavelet analysis and neural networks, Biomed. Signal Process Contr. 8 (6) (2013) 534–541.

[22] S. Chatterjee, S. Dzitac, S. Sen, N.C. Rohatinovici, N. Dey, A.S. Ashour, V.E. Balas, Hybrid modified Cuckoo Search-Neural Network in chronic kidney disease classification, in: 2017 14th International Conference on Engineering of Modern Electric Systems (EMES), IEEE, June 2017, pp. 164–167.

[23] M. Kumar, Prediction of chronic kidney disease using random forest machine learning algorithm, Int. J. Comput. Sci. Mobile Comput. 5 (2) (2016) 24–33.
[24] A.Y. Al-Hyari, A.M. Al-Taee, M.A. Al-Taee, Clinical decision support system for diagnosis and management of chronic renal failure, in: 2013 IEEE Jordan Conference on Applied Electrical Engineering and Computing Technologies (AEECT), IEEE, December 2013, pp. 1–6.
[25] W.H.S.D. Gunarathne, K.D.M. Perera, K.A.D.C.P. Kahandawaarachchi, Performance evaluation on machine learning classification techniques for disease classification and forecasting through data analytics for Chronic Kidney Disease (CKD), in: 2017 IEEE 17th International Conference on Bioinformatics and Bioengineering (BIBE), IEEE, October 2017, pp. 291–296.
[26] A.J. Aljaaf, D. Al-Jumeily, H.M. Haglan, M. Alloghani, T. Baker, A.J. Hussain, J. Mustafina, Early prediction of chronic kidney disease using machine learning supported by predictive analytics, in: 2018 IEEE Congress on Evolutionary Computation (CEC), IEEE, 2018, July, pp. 1–9.
[27] P.F. Halloran, K.S. Famulski, J. Reeve, Molecular assessment of disease states in kidney transplant biopsy samples, Nat. Rev. Nephrol. 12 (9) (2016) 534.
[28] Y. Amirgaliyev, S. Shamiluulu, A. Serek, Analysis of chronic kidney disease dataset by applying machine learning methods, in: 2018 IEEE 12th International Conference on Application of Information and Communication Technologies (AICT), IEEE, October 2018, pp. 1–4.
[29] L. Nie, M. Wang, L. Zhang, S. Yan, B. Zhang, T.S. Chua, Disease inference from health-related questions via sparse deep learning, IEEE Trans. Knowl. Data Eng. 27 (8) (2015) 2107–2119.
[30] L. Jena, N.K. Kamila, Distributed data mining classification algorithms for prediction of chronic-kidney-disease, Int. J. Emerg. Res. Manag. Technol. 9359 (11) (2015) 110–118.
[31] M.J. Huang, M.Y. Chen, S.C. Lee, Integrating data mining with case-based reasoning for chronic diseases prognosis and diagnosis, Expert Syst. Appl. 32 (3) (2007) 856–867.

CHAPTER

13

Enhanced brain tumor detection using fractional wavelet transform and artificial neural network

Bhakti Kaushal[1], *Mukesh D. Patil*[1], *Gajanan K. Birajdar*[2]

[1]Department of Electronics and Telecommunication Engineering, Ramrao Adik Institute of Technology, Navi Mumbai, Maharashtra, India; [2]Department of Electronics Engineering, Ramrao Adik Institute of Technology, Navi Mumbai, Maharashtra, India

1. Introduction

The human body is comprised of many types of cells. To keep the body healthy and working correctly, these cells grow and divide well to make new cells. Some of these cells lose their control over growth and so keep dividing themselves without proper order. This disorderly spread of cells develops a mass of tissue which is termed tumor. If such kind of mass of tissue is present in brain, it is called brain tumor. The proper brain tumor definition is "An abnormal growth of uncontrolled cancerous tissues in the brain" [1].

Depending on whether the brain tumor is harmful or not, it is categorized as benign or malignant. The structure of benign tumor is uniform in nature and its cancer cells are not active. Opposite to this, the structure of malignant tumor is nonuniform and it contains cancer cells, which are active and can spread all over the body. It is more harmful as it may remain untreated. A third type can be explained as a benign tumor when present in a significant area of brain. Even though it does not carry any cancer cells which are active, it can still interfere with the vital functions of brain and hence, it can be considered as malignant [1].

The world health organization uses different scales as a grading system, which starts from grade I to grade IV, for classification of tumor types into benign and malignant. The tumors which are low-level can be graded I and II while the tumors which are high-level can be graded III and IV [2]. The brain tumor can be classified as primary and secondary. The tumor which begins spreading in the brain is named primary brain tumors, whereas the cancerous tissues which are originated in another portion of the body are the cause of secondary brain

https://doi.org/10.1016/B978-0-12-822260-7.00010-8

tumors. The growth of cancerous tissues to various parts of the body is termed as *metastasis*. Let's take an example of metastatic lung cancer, this cancer is originated in lungs but if it is spread to the brain, the tumor cells in brain resemble abnormal lung cancer cells, so it can be a type of metastatic lung cancer [1].

Let us understand about some of the common brain diseases:

- **Glioma:** It is the most frequent class of malignant primitive tumors present in the central nervous system (CNS). As they arise in glial tissue, they are called gliomas. These are categorized as high grade glioma (HGG) and low grade glioma (LGG). The different types are Astrocytomas, Oligodendrogliomas, and Ependymomas. *Astrocytomas* begin from astrocytes and can spread anywhere in brain and spinal cord. A fatty covering which protect nerves is called myelin. *Oligodendrogliomas* arise in the cells which produces myelin and cerebrum. They are slow in growth and do not spread into encompassing brain tissue. *Ependymomas* generally arise in the lining of ventricles and might also begin in the spinal cord. The most common age spans in which these tumors can develop are childhood and adolescence. However, they can develop at any age [1,3,4]. The presence of tumor appears to be focused in white matter on the basis of left postcentral gyrus as shown in Fig. 13.1A.
- **Meningioma:** Meninges are the membranes which protects the spinal cord and brain just inside the skull. The tumors which emerge from these membranes are called as meningioma and it is as shown in Fig. 13.1B. A small number of Meningiomas are malignant. Before Meningioma cause any symptoms, they grow quite large in size or shape as they develop very slowly. Hence, it becomes easy for brain initially to adjust to their presence. It occurs most often between the ages of 30–50 years especially in women [1,5].
- **Alzheimer's disease:** It is a degenerative disease of cerebral cortex, a sample brain MR image is shown in Fig. 13.1C. It is a very frequent reason of dementia in the old people. The disorder usually typically becomes clinically evident with changes in mood and actions as an insidious deficiency of higher intellectual ability. The symptom in patients often does not occur before the age of 50 years. The rate of occurrence is categorized by age groups. The rates for 65–74 years, 75–84 years, and 85 years or more categories of age are 3%, 19%, and 47%, respectively. Although at least 5%–10% of cases are familiar, the majority of cases are infrequent [6].

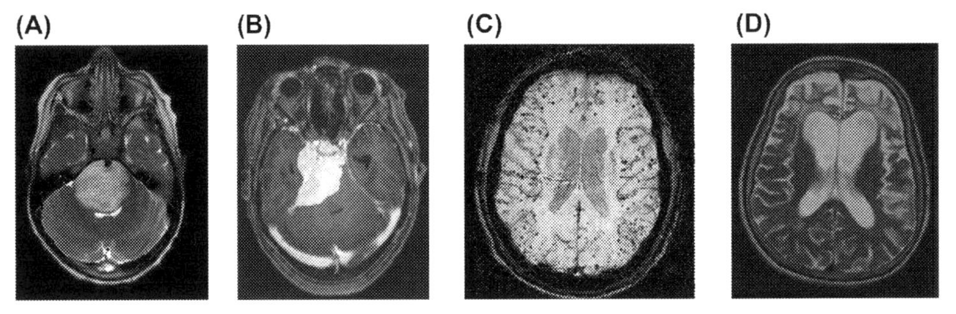

FIGURE 13.1 MR images of some brain diseases (A) Glioma [https://commons.wikimedia.org/wiki/File:Typical_MRI_appearance_of_diffuse_intrinsic_pontine_glioma_(DIPG)_-_Fonc-02-00205-g002.jpg] (B) Meningioma [https://commons.wikimedia.org/wiki/File:Tumor_Meningioma1.JPG] (C) Alzheimer's disease [https://commons.wikimedia.org/wiki/File:Cerebral_amyloid_angiopathy_(CAA)-MRI.png] (D) Pick's disease [https://commons.wikimedia.org/wiki/File:Pick%27s_disease.png].

- **Pick's disease:** It is a progressive but very uncommon dementia that is medically characterized by early beginning of changes in behavior along with changes in personality and language disturbances. The brain always show clearly visible sparing of the posterior two thirds of the superior temporal gyrus and noticeable shrinkage of the temporal and frontal lobes and an uncommon parietal or occipital lobe involvement as shown in Fig. 13.1D. The atrophy is sometimes very intense and it reduces the gyri to a thin wafer like structure (edge of the knife appearance). It is often observed that the putamen and caudate nucleus always have bilateral atrophy [6].
- **Sarcoma:** or Ewing's sarcoma is a tumor family with new and abnormal growth in a shape of small round cells which occur in group. It includes Ewing's sarcoma (EWS), Askin's tumor, primitive neuroectodermal tumor (PNET), PNET of the bone, and also, Extraosseous Ewing's sarcoma (ESS). The origin is believed to be neuroectodermal. In early stages of embryo, out of three germ layers, ectoderm is one of the main germ layers and neuroectoderm is made up of cells derived from it. The inception of neuroectoderm is the initial step in development of nervous system. It gathers hindering signals for bone morphogenetic protein (BMPs) from proteins like noggin which helps in developing the nervous system. Hence, it is frequently present in the soft tissue and bones of children and young adults [7].
- **Huntington's disease:** It is a genetic disorder which is medically characterized by dementia and progressive movement disorders and histological (pathologically) by striatal neuronal degeneration. The patients with progressive movement disorder will have hyperkinetic jerky movements which can affect all the body parts and they might develop parkinsonism in later stage [6].

At any age, people might be affected by brain tumor. Every person might not have the same effect as other. The structure of the human brain is quite intricate which causes the detection of brain tumor a very challenging task in individuals in a timely period. On this basis, it is an important problem in medical science which needs a solution [2].

Various medical imaging modalities are employed for the identification of brain tumor. Some of the different types of modalities are positron emission tomography (PET), magnetic resonance imaging (MRI), single photon emission computed tomography (SPECT) and computed tomography (CT). The CT scan uses X-rays to generate images for diagnosis and exposure to X-ray radiation is harmful for our human body. In PET scan, a radioactive drug which acts as tracer is used. The exposure to radiation risk is low when compared to CT but in some rare cases; the tracer might cause allergic reaction to the patient. SPECT scan uses combination of CT and radioactive material (tracer). These scans are harmful for breast feeding mothers and pregnant women as the material can move to the baby or fetus [8].

MRI is a medical imaging modality which generates images for helpful diagnosis and utilizes nonionizing radiation. The patient is made to lie down inside the MRI scanner which consists of huge and strong magnet. The signals are then made to transmit through patient's body with the help of a radio wave antenna and the transmitted signals are then received by it. The MRI signals received are converted back into images. The MR images are broadly categorized as T1 and T2 images named after the two types of relaxation time measures, i.e., spin-lattice relaxation (T1) and transverse relaxation (T2) time. The two sample images are as shown in Fig. 13.2. It is one such imaging modality which reveals the exact location and

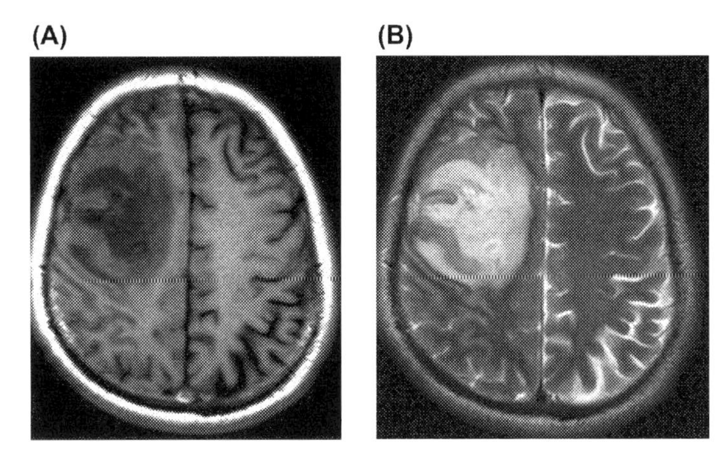

FIGURE 13.2 The two types of MR Images taken in axial plane (A) T1 (B) T2 [https://commons.wikimedia.org/wiki/File:CT_and_MRI_scan_of_the_brain_with_melioidosis.jpg].

extent of the tumor due to its ability to get images in many planes. It does not use any harmful radiation or radioactive material. The MR images illustrate the soft tissue contrast superiorly over other imaging techniques. This quality makes it an ideal imaging technique for the examination of body parts consisting of soft tissues. Due to all these advantages of MRI over the drawbacks of other imaging modalities, MRI is used extensively for brain tumor detection [9]. This imaging scan let the doctors provide an accurate diagnosis by monitoring and tracking the growth and development of regions which are affected by tumor at different stages [2].

Researchers are coming up with new techniques for brain tumor detection, but the most significant problem is its detection at an early stage. Early diagnosis can help in determining the most suitable therapy, radiation, surgery or chemotherapy such that the possibility of survival can be increased [2].

The major contributions of the work presented in this chapter are as follows:

- The early detection of brain tumor with high accuracy, sensitivity and specificity using large dataset of T2-weighted MRI scans.
- This method employs the fractional wavelet transform (FrDWT) for feature extraction which is a cascade of discrete wavelet transform (DWT) and fractional Fourier transform (FrFT) with fractional angle α. To lower the dimension of features, PCA is adopted. To classify the brain MRI into normal or abnormal, ANN is adopted.
- The results of the experiment performed are evaluated using the performance evaluation metrics namely precision, specificity, accuracy, recall, and F1 score which indicate the proposed method is better when compared to other existing techniques.

This chapter is organized as follows: Section 2 presents the literature survey. Section 3 explains about fractional wavelet transform (FrDWT). Section 4 explains principal component analysis (PCA). Section 5 describes artificial neural network (ANN) classifier. Section 6 talks about the proposed methodology. Section 7 analyzes the experimental results. Lastly, Section 8 draws the conclusion of the chapter.

2. Literature survey

The literature contains a lot of approaches to detect and classify brain tumors and a vast amount of research is still happening in this area. Some common methods which are employed for extracting features and classification are gray level cooccurrence matrix (GLCM), DWT, FrFT, ANN, and support vector machine (SVM).

Chaplot et al. [10] suggested a method for binary classification using T2-weighted brain MRI scans classifying them into abnormal or normal. The total numbers of images used are 52, out of which 6 are normal and 46 are abnormal. It utilizes 3-level approximation coefficients of discrete wavelet transform to obtain the features. The self-organizing map provided an accuracy of 94% and support vector machine (SVM) provided 98% accuracy. Maitra and Chatterjee [11] used a different transform i.e., Slantlet transform (ST) and back propagation neural network (BPNN) to extract features and classify brain MR images into normal and person with Alzheimer's disease. In their method, one-dimensional intensity histogram was obtained from brain MR images and ST was applied on the histograms to obtain six relevant features. So, the BPNN was implemented using six inputs and one output. They have accomplished 100% success rate for this proposed work. El-Dahshan et al. [12] used an approach which classifies T2 weighted MRI scans of brain into either normal or abnormal by employing third-level DWT for feature extraction with 1024 features. The PCA method is used for dimensionality reduction to get seven numbers of features. They used two classifiers in their proposed method that are feed forward artificial neural network (FP-ANN) as well as k-nearest neighbor (k-NN). The performance of both the methods are evaluated with the help of accuracy, sensitivity and specificity and the first method produced 98%, 100% and 90% and the second method produced 97%,98.3% and 90% respectively. The texture analysis and classification of Multiple Sclerosis in brain MRI scans using Multiscale Amplitude-Modulation Frequency-Modulation (AM-FM) and SVM was suggested by Loizou et al. [13]. The AM-FM features were based on univariate statistical analysis and measures. The MRI scans of 38 subjects are used within a 2 years period and they were scanned twice with a range of 6−12 months. The classification rate obtained from this proposed work is 86%.

Jaffar et al. [14] employed a technique for segmentation of brain MR image. The removal of noise is done using fast discrete curvelet transform and the results are evaluated using PSNR and RMSE and compared with median and wiener filters. The spatial fuzzy C-means algorithm is utilized on noiseless images for segmenting them into three tissues classes. Saritha et al. [15] proposed a method that employs spider web plot (SWP) which is based on Wavelet entropy (WE) for feature extraction. They used Haar and Daubechies4 wavelet and used entropy to construct spider web plots. Probabilistic neural network was utilized for classification of normal and pathological brain MR images which earned higher success rate with lesser dimension of features i.e., three on small dataset of 75 images. For identification and segmentation of tumors present in brain, Islam et al. [16] proposed combination of multifractal dimension and intensity features for feature extraction and modified AdaBoost algorithm for binary classification of normal and tumorous tissues. The performance is evaluated utilizing Receiver operating characteristic curves for the determination of parameters such as sensitivity and specificity of classifier and some similarity coefficients like Jaccard, Dice, Sokal and Sneath, and Roger and Tanimoto coefficients. Multiclass segmentation as well as classification of tumors present in brain is proposed by Sachdeva et al. [17] using intensity and texture features, PCA with neural network (PCA-ANN) approach. They worked on multiclass primary and secondary brain tumor classification with six classes including normal and used segmented regions of interest (SROI) to extract 218 features. With three set of experiments, the accuracy of 75%–90% is achieved.

El-Dahshan et al. [18] recommended a technique which is based on following methods: the feedback pulse-coupled neural network, DWT, PCA, the feed forward back-propagation neural network for image segmentation, features extraction, reducing the dimensionality, and classification, respectively. The accuracy obtained from training as well as test data proposed from this work is 99%. Karimaghaloo et al. [19] suggested a classifier based on conditional random field (CRF). They have developed temporal hierarchical adaptive texture CRF (THAT-CRF) for segmentation of gad enhancing lesions in brain MR images of patients suffering from multiple sclerosis. They used a very large data of 2380 scans. The performance is evaluated using sensitivity and false discovery rate which came 95% and 20% respectively. The MR image classification of healthy or pathological brain using SVM and its two variants and weighted-type fractional Fourier transform (WFRFT) is proposed by Zhang et al. [20]. The alpha α values of fractional Fourier transform is taken from 0.6 to 1. The two forms of SVM utilized are Generalized Eigenvalue Proximal (GEP) as well as twin SVM. The accuracy obtained from WFRFT + PCA + GEPSVM is 99.11% which is comparably better than other proposed variants.

Nayak, Dash, and Majhi [21] proposed a method which is based on DWT for extracting features, probabilistic PCA for lowering the feature dimension, and AdaBoost algorithm using random forest for classification. They used 3-level wavelet decomposition with 1024 features and used 13 number of principal components. The accuracy obtained by the proposed method is 99.53%. To classify and segment brain tumors, Anitha and Murugavalli [22] proposed the adaptive pillar K-means algorithm with a cascade of two classifiers. The training and testing procedure is achieved in two phases. The DWT is applied and a map neural network which is self-organizing is adopted for training of features. For the second phase, K-nearest neighbor is utilized. The performance is evaluated using sensitivity, accuracy, specificity, Mathews correlation coefficient, negative predictive value, false positive rate, positive predictive value, false discovery rate and it is noted that the method proposed by them performed better when compared than SVM. Bahadure et al. [23] suggested a method which is built using Berkeley Wavelet Transformation to segment tumors and SVM to classify RI scans of brain to detect glial tumor from 15 patients. They used skull-stripping algorithm dependent on threshold method for improvement. The features utilized were intensity and histogram based features. In this method, 96.51% accuracy is achieved for identification of normal as well as abnormal brain MRI. To detect tumors of brain and classify them, Amin et al. [24] proposed an automated method using three variants of SVM and hybrid features set like shape, texture, and intensity. The average accuracy achieved by this proposed method is 97.1%.

Regions of interests (ROIs) based classification is used by Liu et al. [25] to classify Alzheimer's disease (AD). They have employed a whole brain hierarchical network method and regions are divided on the basis of Automated Anatomical Labeling atlas. The Pearson's correlation coefficient of two ROIs is obtained as classification feature and multiple kernel boosting algorithm is utilized for classification of AD. The accuracy of 94.65% is achieved from this proposed work. Mohsen et al. [26] employed a method using deep learning neural networks (DNN) and discrete wavelet transform (DWT) for multiclass brain tumor classification with three kinds of tumors namely glioblastoma, sarcoma, and metastatic bronchogenic carcinoma. The parameters used for evaluating the performance were average recall, average area under the ROC curve, average precision, average F-Measure, and average classification rate. The classification rate obtained from this proposed work is 96.97%.

Latif et al. [27] suggested an enhanced system to classify high-grade and low-grade glioma employing first-order as well as second-order statistical features and also using DWT. The multilayer perceptron (MLP) classifier is then used for classification. The accuracy produced from this proposed work is 96.72% and 96.04% for HGG and LGG, respectively. For the automatic detection of gliomas, Song et al. [28] proposed a noninvasive method which is based upon hybrid features, which includes GLCM, modified CLBP, pyramid HOGs and intensity-based features, and the combination of two classifiers for training. The first one is particle swarm optimization and the second is Kernel SVM. The accuracy reached from this proposed work is 98.36%. Table 13.1 briefs the comparison of existing techniques in literature for brain tumor detection.

3. Fractional wavelet transform

The fractional wavelet transform (FrDWT) was introduced by Mendlovic et al. [29]. After reconstruction, it showed the improved performance as it has the ability to produce minimum reconstruction error compared to traditional wavelet transform. The mathematical definition for the fractional wavelet transform (FrDWT) had been suggested and is defined as the fractional Fourier transform (FrFT) is applied over the complete input signal with an optimal fractional angle α and then the traditional wavelet decomposition is performed [29].

It is used for adapting the localization present in FrFT to the localization present in components of wavelet. While reconstructing back to input function's plane, the traditional inverse wavelet transform is used and then the FrFT is carried out with the negative fractional angle ($-\alpha$). The flowchart is presented in Fig. 13.3. The FrDWT is used in many researches, some of the applications discussed in various papers are image watermarking [30,31], image encryption [32,33], fingerprint security under biometrics domain [34].

The mathematical equation of FrDWT [29] can be given by Eq. (13.1) as:

$$W^{(\alpha)}(a,b) = \int_{-\infty}^{\infty}\int_{-\infty}^{\infty} B_\alpha(x,x')f(x')h_{ab}^{*}(x)dx'dx \tag{13.1}$$

where $W^{(\alpha)}(a,b)$ = FrDWT, $B_\alpha(x,x')$ = kernel of the transformation, a = scale parameter, b = shift amount and $h_{ab}(x)$ = mother wavelet and its equation [29] is given by Eq. (13.2) as:

$$h_{ab}(x) = \frac{1}{\sqrt{a}}h\left(\frac{x-b}{a}\right) \tag{13.2}$$

TABLE 13.1 Comparison of feature extraction, feature selection/reduction, classification and accuracy of various existing techniques present in literature.

Author	Feature extraction	Feature selection/ reduction	Classification	Accuracy
Chaplot et al. [10]	DWT	–	Self-organizing map and SVM	94% and 98%
Maitra and Chatterjee [11]	Slantlet transform	–	BPNN	100%
El-Dahshan et al. [12]	DWT	–	k-NN and FP-ANN	98% and 97%
Loizou et al. [13]	AM-FM	–	SVM	86%
Sachdeva et al. [17]	Intensity and texture features	PCA	ANN	75% −90%
El-Dahshan et al. [18]	DWT	PCA	Feed forward BPNN	99%
Zhang et al. [20]	Weighted-type FrFT	–	SVM	99.11%
Nayak, Dash and Majhi [21]	DWT	PCA and PPCA	Adaboost with random forest on three databases DS-66, DS-160, DS-255	100%, 100% and 99.53%
Amin et al. [24]	Hybrid features set like shape, texture, and intensity	–	Variants of SVM	97.1%
Liu et al. [25]	Pearson's correlation coefficient of two ROIs	–	Multiple Kernel Boosting (MKBoost)	94.65%
Mohsen et al. [26]	DWT	–	DNN	96.97%
Latif et al. [27]	First-order and second-order statistical features and DWT	–	Multilayer perceptron (MLP)	96.72% (HGG) and 96.04% (LGG)
Song et al. [28]	GLCM, modified CLBP, pyramid HOG and intensity-based features	–	PSO-KSVM	98.36%

The different families of wavelet [35] are mentioned below:

(1) Haar wavelet: Fig. 13.4A shows the "**haar**" wavelet. It is mathematically defined by Eq. (13.3) as given below [35]:

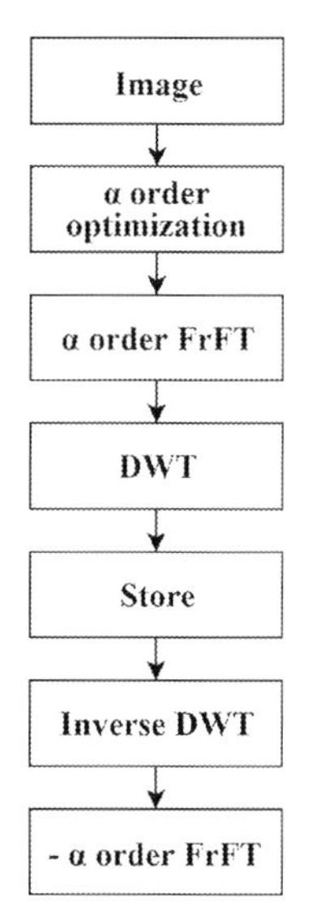

FIGURE 13.3 The block diagram of fractional wavelet transform.

$$W(t) = \begin{cases} -1, & 1/2 \leq t < 1 \\ 1, & 0 \leq t < 1/2 \\ 0 & \text{otherwise} \end{cases} \tag{13.3}$$

(2) Daubechies wavelet (dbN): where N are the number of vanishing moments. They are compactly supported and also the extent of them is finite. So, the value other than their finite interval is zero. Fig. 13.4B shows the Db4 wavelet.

(3) Morlet wavelet: It is an infinite duration wavelet which is obtained by modulating the Gaussian function with sinusoidal function. But the concentration of energy is present for finite interval. Fig. 13.4C shows the Morlet wavelet. It is defined by Eq. (13.4) as given below [35]:

$$w(t) = e^{-t^2} \cos\left(\pi \sqrt{\frac{2}{\ln 2}} t\right) \tag{13.4}$$

(4) Mexican Hat wavelet: The plot is as shown in Fig. 13.4D. The negative Gaussian function is considered and its second derivative is taken to obtain this wavelet. Mathematically, it is represented by Eq. (13.5) as [35]:

$$w(t) = \left(1 - 2t^2\right)e^{-t^2} \tag{13.5}$$

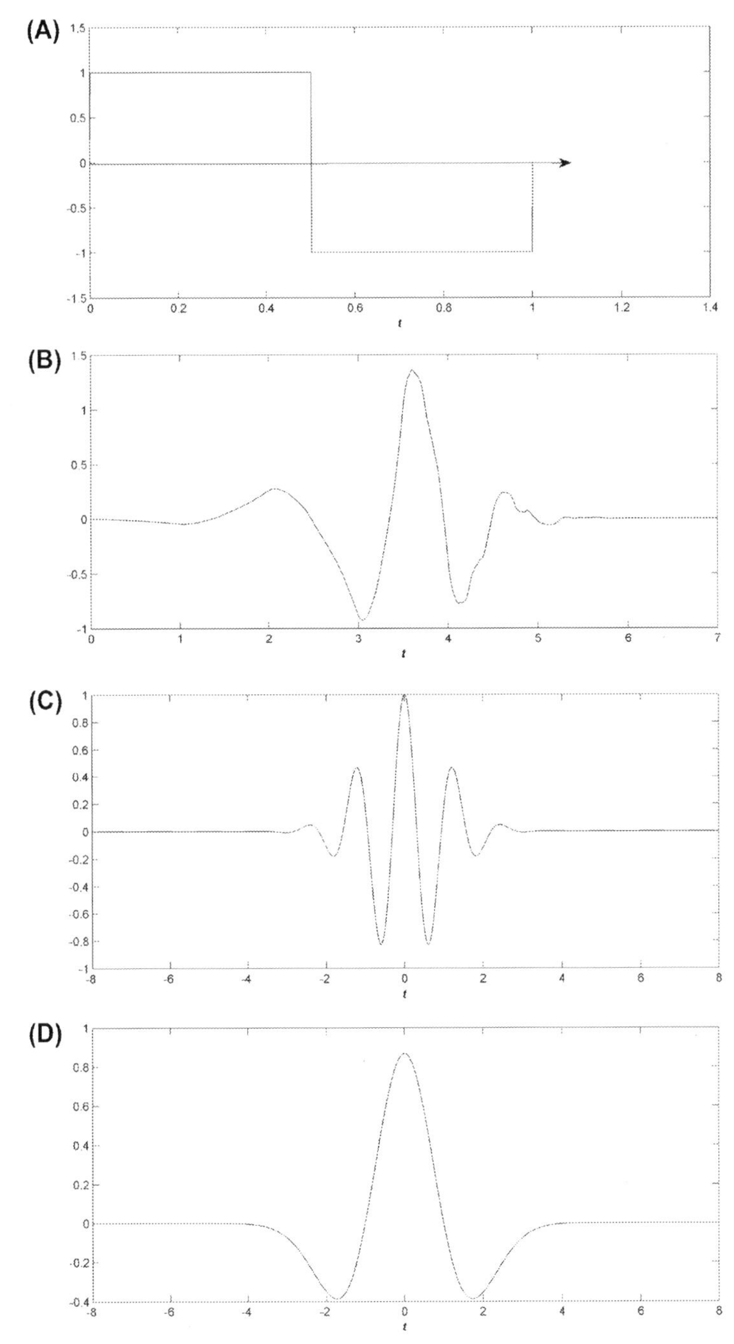

FIGURE 13.4 Wavelet families [35] (A) haar wavelet (B) Db4 wavelet (C) Morlet wavelet (d) Mexican Hat wavelet.

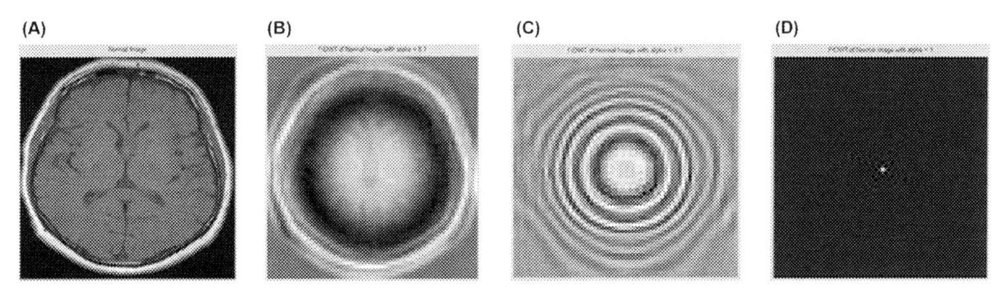

FIGURE 13.5 (A) Normal MR image of brain of size 512 × 512 and its 3-level FrDWT with (B) $\alpha = 0.1$ (C) $\alpha = 0.5$ (D) $\alpha = 1$.

Fig. 13.5 shows the fractional wavelet transform of a normal brain MR image with alpha (α) values 0.1, 0.5 and 1.

3.1 Discrete wavelet transform

DWT is used for simultaneous localization of signal in scale and time for image processing applications when compared with DFT and DCT which can only localize signal in frequency domain. A number of digital filters are present in series at various scales, an image is passed through them to obtain the DWT by filtering. The resolution of the image is changed using subsampling process for scaling operation. The input image is disintegrated into low-pass and high-pass subbands. When compared with original sequence, each subband consists of half of the original samples [36]. Fig. 13.6 shows an input image and its subbands after going through DWT at two different decomposition levels.

It is carried out using filter bank and two-channels are used to implement subband coding. The process of subband coding involves dividing the input image into different frequency bands. For DWT, the filter bank have a common input in the form of an image, hence, they form analysis bank. For inverse transform, they form synthesis bank to reconstruct the image back from its components [36].

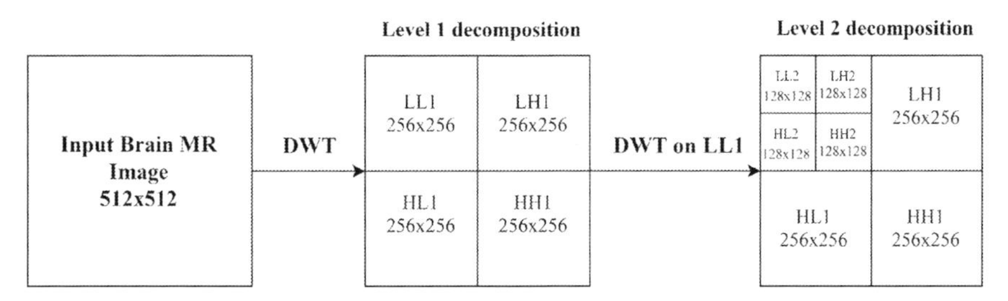

FIGURE 13.6 The sizes of each sub-band of DWT with 2-level decomposition of a brain MR image.

The image is moved through an analysis filter bank and then decimation operation is performed. At each decomposition stage, the image splits into two equal frequency bands as it consists of low as well as high-pass filter. The coarse information of the image is taken out by the low-pass filter, this operation corresponds to averaging and the detailed information of the image is taken out by high-pass filter, this operation corresponds to differencing. After filtering, the output is decimated by two. The reason for performing down-sampling operation is that the sampling frequency of the image signal is too high after filtering. So, for discarding half of the samples down-sampling operation is performed [36].

For image signals, a two-dimensional (2D) transform is required which can be accomplished by performing one-dimensional transform twice separately. Firstly, the filtering operation is done along row on an image and then it is decimated by two. After that, filtering operation is carried along column on the subimage and then it is again decimated by two. The above process splits an image "I[i,j]" into four various subbands which are LL with approximation coefficients, LH with horizontal detail, HL with vertical detail and HH with diagonal detail respectively as represented in Fig. 13.7. Further levels of decomposition are carried on the LL subband back to back and the resulting image can be divided in further four different bands [36].

3.2 Fractional Fourier transform

In 1980, a general type of Fourier transform with fractional powers was rediscovered by NAMIAS [37] in quantum mechanics which is called as the FrFT but this idea in literature appears too early in 1929. It became very popular in optics [38,39] and signal processing communities [40].

A plane is utilized with two orthogonal axes for time-frequency representation, one corresponds to time "t" and other corresponds to frequency "ω," respectively, as shown in Fig. 13.8. When the signal along the time "t" axis be $x(t)$ and its Fourier transform along the frequency "ω" axis be $X(\omega)$, then the signal's Fourier transform operator $\boldsymbol{F}$ is said to be a counterclockwise rotation of $\pi/2$ which corresponds to the change in signal representation. The above statement is consistent with the fact that if $\boldsymbol{F}$ operator is applied repeatedly it

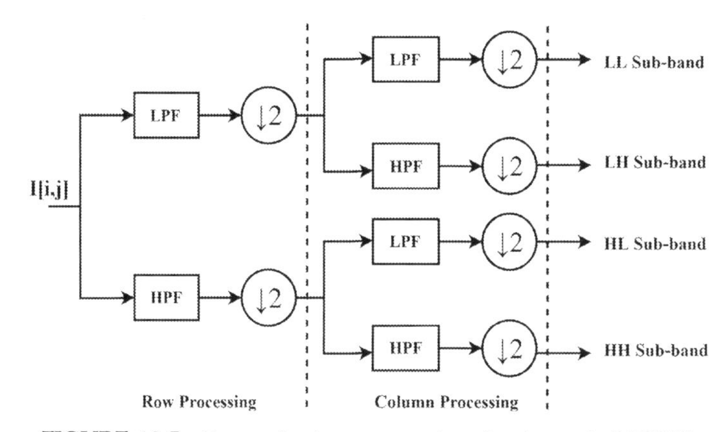

FIGURE 13.7 Row and column processing of an image in 2d-DWT.

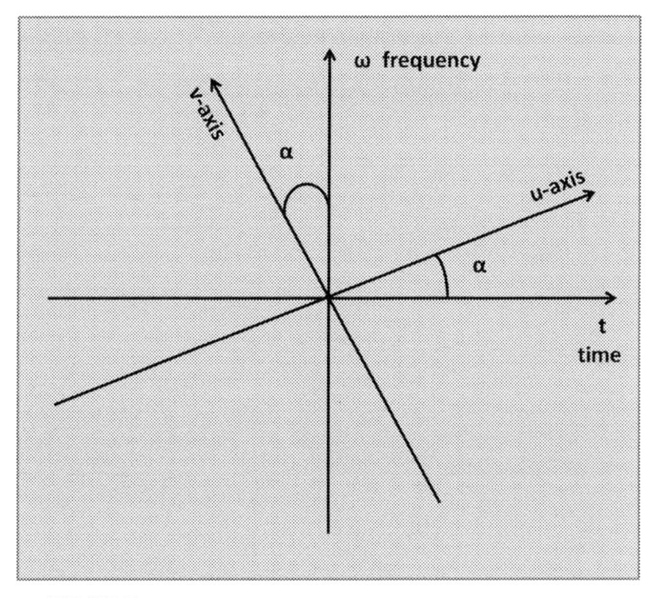

FIGURE 13.8 The time-frequency representation of FrFT.

results in successive rotations of time "t" axis by $\pi/2$. For example, if twice the Fourier operator $\boldsymbol{F}$ is applied, then twice the rotations of time "t" axis by $\pi/2$ successively can be interpreted as an axis which is directed along negative time "$-t$" axis [40]. It is given by Eq. (13.6) as:

$$\boldsymbol{FF}z(t) = x(-t) \tag{13.6}$$

A linear operator when rotated by an angle α (not a multiple of $\pi/2$) or a signal representation along "u" axis with an angle α with respect to time axis as shown in Fig. 13.8 is called as fractional Fourier transform (FrFT) operator which can be represented by $\boldsymbol{R}^{\alpha}$. Consider, a signal "x" with angle α, then the fractional Fourier transform of the signal $\boldsymbol{R}^{\alpha}x = X_{\alpha}$ [37,41] is given by the following Eq. (13.7) as:

$$X_{\alpha}(u) = \int_{-\infty}^{\infty} x(t)K_{\alpha}(t,u)dt$$

$$= \begin{cases} \sqrt{\dfrac{1-j\cot\alpha}{2\pi}}e^{j\frac{u^2}{2}\cot\alpha} \displaystyle\int_{-\infty}^{\infty} x(t)e^{j\frac{t^2}{2}\cot\alpha}e^{jut\csc\alpha}dt, & \text{if } \alpha \text{ is not a multiple of } \pi \\ x(t), & \text{if } \alpha \text{ is a multiple of } 2\pi \\ x(-t), & \text{if } \alpha+\pi \text{ is a multiple of } 2\pi \end{cases} \tag{13.7}$$

Where $K_\alpha(t,u)$ is transformation kernel given by Eq. (13.8) as:

$$K_\alpha(t,u) = \begin{cases} \sqrt{\dfrac{1-j\cot\alpha}{2\pi}}e^{j\frac{t^2+u^2}{2}\cot\alpha - jut\csc\alpha}, & \text{if } \alpha \text{ is not a multiple of } \pi \\ \delta(t-u), & \text{if } \alpha \text{ is a multiple of } 2\pi \\ \delta(t+u), & \text{if } \alpha+\pi \text{ is a multiple of } 2\pi \end{cases} \tag{13.8}$$

4. Principal component analysis

Component analysis is a method to find the right features in an unsupervised way from the data. The PCA is one of the leading dimensionality reduction methods. The main idea is to reduce the data with dimension "*d*" comprising of interrelated variables in large number by a lower dimension compared to "*d*." This is accomplished by transforming the data into a new set of principal components (PCs), which are nothing but a set of uncorrelated variables. They are arranged in such a manner that most of the variation existing in the original is kept by the starting few of the PCs [42].

Let us assume x is a vector of n random variables. For understanding PCAs, the concern is about variance of n random variables and also about how the covariance or correlation between these random variables is structured. If the value of n is very small, then looking at all the variances will be feasible. But if n is not small, then it is unnecessary to check all the n number of variances. So, the new approach is to check a small number of new set of variables compared to n ($<<n$) which retain most of the variance. The PCAs give more concentration on variances as compared to correlation or covariance [42].

The linear function $\alpha_1'x$ meeting maximum variance with elements of x is found and given by the following Eq. (13.9) [42]:

$$\alpha_1'x = \alpha_{11}x_1 + \alpha_{12}x_2 + \ldots + \alpha_{1n}x_n = \sum_{j=1}^{n} \alpha_{1j}x_j \tag{13.9}$$

where α_1 is a vector of n constants i.e., $(\alpha_{11}, \alpha_{12}\alpha_{13}, \ldots\alpha_{1n})$ and α_1' is its transpose.

Similarly, all the linear functions with maximum variance with elements of x are found, next being $\alpha_2'x$ till kth function, i.e., $\alpha_k'x$ which is the kth principal component. It is said that PCs can be found up to nth stage. But there are m PCs in which maximum variance in x is accounted where ($m << n$), hence, the dimensionality is reduced from n to m as, it is observed that with m number of PCs also efficient results are obtained. Due to this reason, it is a very common dimensionality reduction method [42].

5. Artificial neural network

The foundation of neural networks is based on the idea that our human brain works in a completely different way as compared to a computer. Some computations like pattern recognition performed by the human brain are much faster than a computer due to its complex, nonlinear and parallel computer like structure [43]. The ANN has a minimum of three layers, i.e., input, hidden and output layer as shown in Fig. 13.9. There can be more than one hidden layer depending on the network requirement. The features are fed as inputs to input layer. The output of previous layer is fed as input to the layer present next to it, as they are connected with each other via neurons. Finally, the output can be gathered from the output layer.

Just like the human brain consist of neurons, the ANN is also made up of artificial neurons. The neuron is made up of three main components which are a set synaptic weight, an adder, and an activation function. A sample model of the neuron is as shown in Fig. 13.10. Let "m" number of input signals x_i are linked to neuron "k" and the product of input signals with its respective synaptic weight "w_{ki}" is taken. The total input of activation function, $\phi(\bullet)$, can be controlled by an externally applied bias "b_k," which can either increase or decrease it. The output signal of the neuron is given by "y_k." The neuron can be expressed mathematical by the following equations Eqs. (13.10–13.12) [43];

$$y_k = \phi(v_k) \tag{13.10}$$

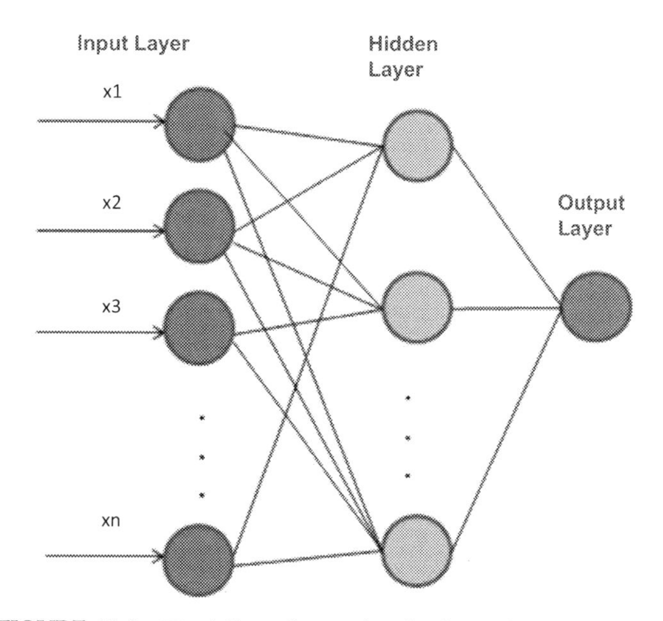

FIGURE 13.9 The different layers of artificial neural network (ANN).

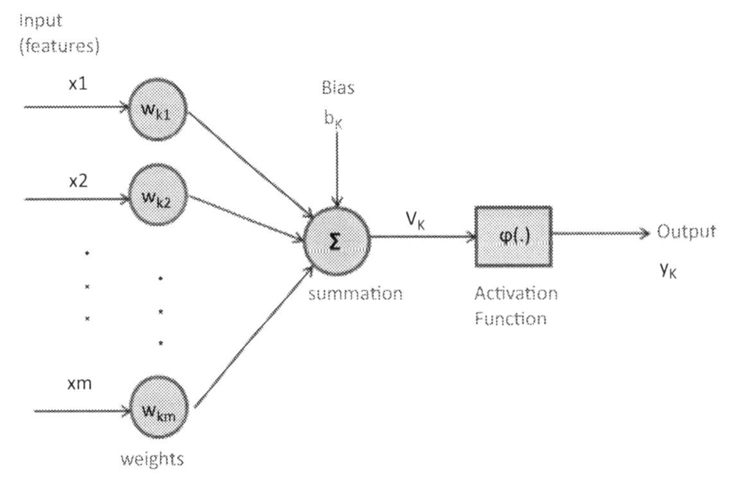

FIGURE 13.10 The model of neuron.

where

$$v_k = u_k + b_k \tag{13.11}$$

and

$$u_k = \sum_{i=1}^{m} w_{ki} x_i \tag{13.12}$$

The activation function can be broadly classified by two types, i.e., threshold function and sigmoid function [43]. The most widely used activation function is sigmoid. The ANN is a huge parallel distributed processor which can store information and can use it later. This information is gathered through a learning process which can either be supervised or unsupervised. In supervised learning, the ANN is fed with a labeled data along with input features. The network is trained with the help of this labeled data. In unsupervised learning, the labeled data is not provided along with the input and the network learns via self-organized learning. Various applications of ANN can be designed with some common techniques of machine learning which includes classification, pattern recognition, regression, clustering, supervised learning and unsupervised learning [44].

6. Proposed method

The important stages of our proposed method are application of fractional wavelet transform (FrDWT), PCA, and neural network classifier. The block diagram is as shown in Fig. 13.11. The following section provides an explanation of each of the blocks present in the block diagram.

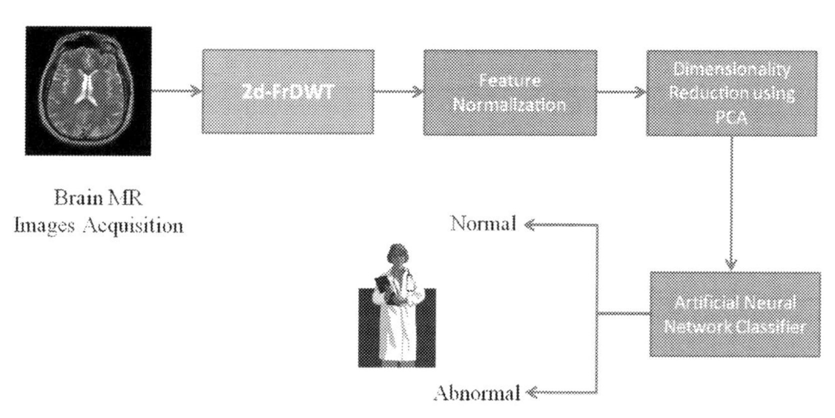

FIGURE 13.11 The block diagram of method proposed.

6.1 Brain MR image acquisition

The first block of our proposed block diagram talks about the acquisition of brain MR image dataset. The MR images used in this experiment are collected from the Harvard Medical School website [45], BRATS 2015 [3,4], e-health laboratory [13] dataset. These are all standard MRI dataset which had been used in various research.

6.2 Feature extraction based on fractional wavelet transform

The second block talks about fractional wavelet transform or FrDWT. The definition says that it is cascade of FrFT and DWT but in this approach FrDWT construction is developed with a slight difference as proposed by Ref. [30]. First, 2D-DWT is applied with 3-level decomposition on input image dataset and then FrFT is applied on LL3 image. This result in a coefficient matrix which is a combination of both and has an added advantage compared to DWT or FrFT alone. The subbands which are obtained at each level are LL, LH, HL and HH. The first level decomposition produced four subband images (LL1-approximation image, LH1-horizontal detail image, HL1-vertical detail image, and HH1-diagonal detail image). The LL1 can be further used to produce second level 2D-DWT which results in another four subband images (LL2, LH2, HL2, and HH2), and the LL2 is further used to obtain third level 2D-DWT which produces four subband images (LL3, LH3, HL3, and HH3). The detailed explanation is provided in Section 3.

In this work, a Haar wavelet is used as it is simplest, very fast, orthogonal, and symmetric, and in the presence of noise, provides better results. In the next step, 2D-FrFT is applied on third-level approximation image (LL3) subband with alpha value ranging from [0, 1]. As FrFT is complex in nature, so, the absolute value of the resulting image is taken and the coefficients are placed in a primary feature vector FV. A matrix is then created by combining the entire feature vector FV with M number of features and of N images named feature matrix FM (NxM) from dataset as explained in the step "a" and "b" of Algorithm 1 of the proposed method.

6.3 Feature normalization

The third block talks about the feature normalization. The normalization of FM is done by applying unit variance and zero mean before the dimensionality reduction step. As a result, a new normalized feature matrix named NFM is procured to increase the efficiency. This new matrix is then given to the next block which is PCA. The step "c" of proposed method given in Algorithm 1 explains the above process.

6.4 Dimensionality reduction using PCA

The dimensionality reduction step is important as it calculates a lower dimensional feature matrix (LFM) compared to the original feature matrix. In our proposed method, the PCA is utilized for decreasing the feature matrix dimension. The new matrix (LFM) obtained from PCA is computationally efficient as one can utilize only the first few PCs from feature matrix in a required situation, like, classification problems. It is given by the step "d" presented in Algorithm 1.

Algorithm1. The Algorithm of proposed method

Input -
N = total number of images in dataset of size k x k
M = k/8
α = fractional angle = [0.1-1]
P = no of PCs
Output –
accuracy, sensitivity, specificity, precision, F1 score
for $\alpha \leftarrow$ 0.1 to 1 **do**
for n $\leftarrow$ 1 to N **do**
a) LL3 [1:M, 1:M] $\leftarrow$ compute fractional wavelet transform (FrDWT) at level 3 of LL sub-band
b) FM[1(M x M), N] $\leftarrow$ LL3
c) NFM $\leftarrow$ normalize FM with zero mean and unit variance
d) LFM[1:P, 1:N] $\leftarrow$ compute PCA of NFM
end for
e) Confusion matrix $\leftarrow$ provide the neural net with LFM and labeled set for training
f) accuracy, sensitivity, specificity, precision, F1 score $\leftarrow$ compute the performance evaluation
parameters from confusion matrix
end for

6.5 Classification using artificial neural network classifier

The last block explains the ANN as a classifier in the proposed method and the steps "e" and "f" are presented in the Algorithm 1. From the literature survey, it has been observed that it provides better accuracy and unmatched results when compared to other classification techniques. Like the human brain, the neural network also has a layered structure. An

ANN with desired hidden layers as well as a number of neurons can be used which helps to provide efficient classification results. The new feature matrix (LFM) is split into training set as well as testing set with 70% and 30% data respectively. The training samples along with labeled dataset of "normal" or "abnormal" labels are given as input to train the feed forward neural network. The desired number of hidden layers and neurons used in hidden layer are provided as input. The test dataset is used to analyze the performance of the trained classifier using performance evaluation parameters such as precision, recall, specificity, sensitivity and F1 score. It is expected to render high classification accuracy.

7. Experimental results

The experiment is performed using *"MATLABR2014a"* on a personal computer with a configuration of 2.30 GHz Pentium(R), a dual core processor, 2 GB RAM, and an operating system with 64-bit Windows 7. The results of our proposed method are evaluated using some performance parameters and they are computed as shown in Eqs. (13.13)–(13.17) [21]:

$$\text{Accuracy} = \frac{\text{TN} + \text{TP}}{\text{TP} + \text{TN} + \text{FP} + \text{FN}} \tag{13.13}$$

$$\text{Recall(R)} = \frac{\text{TP}}{\text{FN} + \text{TP}} \tag{13.14}$$

$$\text{Specificity} = \frac{\text{TN}}{\text{FP} + \text{TN}} \tag{13.15}$$

$$\text{Precision(P)} = \frac{\text{TP}}{\text{TP} + \text{FP}} \tag{13.16}$$

$$\text{F1 score} = 2 * \frac{\text{P} * \text{R}}{\text{P} + \text{R}} \tag{13.17}$$

where.

True Positive (TP) = correct classified normal samples.
True Negative (TN) = correct classified abnormal samples.
False positive (FP) = incorrect classified normal samples.
False Negative (FN) = incorrect classified abnormal samples.

7.1 Database

The proposed method is implemented on two datasets named "db1" and "db2." The images of "db1" are procured from Harvard Medical School website [45] (http://www.med.harvard.edu/AANLIB/home.html). The images of abnormal brain in dataset "db2" are taken from the

TABLE 13.2 Shows the number of normal, abnormal and total images used in db1 and db2.

Dataset	Normal	Abnormal	Total
Db1	18	48	66
Db2	60	60	120

BraTS 2015 challenge for brain tumor segmentation website [3,4] (braintumorsegmentation.org) and the normal brain images are taken from the website of the Laboratory of e-Health [13] (http://www.medinfo.cs.ucy.ac.cy).

The first dataset "db1" has a total of "66" images out of which "18" are labeled "normal" and "48" are labeled "abnormal" as shown in Table 13.2. The abnormal images are of various brain diseases which are meningioma, glioma, Huntington's disease, AD, Pick's disease, sarcoma, and AD plus visual agnosia.

The second dataset "db2" has a total of "120" images out of which "60" are labeled "*normal*" and "60" are labeled "*abnormal*" as shown in Table 13.2. The abnormal images are of diseases HGG as well as LGG. The images are T2-weighted and are axial in plane.

7.2 Feature extraction using fractional wavelet transform

In the preprocessing step, all the images are resized to 256 $\times$ 256. In the next step, fractional wavelet transform (FrDWT) is applied which is a cascade of 2D-DWT and FrFT for feature extraction. Three-level wavelet decomposition is performed, so the size of the LL3 image is reduced to 32 $\times$ 32 (256/(23) = 32). On this LL3 image, fractional Fourier transform is applied with α ranging from [0.1−1]. Fig. 13.12 shows the four different subband images of

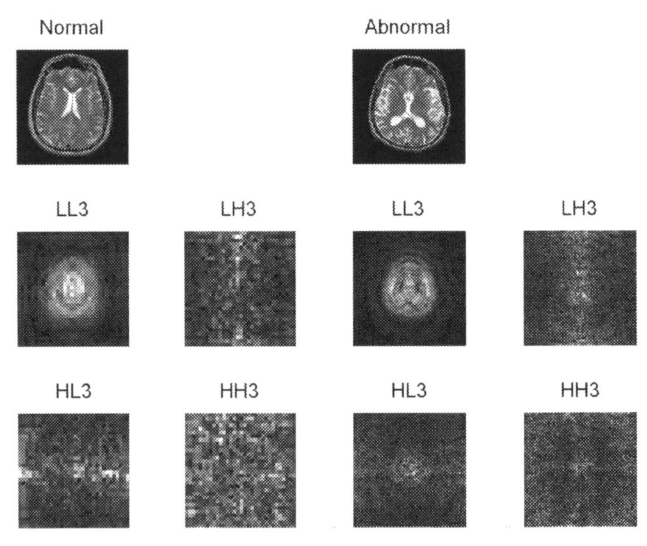

FIGURE 13.12 The four different sub-bands of 3-level fractional wavelet transform at α = 0.5 of normal and abnormal brain MR images.

3-level fractional wavelet transform at $\alpha = 0.5$ applied on normal and abnormal brain MR image. The visual difference can be seen in the LL3 subband image of normal and abnormal brain. The size of the image remains same, i.e., $32 \times 32 = 1024$, these coefficients are used as features.

7.3 Dimensionality reduction using PCA

As the size of the features is very large, PCA is implemented to lower the dimension of feature matrix. The normalization of the features is required before applying PCA, so the normalization with zero mean and unit variance is performed. The normalized features can now be reduced using PCA. Fig. 13.13 shows all the plots of cumulative variance (%) versus number of PCs. Fig. 13.13A shows the plot for DWT and dataset db1, Fig. 13.13B shows the plot for FrDWT and dataset db1, Fig. 13.13C shows the plot for DWT and dataset db2 and Fig. 13.13D shows the plot for FrDWT and dataset db2. It is evident from these plots that **"48"** number of PCs, for all α values ranging from [0.1−1] of FrDWT, can **retain 95% of total variance** in both the datasets. Hence, "48" number of PCs are selected for both the datasets. The new feature matrix is reduced from **"1024" to "48.**" The selected number of PCs are the result of trial and error method.

7.4 Classification using ANN and performance evaluation

The reduced features with the labeled data are given as input to the feed forward neural network as it is a supervised learning. One hidden layer with two numbers of neurons are used. In the proposed method the feature samples are divided into training by 70% ($0.7 \times N$), validation and testing by 30% ($0.3 \times N$) where N is the total number of images present in the dataset, that is, $\mathbf{N_{db1} = 66}$ and is $\mathbf{N_{db2} = 120}$. The scaled conjugate gradient (SCG) method is used as the training algorithm. The **"tansig"** activation function is used for hidden layer. Table 13.3 explains all the parameters used to set up ANN.

The test dataset is used to analyze the performance of the trained classifier. The confusion matrices of test data with $\alpha = 0.1, 0.5$, one of both the datasets are as shown in Fig. 13.14A−F. The comparison between DWT and FrDWT with different α values on the basis of sensitivity, specificity, accuracy, precision and F1 score on "db1" and "db2" is as presented in Table 13.4.

From Table 13.4, it is seen that the FrDWT accuracy on Db1 and Db2 is 100% near $\alpha = 0.5$. At $\alpha = 1$, FrDWT matches DWT for Db2 but for Db1, there is some difference, as the images in Db1 is less in number when compared with a large dataset "Db2." The performance is experimented by varying the number of principal components and it is observed that with 48 PCs the systems performs well.

The performance of the proposed system on db1 at $\alpha = 0.5, 0.6, 0.7, 0.8$, and 1 is good. As the db1 is comparatively smaller in size than db2, thus, the results of DWT are better. But when the size of the dataset increases in db2, the performance of the proposed system

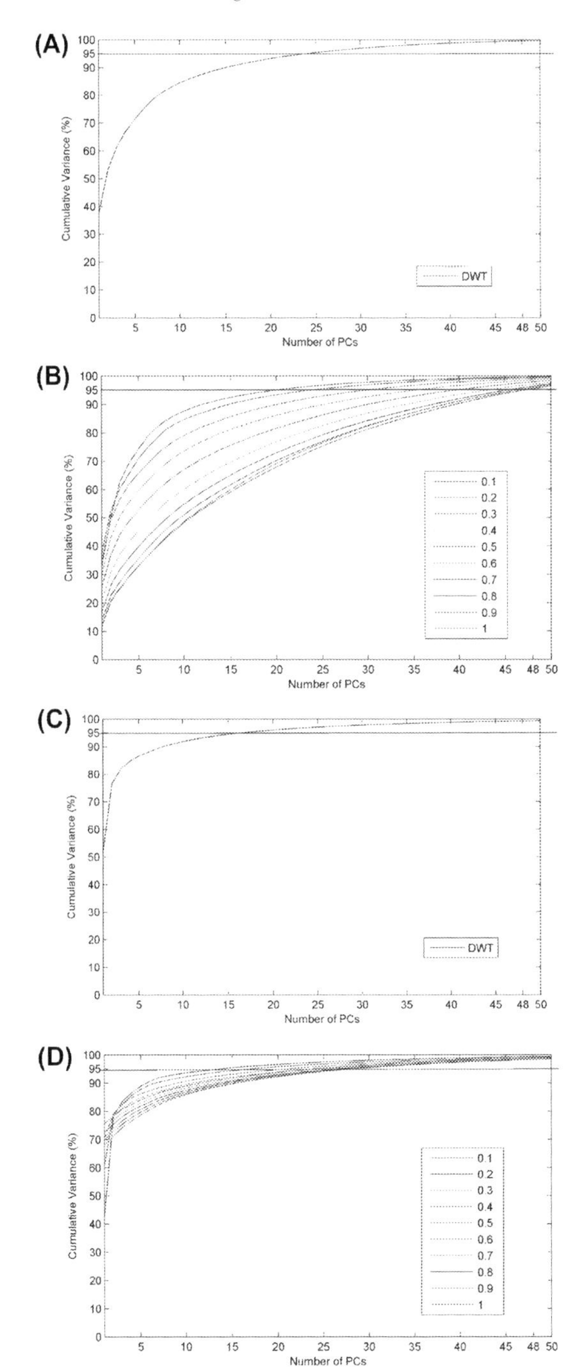

FIGURE 13.13 Plot of cumulative variance versus number of PCs: (A) DWT on 'db1', (B) FrDWT with α ranging from [0.1–1] on 'db1', (C) DWT on db2 and (D) FrDWT with α ranging from [0.1–1] on db2.

TABLE 13.3 shows the parameter setup of ANN.

Parameters	Value
No. of hidden layers	1
No. of neurons in hidden layer	2
No. of features input to network	48
Output neurons	1
Training method	Scaled conjugate gradient (SCG)
Activation function for hidden layer	Tan sigmoid
Activation function for output layer	Soft max
Performance function	Cross-entropy

near $\alpha = 0.5$ of FrDWT provides good results and at $\alpha = 1$ matches with conventional DWT. It is evident from the results that the FrDWT is better compared to conventional DWT for larger datasets in terms of sensitivity, specificity, accuracy, precision, and the F1 score shows improvement near $\alpha = 0.5$.

8. Conclusion

In this chapter, a brain tumor detection technique is proposed based on fractional wavelet transform (FrDWT) which is a generalization of conventional DWT. It is implemented using PCA and ANN. The use of fractional wavelet transform in brain tumor detection is explored in this approach. We have extended the approach of 2D-DWT to 2D-FrDWT. The features are reduced using PCA and to classify brain MR images, ANN is employed.

The performance evaluation of proposed method is carried out using recall, accuracy, precision, specificity, F1 score and various confusion matrix. The Harvard Medical School dataset, BraTS 2015 dataset and e-health laboratory dataset for brain MRI scans are employed for performing the experiment. The results of FrDWT are compared with conventional DWT, and it is observed that FrDWT is better in terms of conventional DWT. The FrDWT results show significant improvement near $\alpha = 0.5$. The presented algorithm can further apply on diseases other than brain tumors, for example, lung cancer.

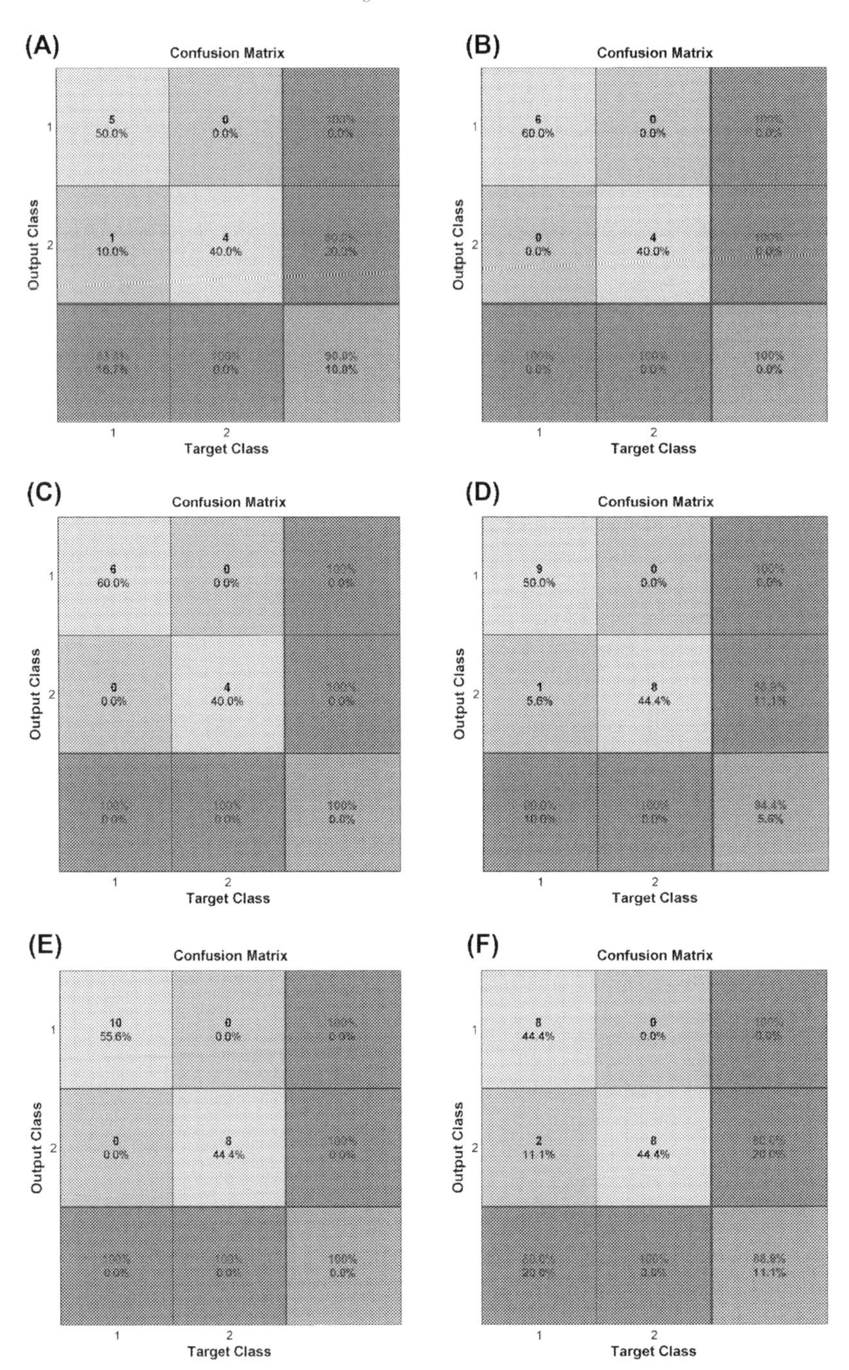

FIGURE 13.14 Test confusion matrix (A)–(C) for db1 with $\alpha = 0.1, 0.5, 1$ (D)–(F) for db2 with $\alpha = 0.1, 0.5, 1$.

TABLE 13.4 The comparison of various performance evaluation parameters between conventional DWT and FrDWT with α [0.1−1] for both the datasets "db1" and "db2."

FrDWT on Db1					
Alpha(α)	**Sensitivity**	**Specificity**	**Accuracy(%)**	**Precision**	**F1 score**
0.1	1	0.8	90	0.833	0.909
0.2	0.6	0	60	1	0.75
0.3	1	0.8	90	0.833	0.909
0.4	0.75	1	80	1	0.85
0.5	1	1	100	1	1
0.6	1	1	100	1	1
0.7	1	1	100	1	1
0.8	1	1	100	1	1
0.9	0.6	0	60	1	0.75
1	1	1	100	1	1
DWT	1	0.8	90	0.833	0.909

FrDWT on Db2					
Alpha(α)	**Sensitivity**	**Specificity**	**Accuracy(%)**	**Precision**	**F1 score**
0.1	1	0.889	94.44	0.9	0.947
0.2	1	1	100	1	1
0.3	1	1	100	1	1
0.4	1	0.8	88.89	0.8	0.889
0.5	1	1	100	1	1
0.6	1	0.889	94.44	0.9	0.947
0.7	1	1	100	1	1
0.8	1	0.889	94.44	0.9	0.947
0.9	1	1	100	1	1
1	1	0.8	88.89	0.8	0.889
DWT	1	0.8	88.89	0.8	0.889

References

[1] National Cancer Institute, Brain Tumors, 2019 [viewed 21-March-2019], Available from: https://www.cancer.gov/types/brain.

[2] N. Varuna Shree, T.N.R. Kumar, Identification and classification of brain tumor MRI images with feature extraction using DWT and probabilistic neural network, Brain Inform. 5 (1) (2017) 23–30.

[3] Kistler, et al., The virtual skeleton database: an open access repository for biomedical research and collaboration, J. Med. Internet Res. 15 (11) (2013) e245.

[4] Menze, et al., The multimodal brain tumor image segmentation benchmark (BRATS), IEEE Trans. Med. Imaging 34 (10) (2015) 1993–2024, https://doi.org/10.1109/TMI.2014.2377694.

[5] J. Watts, G. Box, A. Galvin, P. Brotchie, N. Trost, T. Sutherland, Magnetic resonance imaging of meningiomas: a pictorial review, Insights Imaging 5 (1) (2014) 113–122.

[6] V. Kumar, A. Abbas, J. Aster, Robbins Basic Pathology, Elsevier, 2003.

[7] K.B. Choudhury, S. Sharma, R. Kothari, A. Majumder, Primary extraosseous intracranial ewing's sarcoma: case report and literature review, Indian J. Med. Paediat. Oncol. 32 (2) (2011) 118–121.

[8] P. Sprawls, Physical Principles of Medical Imaging, An Aspen publication, Aspen Publishers, 1993.

[9] R.A. Pooley, Fundamental physics of MR imaging, Radiographics 25 (4) (2005) 1087–1099.

[10] S. Chaplot, L. Patnaik, N. Jagannathan, Classification of magnetic resonance brain images using wavelets as input to support vector machine and neural network, Biomed. Signal Process Contr. 1 (1) (2006) 86–92.

[11] M. Maitra, A. Chatterjee, A Slantlet transform based intelligent system for magnetic resonance brain image classification, Biomed. Signal Process Contr. 1 (4) (2006) 299–306.

[12] E.A. El-Dahshan, T. Hosny, A.M. Salem, Hybrid intelligent techniques for MRI brain images classification, Digit. Signal Process. 20 (2) (2010) 433–441.

[13] C.P. Loizou, V. Murray, M.S. Pattichis, I. Seimenis, M. Pantziaris, C.S. Pattichis, Multiscale Amplitude-modulation frequency-modulation (AM-FM) texture analysis of multiple sclerosis in brain MRI images, IEEE Trans. Inf. Technol. Biomed. 1 (2011) 119–129.

[14] M.A. Jaffar, Q. Ain, T.S. Choi, Tumor detection from enhanced magnetic resonance imaging using fuzzy curvelet, Microsc. Res. Tech. 75 (4) (2012) 499–504.

[15] M. Saritha, K.P. Joseph, A.T. Mathew, Classification of MRI brain images using combined wavelet entropy based spider web plots and probabilistic neural network, Pattern Recogn. Lett. 34 (16) (2013) 2151–2156.

[16] A.S. Islam, M.S. Reza, K.M. Iftekharuddin, Multifractal texture estimation for detection and segmentation of brain tumors, IEEE (Inst. Electr. Electron. Eng.) Trans. Biomed. Eng. 60 (2013) 3204–3215.

[17] J. Sachdeva, V. Kumar, I. Gupta, N. Khandelwal, C.K. Ahuja, Segmentation, feature extraction, and multiclass brain tumor classification, J. Digit. Imaging 26 (2013) 1141–1150.

[18] E.A. El-Dahshan, H. Mohsen, M.K. Revett, A.M. Salem, Computer-aided diagnosis of human brain tumor through MRI: a survey and a new algorithm, Expert Syst. Appl. 41 (11) (2014) 5526–5545.

[19] Z. Karimaghaloo, H. Rivaz, D.L. Arnold, D.L. Collins, T. Arbel, Temporal hierarchical adaptive texture CRF for automatic detection of gadolinium-enhancing multiple sclerosis lesions in brain MRI, IEEE Trans. Med. Imaging 34 (2015) 1227–1241.

[20] Y. Zhang, S. Chen, S. Wang, J. Yang, P. Phillips, Magnetic resonance brain image classification based on weighted-type fractional Fourier transform and nonparallel support vector machine, Int. J. Imaging Syst. Technol. 25 (4) (2015) 317–327.

[21] D.R. Nayak, R. Dash, B. Majhi, Brain MR image classification using two-dimensional discrete wavelet transform and AdaBoost with random forests, Neurocomputing 177 (2016) 188–197.

[22] V. Anitha, S. Murugavalli, Brain tumour classification using two-tier classifer with adaptive segmentation technique, IET Comput. Vis. 10 (1) (2016) 9–17.

[23] N.B. Bahadure, A.K. Ray, H.P. Thethi, Image analysis for MRI based brain tumor detection and feature extraction using biologically inspired BWT and SVM, Int. J. Biomed. Imaging (2017) 1–12.

[24] J. Amin, M. Sharif, M. Yasmin, S.L. Fernandes, A Distinctive Approach in Brain Tumor Detection and Classification Using MRI, Pattern Recognition Letters, 2017. ISSN 0167-8655.

[25] J. Liu, M. Li, W. Lan, F. Wu, Y. Pan, J. Wang, Classification of Alzheimer's disease using whole brain hierarchical network, IEEE ACM Trans. Comput. Biol. Bioinf. 15 (2018) 624–632.

[26] H. Mohsen, E.A. El-Dahshan, E.M. El-Horbaty, A.M. Salem, Classification using deep learning neural networks for brain tumors, Future Comput. Inform. J. 3 (1) (2018) 68–71.

[27] G. Latif, D.N.F.A. Iskandar, J.M. Alghazo, N. Mohammad, Enhanced MR image classification using hybrid statistical and wavelets features, IEEE Access 7 (2019) 9634–9644.
[28] G. Song, Z. Huang, Y. Zhao, X. Zhao, Y. Liu, M. Bao, J. Han, P. Li, A noninvasive system for the automatic detection of gliomas based on hybrid features and PSO-KSVM, IEEE Access 7 (2019) 13842–13855.
[29] D. Mendlovic, Z. Zalevsky, D. Mas, J. García, C. Ferreira, Fractional wavelet transform, Appl. Optic. 36 (1997) 4801–4806.
[30] C. Pujara, A. Bhardwaj, V.M. Gadre, S. Khire, Secure watermarking in fractional wavelet domains, IETE J. Res. 53 (6) (2007) 573–580.
[31] E.H. Elshazly, O.S. Faragallah, A.M. Abbas, M.A. Ashour, E.S.M. El-Rabaie, H. Kazemian, S.A. Alshebeili, F.E. Abd El-Samie, H.S. El-sayed, Robust and secure fractional wavelet image watermarking, Signal Image Video Process. 9 (2015) 89–98.
[32] N. Taneja, B. Raman, I. Gupta, Selective image encryption in fractional wavelet domain, AEU – Int. J. Electron. Commun. 65 (4) (2011) 338–344.
[33] G. Bhatnagar, Q.J. Wu, B. Raman, Discrete fractional wavelet transform and its application to multiple encryption, Inf. Sci. 223 (2013) 297–316.
[34] G. Bhatnagar, Q.M.J. Wu, B. Raman, A new fractional random wavelet transform for fingerprint security, IEEE Trans. Syst. Man Cybern. Syst. Hum. 42 (2012) 262–275.
[35] M.R. Raghuveer, A.S. Bopardikar, Wavelet Transforms: Introduction to Theory and Applications, 1998.
[36] S. Jayaraman, T. Veerakumar, S. Esakkirajan, Digital Image Processing, Tata McGraw Hill Education, 2009.
[37] V. Namias, The fractional order Fourier transform and its application to quantum mechanics, IMA J. Appl. Math. 25 (1980) 241–265.
[38] D. Mendlovic, H.M. Ozaktas, Fractional Fourier transforms and their optical implementation: I, J. Opt. Soc. Am. A 10 (1993) 1875–1881.
[39] H.M. Ozaktas, D. Mendlovic, Fractional Fourier transforms and their optical implementation: II, J. Opt. Soc. Am. A 10 (1993) 2522–2531.
[40] L.B. Almeida, The fractional Fourier transform and time-frequency representations, IEEE Trans. Signal Process. 42 (11) (1994) 3084–3091.
[41] A.C. McBride, F.H. Kerr, On Namias's fractional Fourier transforms, IMA J. Appl. Math. 39 (1987) 159–175.
[42] I.T. Jolliffe, Principal Component Analysis, Springer, 2002.
[43] S. Haykin, Neural Networks and Learning Machines, Pearson, 2009.
[44] K. Faezehossadat, S.M. Jamal, Predicting the 28 days compressive strength of concrete using artificial neural network, I-Manager's J. Civ. Eng. 6 (2) (2016) 1–7, https://doi.org/10.26634/jce.6.2.5936.
[45] K.A. Johnson, J.A. Becker, The Whole Brain Atlas, 1995 [viewed 21-October-2018], Available from: http://www.med.harvard.edu/aanlib/home.html.

CHAPTER

14

A study on smartphone sensor-based Human Activity Recognition using deep learning approaches

Riktim Mondal[1], Dibyendu Mukhopadhyay[1], Sayanwita Barua[1], Pawan Kumar Singh[2], Ram Sarkar[1], Debotosh Bhattacharjee[1]

[1]Department of Computer Science and Engineering, Jadavpur University, Kolkata, West Bengal, India; [2]Department of Information Technology, Jadavpur University, Kolkata, West Bengal, India

1. Introduction

Human Activity Recognition (HAR) plays a paramount role in our circadian life with a wide range of applications. HAR can be performed with different types of inputs like sensor-based data, images [1], and RGB videos [2]. It has several applications in different areas such as (a) building smart ubiquitous devices for the conservation of energy [3], (b) in the medical field, namely, medical imaging, translational bioinformatics, medical informatics, and public health [4], (c) enabling the smart technologies for the Active and Assisted Living (AAL) and Ambient Intelligence [5], (d) in antiterrorism and cyber law and securities [6], and (e) as a cost-effective solution for healthcare, well-being, and sports using inertial sensor data [7]

Wearable sensors and smartphones are transforming our society at an increasing pace, which, in turn, result in excellent opportunities for creating knowledge by extracting meaningful information from new raw data sources. For example, we would be able to know about the detailed information of the activity of a person from the movement of the sensors attached with mobile devices whose data can be obtained in the raw format [8,9].

The role of health informatics has been increased rapidly in the last decade. Among the various strategies of deep learning methods, Convolution Neural Network (CNN) has the

Handbook of Computational Intelligence in Biomedical Engineering and Healthcare
https://doi.org/10.1016/B978-0-12-822260-7.00006-6

greatest impact in the research field of healthcare and health informatics. With the information obtained from the sensors, one cannot only predict the state of a person based on the activity but also visualize the raw sensor data. As a result, one can even decipher whether a person is in a good or bad state of health, which can further help in medical analysis to maintain a healthy lifestyle. Several embedded medical devices for personalized care have been developed to tackle such issues [10]. With the availability of massive data and several types of deep learning-based classification models, it has become much easier to identify the kind of activities a user is performing. Deep learning-based models can directly extract the required information from the input data without any additional handcrafted feature engineering, which is a prerequisite of typical machine learning-based methods. Thus, these deep learning methods are our first choice for activity recognition from sensor data to build an efficient HAR model.

HAR is recognized as one of the hottest research topics in the field of computer vision. In our present work, we concentrate on smartphone device-based sensor data in order to detect human activities using different CNN and Recurrent Neural Network (RNN) based models. Our main objective is to study several activity recognition models based on smart device sensors that generate impressive results in real-time scenarios. Fig. 14.1 shows the overall architecture of the generalized deep learning-based HAR model. For the model to learn proper generalization, the dataset is collected in various environmental conditions like outdoor activity, indoor activity, particular controlled environment, etc. [8].

This paper is structurally arranged as follows: In Section 2, a brief literature survey is performed summarizing some of the past methods done for sensor-based HAR, whereas Section 3 discusses the preparation of datasets that have been used in our present work. In Section 4, a detailed description of five different CNN architectures used for solving the HAR problem is provided. In contrast, Section 5 reports a comprehensive evaluation of experimental results achieved using different HAR models. Finally, Section 6 covers the conclusion and future work.

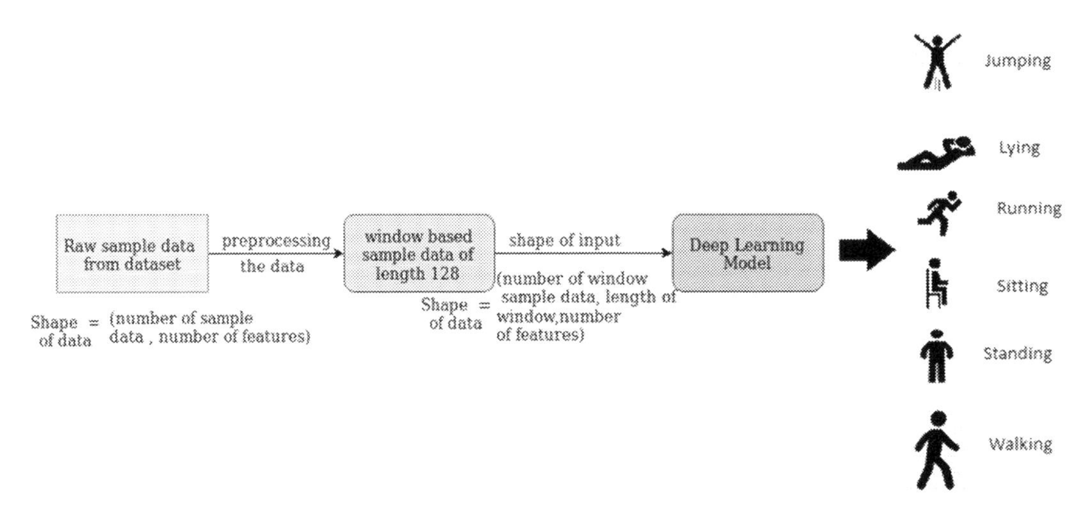

FIGURE 14.1 A generic flowchart describing the flow of our present work, which shows how raw data from sensors are preprocessed and then fed to the deep learning model for identifying human activities.

2. Literature survey

Activity recognition plays a vital role in tracking human behavior while interacting with other humans or interacting with one's surroundings. Many researchers have been working on a smartphone sensor based on HAR using machine learning approaches since long [11]. In this section, we present a brief study of some of the recently developed methods used for HAR.

We are now approaching the age of technology where devices like smartphones, fit bands, smartwatches, etc. have become an integral part of our lives. Raw data from sensors are considered as features for identifying the patterns needed for the classification of activity. Some basic preprocessing steps like rounding off raw values, normalization of raw values are usually performed before fitting into the deep learning models for training and testing where the models could easily extract significant information on its own.

Micucci et al. [12] introduced a HAR methodology that was based on five-fold cross-validation and a leave-one-subject-out cross-validation on both Activities of Daily Life (ADL) and datasets by using raw sensor data [13]. Their proposed dataset was divided into four classification tasks such as (a) AF-17, where eight classes for ADL and remaining nine classes for falls, (b) AF-2 in which all ADLs were considered as one class and all falls were considered as another class, (c) A-9 consist of nine classes for ADLs, and (d) F-8 comprised of eight classes for falls. In order to improve the Mean Average Accuracy (MAA) of their proposed model, a leave-one-subject-out cross-validation approach was also implemented. The features were classified using several classifiers like K-Nearest Neighbor (K-NN), Random Forest (RF), and Support Vector Machine (SVM).

In work described in Ref. [14], He et al. extracted the autoregressive coefficients of accelerometer data as the characteristics of HAR, and used SVM to classify the human activities. The average classification performance obtained on a self-prepared dataset consisting of *Running, Standing, Jumping,* and *Walking* activities is found to be 92.25%. A 10-fold cross-validation procedure and Gaussian kernels were followed in their method to obtain satisfying results. Zeng et al. [15] described a CNN-based methodology with additional modified weight sharing technique named as partial weight sharing. It was used with the accelerometer signal to achieve high recognition accuracy. They applied their model over three datasets, *namely* Skoda [16], Opportunity [17], and Antitracker [18] and obtained 88.19%, 76.83%, and 96.88% recognition accuracies on Skoda, Opportunity, and Antitracker, respectively. Hassan et al. [19] extracted features, namely mean, median, autoregressive coefficients, etc. from the raw sensor data. These features were treated through kernel principal component analysis (KPCA) and Linear Discriminant Analysis (LDA) techniques to make those features more robust for the classification of human activities. Finally, Deep Belief Network (DBN) was used to train those robust features to the final prediction.

Ronao et al. [20] used the idea of "ConvNet" based HAR on sensor data, which could extract features automatically with the help of convolutional and pooling layers. Their model was based on the exploitation of in-built characteristics of the activities and the robust raw data. They also showed the effect of temporal fast Fourier transform on HAR datasets for more accurate activity identification.

Matsui et al. [21] proposed the idea of an adaptive model specific to users' data. They have proposed an adaptation based method using Learning Hidden Unit Contribution (LHUC) in their CNN model where two special layers with very less number of free parameters had been embedded between each of the two CNN layers. The free parameters were calculated based on a small amount of users' data.

Rokni et al. [22] developed a personalized HAR model called Transfer Convolutional (TrC), which helped to build personalized activity recognition with little supervision for individual users. Their base model was fined tuned with a small amount of new user data whenever a new user was added to track their activity, which made the model more specific toward separate users based on their individual patterns of activity.

In the year of 2016, Hinton et al. [23] presented the first successful deep-stacked autoencoder (SAE) to strengthen up the accuracy of HAR. Another work done by Almaslukh et al. [24] used this method but with a random weight initialization. Their method was a low-cost SAE model in which the parameters were adjusted continuously until it achieved the best parameters. This was done in order to get results comparable to state-of-the-art models.

Kwapisz et al. [25] used a single 3D accelerometer in a cell phone to recognize six basic activities such as *Walking*, *Jogging*, *Ascending Stairs*, *Descending Stairs*, *Sitting* and *Standing*. Their performance showed that the J48 algorithm could deduce immobile activities like *Standing*, *Sitting*, etc. easily but failed in mobile activities like *Upstairs* and *Downstairs climbing*.

Lee and Cho [26] collected accelerometer data, performed by four volunteer users (subjects), to recognize human activities with hierarchical hidden Markov models (HHMMs). Their outcome showed that there was still confusion between two activities like *Upstairs* and *Downstairs* movements. Still, the authors claimed that HHMMs were more efficient in classifying low and high-level activities than simple HMMs and neural networks.

In a similar way, the work done in Ref. [27] by Ronao et al. proposed a two-stage continuous HMM on different activity datasets. The first level was for coarse classification, which generally had two cases - moving and stationary. The highest probability of activity was moved to the subclass of the particular stage. In contrast, the second level was for fine recognition, where the subclass category from the first level would finally verify and recognize the target activity. Their results also showed that the stationary activities turned out to be more confused while the moving activities gave a good result.

In this paper, we have aimed at a comparative assessment of different CNN models in order to carry out HAR on sensor-based datasets while evaluating their performances on three different benchmark activity datasets.

3. Dataset description

In our present work, we have considered three publicly available standard datasets: (1) UCI HAR dataset (https://archive.ics.uci.edu/ml/datasets/human+activity+recognition+using+smartphones) [28], (2) RealWorld HAR dataset (https://sensor.informatik.uni-mannheim.de/#dataset_realworld) [29], and (3) MHEALTH (Mobile HEALTH) dataset (https://archive.ics.uci.edu/ml/datasets/MHEALTH+Dataset) [30].

Each of the datasets differs with respect to the number of activities to be classified, the number of subjects (users) used to collect the data and the number of samples of each activity of each subject. Each of the three dataset samples, considered here, has been measured with a frequency of 50 Hz. We have sampled all the raw data samples in a window of a fixed length of 128 for both training and testing. Labeling of the activity of each of those window activities is decided based on the majority type activity present in those 128 data samples of a window. For proper training of our model, we have performed some preprocessing like removal of a sample (a sample observation consisting of all features as a vector) with any missing features, normalization of each observed feature by diving the max value from the feature vector to all the values in that feature vector, and rounding off each observed feature values to four decimal places.

We have performed our experimentations on the RealWorld HAR dataset and MHEALTH dataset using two setups. In the first case, we have performed k-fold cross validation–based classification (considering k as 7) with data samples collected from all the 10 users. In the second case, we have considered the data samples from the first nine users as the training set, and the last user's data have been considered as the validation (testing) set.

3.1 RealWorld HAR dataset

The labeling of the raw sensor data has been performed by manually figuring out the activity, the human is performing by watching videos and labeling each time frame based on the activity of the videos frame at each time step. We have considered a total of 10 subjects (users) for our experiments. A set of three features is considered from linear acceleration (x-axis, y-axis, and z-axis), and eight different activities are recorded in this dataset as follows:

A. Running (R0),
B. Jumping (R1),
C. Sitting (R2),
D. Standing (R3),
E. Climb Upstairs (R4),
F. Climb Downstairs (R5),
G. Walking (R6), and
H. Lying (R7).

3.2 MHEALTH (mobile HEALTH) dataset

This dataset is available in raw format, so we have preprocessed the data following the way, as discussed in the previous section. The samples are recorded using a video camera based on which activities are labeled along with a collection of sample features from 23 sensors. The features are linear acceleration from the chest sensor (along the x-axis, y-axis, and z-axis), electrocardiogram signal (lead one and lead two), linear acceleration from the left-ankle sensor (along all three axes), angular acceleration from the left-ankle sensor (along all three axes), magnetometer readings from the left-ankle sensor (along all three axes), linear acceleration from the right-lower-arm sensor (along all three axes), angular acceleration from the right-lower-arm sensor (along all three axes), and magnetometer readings from the right-lower-arm sensor (along all three axes). There are 12 activities collected in this dataset, which are mentioned below:

A. Standing still (L1),
B. Sitting and relaxing (L2),
C. Lying down (L3),
D. Walking (L4),
E. Climbing stairs (L5),
F. Waist bends forward (L6),
G. Frontal elevation of arms (L7),
H. Knees bending (crouching) (L8),
I. Cycling (L9),
J. Jogging (L10),
K. Running (L11), and
L. Jump front and back (L12).

3.3 UCI HAR dataset

The dataset is already available in a preprocessed format where the accelerometer and the gyroscope measurements are filtered using a noise filter. The observed sensor data are converted into a window of data where each window consists of 128 observed sensor data using a sliding window method for training and testing a learning model. The data are collected from 30 subjects who performed six types of activities, as mentioned below:

A. Walking (A0),
B. Walking Upstairs (A1),
C. Walking Downstairs (A2),
D. Sitting (A3),
E. Standing (A4), and
F. Lying (A5).

Their movements are recorded in videos for labeling. Nine types of measurements such as linear movement data using an accelerometer (along the x-axis, y-axis, and z-axis), angular movement data using gyroscope (along all three axes) and total body movement data (using accelerometer and gyroscope data along all three axes) are collected. Each of the raw collected samples is labeled based on the observation from the video. This preprocessed dataset is available as separate training and testing sets where the first 21 subjects have been considered as the training set and the remaining nine subjects as the test set.

4. Architecture of different deep networks

For our study, we have considered five existing deep learning-based classification models for human activity prediction. Here, all the models are of the supervised based learning models where it can recognize the activities based on the domain it has been trained on. These models are described in detail in the following subsections.

4.1 Convolutional Neural Network (1D-CNN)

In general, CNN models are used for image classification using 2D-CNN and video classification using 3D-CNN, where the models try to extract hidden features from the images or videos. It creates its own internal representation of the source image or video using those features and can easily identify the type of an unknown object based on the knowledge it acquires during training. A similar concept is true for datasets, which consists of a 1D sequence of data from sensors like accelerometer and gyroscope. Here the CNN model can directly extract features from the raw time series data without any need of feature engineering and learn to classify the type of activity being performed.

Our network architecture (shown in Fig. 14.2) consists of two 1D-CNN layers with a kernel size of 3, a number of filters as 64, stride 1 with rectified linear unit (ReLU) [31] as an activation function. Then it has a dropout layer with a probability of dropout as 0.5, followed by a 1D max-pooling layer with a pooling size of 2. Next is a fully connected layer, followed by a dense layer with 100 neurons and with ReLU activation. It ends with the final dense layer of several activities as its output channel with Softmax activation.

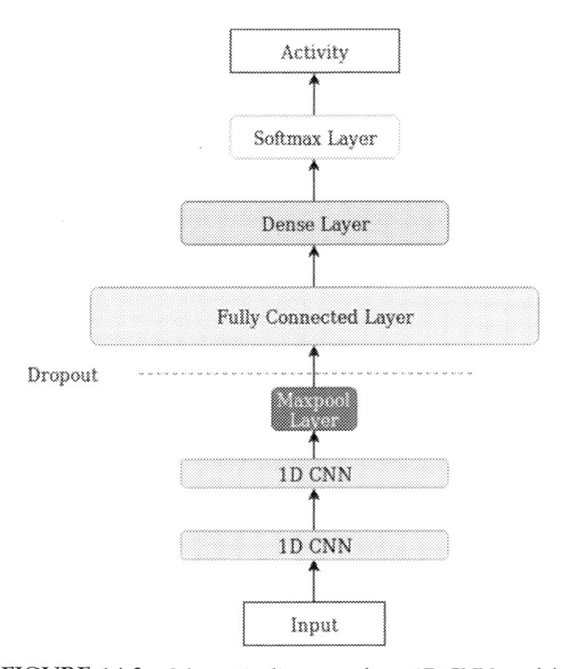

FIGURE 14.2 Schematic diagram of our 1D-CNN model.

4.2 Recurrent Neural Network with Long-Short Term Memory (RNN-LSTM)

LSTM is the type of RNN that can remember a long sequence of input data. They were introduced as a remedy toward the vanishing gradient problem of RNN. LSTM network [32] typically consists of a cell, an input gate, forget gate and the output gate. The cell is used to remember the information over a long interval of time steps and the three gates in regulating information flow across the cells, which make a suitable model for our task.

Here, our LSTM model (illustrated in Fig. 14.3) consists of a single LSTM layer with 200 neurons, followed by a dropout layer with a probability of dropout as 0.5. It is followed by a dense layer consisting of 200 neurons and lastly the final Softmax prediction layer.

4.3 CNN-LSTM

Here, we have considered a combination of 1D CNN and LSTM architectures. Here the window of 128 time steps has been divided into 4 equal subsequent blocks containing 32 time steps. Each of the four subsequent blocks is fed to a CNN model that reads an input that is none other than a sequence of data of each 32 time steps and contains the specified number of features as obtained from the dataset. Our CNN-LSTM network layers (similar to Fig. 14.2) are defined layers of 1D-CNN and RNN-LSTM networks. The overall architecture is shown in Fig. 14.4.

4.4 ConvLSTM

In this model, unlike CNN-LSTM, where the outputs from the CNN layers are passed to the LSTM layer for classification, while in ConvLSTM, besides reading the time step data into

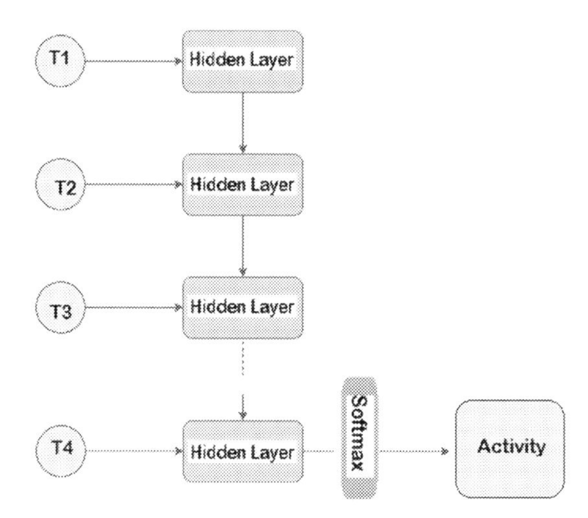

FIGURE 14.3 A block diagram of an LSTM based RNN model where T1, T2, T3, and T4 are the samples of the four time step we have considered from the window provided in sequence to predict the activity.

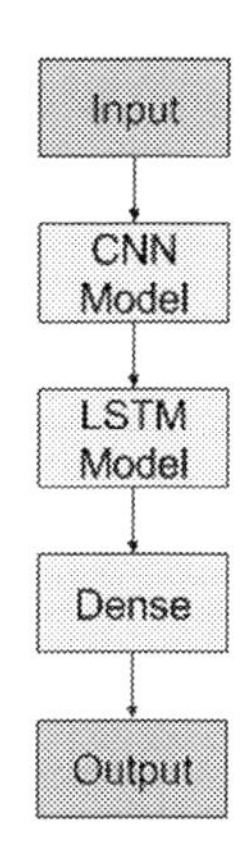

FIGURE 14.4 A schematic diagram of our CNN-LSTM model.

LSTM units, convolution is also applied at the same time. So, each input time step has been considered as a 2D image of sample points where the number of time steps is similar to several subsequent blocks (as in CNN-LSTM), i.e., four and the dimensionality of the data is (1, 32) where the number of rows and columns are taken as 1 and 32 respectively.

The architecture (presented in Fig. 14.5) consists of ConvLSTM2D [33] with a kernel size of (1, 3) having 64 filters, followed a dropout layer with 0.5 probability. Then, a dense layer consisting of 100 neurons with ReLU as activation is present with a Softmax prediction layer at the end.

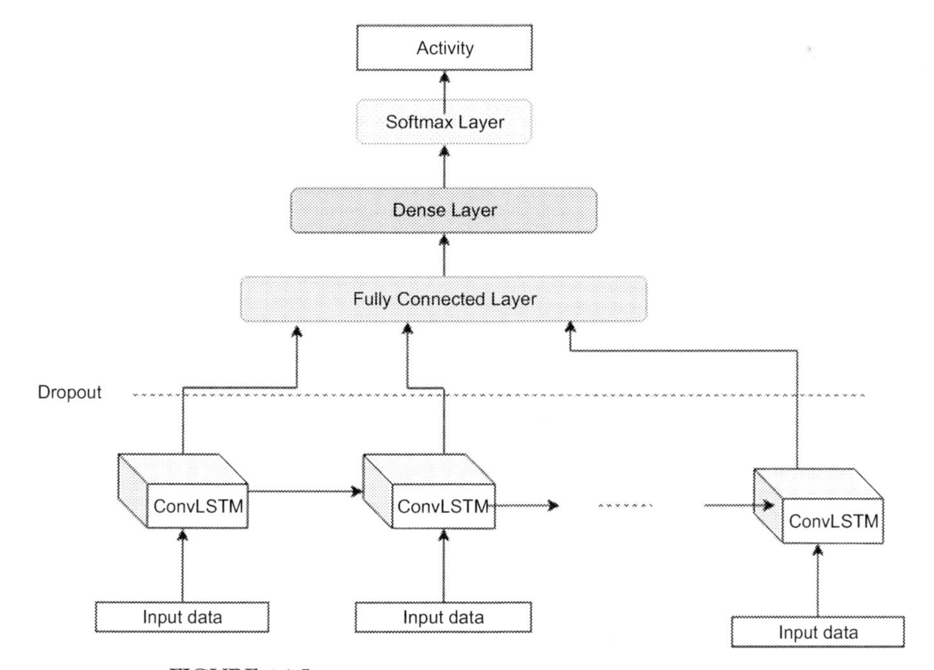

FIGURE 14.5 A schematic diagram of our ConvLSTM model.

4.5 Stacked-CNN

In the Stacked-CNN model, we have considered a stack of three 1D CNN models where each of three separate models reads data from the input time series using different kernel sizes of 3, 5, and 11, respectively. Each of the models extracts features from a different number of samples present in the window due to varying kernel size and finds their own internal representation. Each of the internal representations is considered as a latent vector (usually represented as a fully connected layer). Finally, they are concatenated in the same model, followed by another fully connected layer before the final Softmax prediction layer.

Here, each of the individual CNN models consists of a single 1-D CNN layer with k as kernel size (which is set to 3, 5, and 11 respectively), a number of filters as 64, stride 1 with ReLU as activation. Then, a dropout layer with a probability of dropout as 0.5 is provided, followed by a max-pooling layer with a pooling size of 2. The next layer is fully connected. Each of the three fully connected layers is concatenated together, which is then followed by a dense layer having 100 neurons and with ReLU activation. The Softmax prediction layer ends the network with its number of channels equal to the number of activity classes present in the dataset. The model is illustrated in Fig. 14.6.

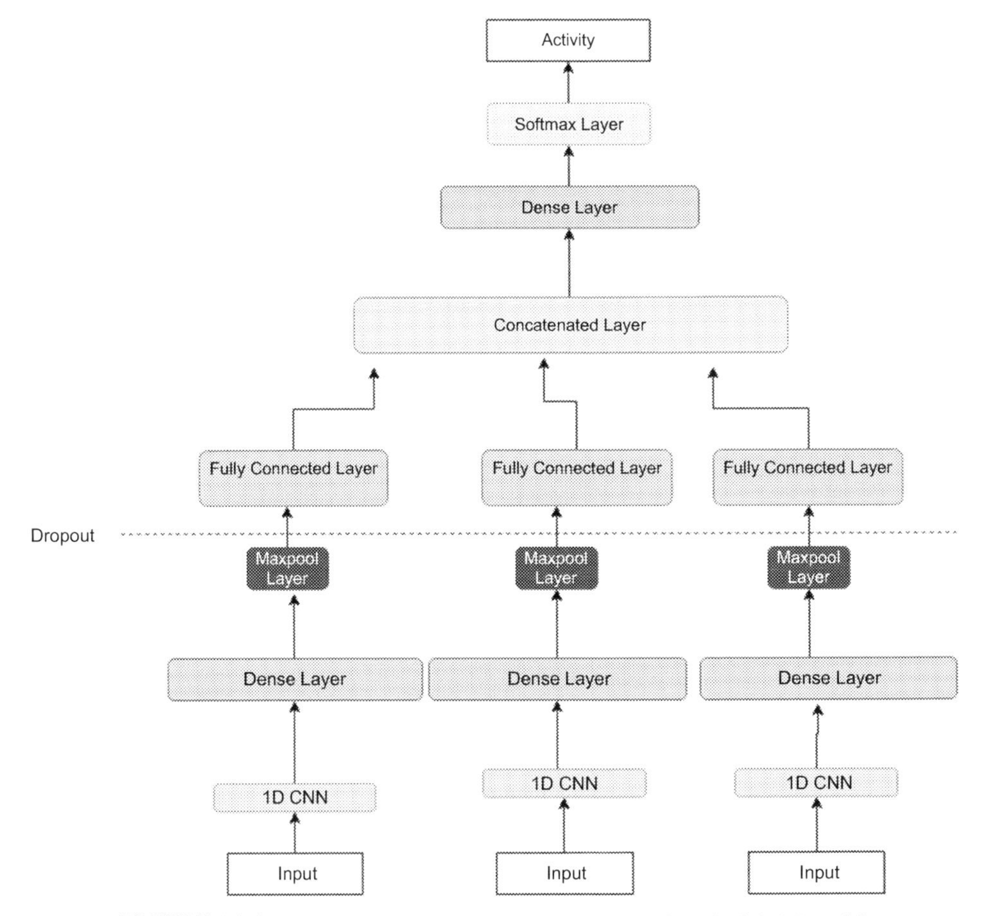

FIGURE 14.6 A schematic diagram of our multi-headed stacked-CNN model.

5. Results and discussion

As mentioned in Section 3, for our study, we have performed two types of experiments on the RealWorld HAR dataset and MHEALTH dataset and a single type of experiment on the UCI HAR dataset, which we have described below. The training of our models has been performed for 20 epochs with Adam [34] as an optimizer having a learning rate of 0.001. The values of the parameters such as beta_1 and beta_2 are taken as 0.9 and 0.999, respectively. The batch size has been considered as 32 in the present work. For the evaluation of our models, we have provided the overall accuracy of correctly identifying human activities.

$$\text{Overall Accuracy}(\%) = \frac{\text{number of correctly classified activities}}{\text{total number of activities present} \in \text{the dataset}} \times 100 \quad (14.1)$$

We have also provided the Precision, Recall, and F1-score along with the confusion matrix of the best performing model in each experiment for every dataset. The confusion matrix is a general metric of classification problem where the predicted classes are compared against actual classes.

As illustrated in Fig. 14.7, TP is the true positive where labels of both the actual and the predicted classes are 1; FP is the false positive where the label of the predicted class is 1 but the label of the actual class is 0; FN is the false negative where the label of the predicted class is 0; and the label of the actual class is 1. Finally, TN is the true negative where the label of the predicted class is 0, and the label of the actual class is 0.

The metric df*Precision* can be defined as:

$$\text{Precision} = \text{TP}/(\text{TP} + \text{FP}) \quad (14.2)$$

which means out of all positive classes which have been predicted correctly, how many are actually positive.

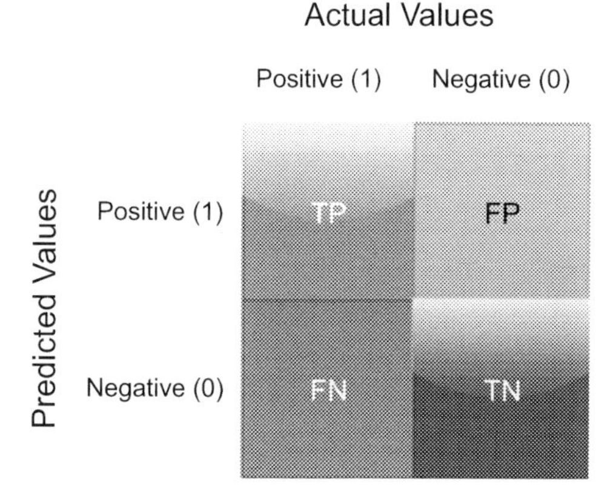

FIGURE 14.7 A confusion matrix for a two-class classification problem ("0" and "1").

The metric *Recall* can be defined as:

$$\text{Recall} = \text{TP}/(\text{TP} + \text{FN}) \tag{14.3}$$

which means out of all positive classes how many have been predicted correctly.

F1-score is a metric which is defined as follows:

$$F1 - score = 2 * \text{Precision} * \text{Recall} / (\text{Precision} + \text{Recall}) \tag{14.4}$$

5.1 RealWorld HAR dataset

(a) Case 1: For the first kind of experiments, *k*-fold based training and validation have been considered with a value of *k* as 7 (Fig. 14.8.). The dataset has been split into 14,756 window samples where each window contains 128 samples with three features for training (consisting of six folds data) and 2460 window samples where each window contains 128 samples with three features for validation (1 fold data). Since we have considered shuffling of seven-fold cross-validation, every time the validation set (2460 window samples) becomes unique with no overlapping. The mean accuracy of the seven-fold cross-validation is considered as the final accuracy. The graph of training and validation for seven-fold experiments on the RealWorld HAR dataset is shown in Fig. 14.9.

Here, we have found that the 1D-CNN model performs the best in this experimental setup with an overall accuracy of **92.07%**, followed by CNN-LSTM (91.35%), Stacked CNN (91.28%), ConvLSTM (90.36%). RNN-LSTM performs the worst with an overall accuracy of 80.88%. Table 14.1 shows the activity-wise performance measures for the best case, whereas Fig. 14.10 shows the confusion matrix for the same.

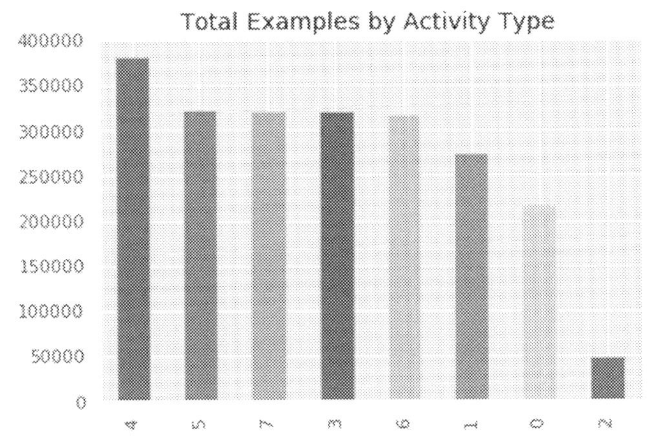

FIGURE 14.8 Illustration of the count of a total number of preprocessed window samples for each type of activity in the RealWorld HAR dataset where the activities are labeled as 0-ClimbDownstairs, 1-ClimbUpstairs, 2-Jumping, 3-Lying, 4-Running, 5-Sitting, 6-Standing, and 7-Walking.

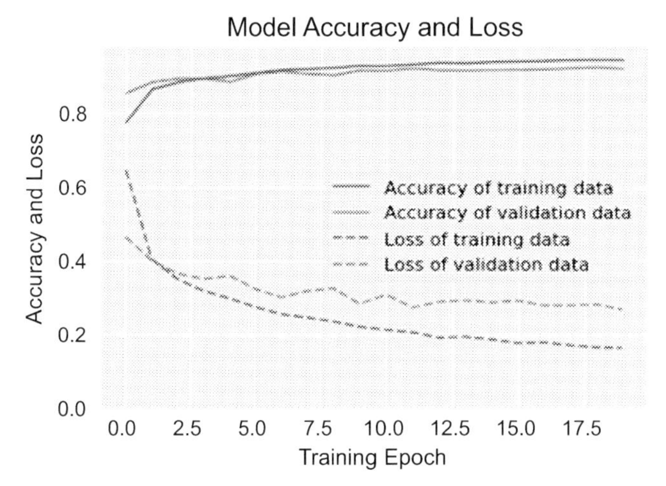

FIGURE 14.9 Training graph of 1D-CNN model for RealWorld HAR dataset for 7-fold cross-validation experiment. Here, the graph corresponds to statistics from a single iteration out of 7.

TABLE 14.1 Performance measures in terms of Precision, Recall, and F1-score of each activity class using the 1D-CNN model on RealWorld HAR dataset (*Support is the number of samples in each activity).

Activity class	Precision	Recall	F1-score	Support
Climb downstairs	0.90	0.91	0.91	246
Climb upstairs	0.93	0.90	0.91	309
Jumping	0.98	0.84	0.90	50
Lying	0.99	0.97	0.98	350
Running	0.95	0.81	0.87	423
Sitting	0.96	0.95	0.95	358
Standing	0.79	0.97	0.87	363
Walking	0.96	0.95	0.95	360
Mean average	0.93	0.92	0.92	2459

(b) Case 2: In this case, the training set (shown in Fig. 14.11) consists of data taken from the first nine subjects (users), whereas the test set (validation set) (illustrated in Fig. 14.12) consists of the data from the 10th subject. A total of 15,565 window samples are present in the training set, whereas 1651 window samples are present in the testing set (validation set). Both the sets contain 128 sample data in each window with three features. The graphs of training and validation on the RealWorld HAR dataset for the second case are shown in Fig. 14.13.

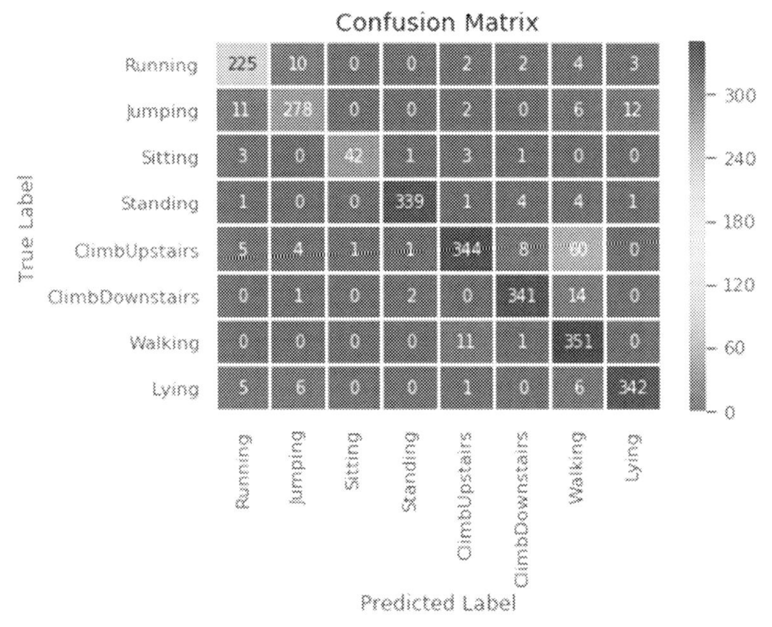

FIGURE 14.10 Confusion matrix of the last test fold set of RealWorld HAR dataset for 1D-CNN model where the values in the diagonal show correctly classified data and the values outside the diagonal show classes predicted as wrong categories.

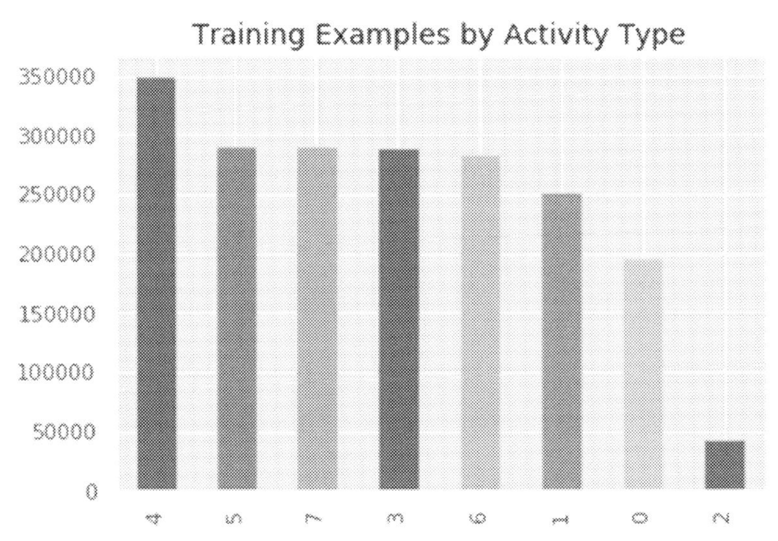

FIGURE 14.11 A chart is displaying the number of preprocessed window samples considered in the training phase for each type of activity in the RealWorld HAR dataset, where the activities are labeled same as in Fig. 14.8.

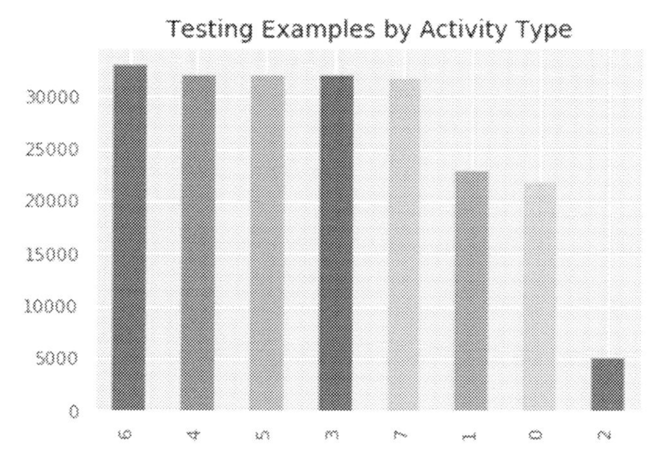

FIGURE 14.12 A chart is displaying the number of preprocessed window samples considered in the testing/validation phase for each type of activity in the RealWorld HAR dataset, where the activities are labeled same as in Fig. 14.8.

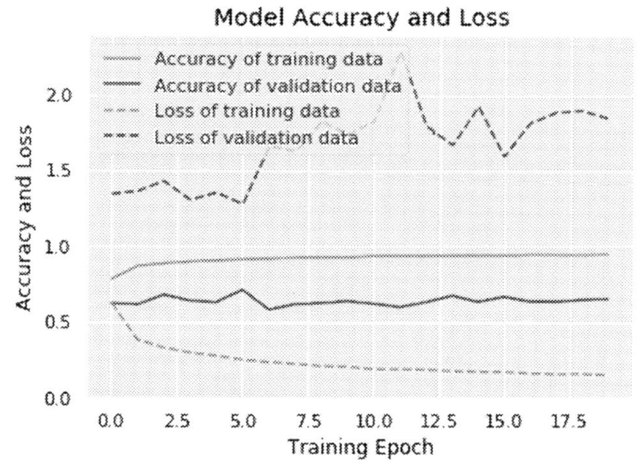

FIGURE 14.13 Training graph of 1D-CNN model for RealWorld HAR dataset considering training with nine subjects and testing with one subject.

It is observed that the 1D-CNN model also gives the best result among the other models, but its overall accuracy is quite low compared to *k*-fold based results. This is because the test set consists of data from only one subject whose pattern might differ from other train set subjects. For example, for one subject, *running* activity can be considered as *walking* for another subject due to age, gender, and other varied factors. The overall accuracy of 1D-CNN is 65.47%, followed by Conv-LSTM (63.77%), Stacked CNN (63.23%), CNN-LSTM (60.87%), and RNN-LSTM (54.39%). Table 14.2 shows the activity-wise performance measures for the best case, whereas Fig. 14.14 shows the confusion matrix for the same.

TABLE 14.2 Performance measures in terms of Precision, Recall, and F1-score of each activity class using the 1D-CNN model on the RealWorld HAR dataset.

Activity class	Precision	Recall	F1-score	Support
Climb downstairs	0.92	0.06	0.12	171
Climb upstairs	0.36	0.94	0.52	180
Jumping	1.00	0.93	0.96	40
Lying	1.00	0.96	0.98	250
Running	0.76	0.97	0.85	252
Sitting	0.62	0.87	0.72	251
Standing	0.71	0.50	0.59	259
Walking	1.00	0.13	0.23	248
Mean average	0.78	0.65	0.61	1651

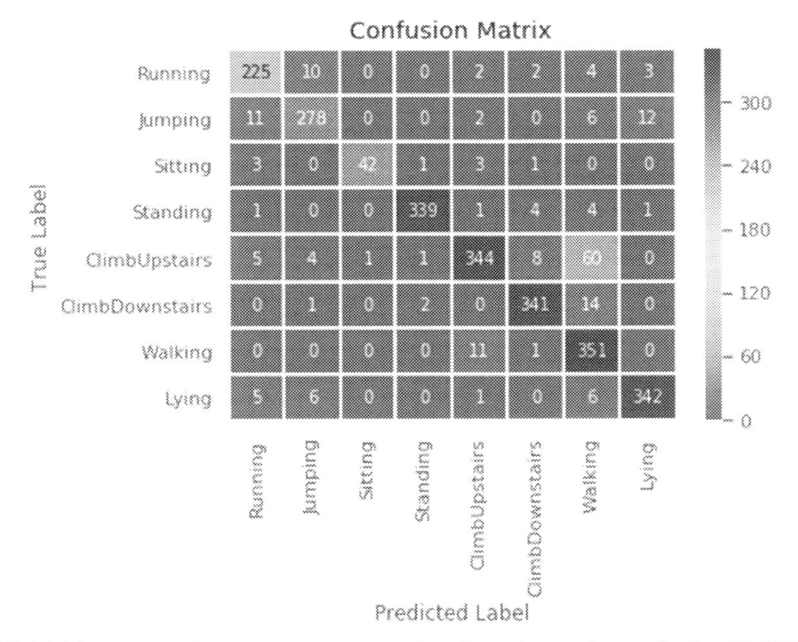

FIGURE 14.14 Confusion matrix of the test set of RealWorld HAR dataset for the 1D-CNN model.

5.2 MHEALTH dataset

(a) Case 1: In the first case, the value of k in k-fold cross-validation has also been considered as 7. The dataset has been split into 2298 window samples, where each window contains 128 samples with 23 features for training (consisting of six-folds of data) and 383 window samples for testing. Here, each window contains 128 samples with 23 features for validation (one-fold data), which is shown in Fig. 14.15. Similar to the RealWorld HAR dataset, here also we have considered shuffling of seven-fold cross-validation every time such that the validation fold data (383 window samples) become unique with no overlapping. The mean accuracy of the seven-fold iteration is considered as the overall accuracy. The graphs of training and validation for the seven-fold experiment on the MHEALTH dataset are shown in Fig. 14.16.

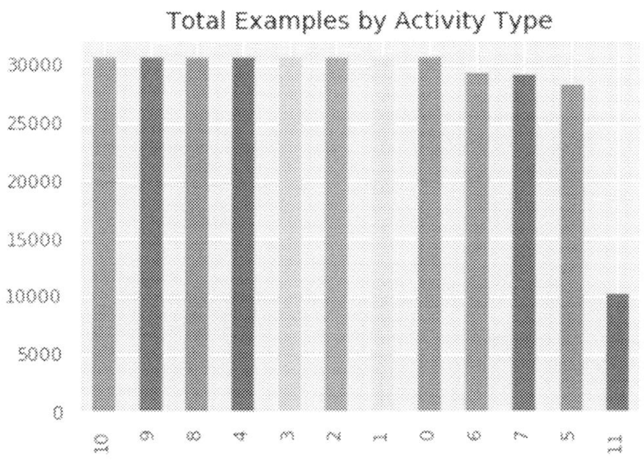

FIGURE 14.15 A total number of preprocessed window samples for each type of activity in MHEALTH dataset where the activities are labeled as 0 - Standing still (L1), 1 - Sitting and relaxing (L2), 2 - Lying down (L3), 3 - Walking (L4), 4 - Climbing stairs (L5), 5 - Waist bends forward (L6), 6 - Frontal elevation of arms (L7), 7 - Knees bending (crouching) (L8), 8 - Cycling (L9), 9 - Jogging (L10), 10 - Running (L11), and 11 - Jump front and back (L12).

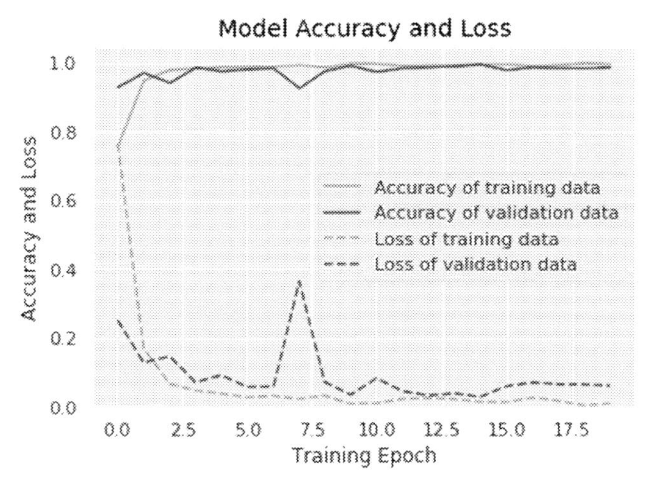

FIGURE 14.16 Seven-fold cross-validation training graph of the 1D-CNN model for the MHEALTH dataset. Here the graph corresponds to statistics from single iteration out of seven.

It has been found that the 1D-CNN model performs the best with an overall accuracy of 99.50%, followed by CNN-LSTM (98.58%), ConvLSTM (98.09%), and Stacked CNN (97.39%). Finally, RNN-LSTM scores the lowest overall accuracy of 58.15%. Table 14.3 shows the activity-wise performance measures for the best case, whereas Fig. 14.17 shows the confusion matrix for the same.

(b) Case 2: In the second case, the division of train set (shown in Fig. 14.18) and test set (see Fig. 14.19) are similar to that of the RealWorld HAR dataset (in case 2) with the splitting of nine subjects for training and one subject for testing. A total number of 2418 window samples have been considered in the training set and 263 window samples in the validation (testing) set. Both the sets contain 128 sample data in each window with 23 features. The graphs of both training and validation on the MHEALTH dataset for the second case are shown in Fig. 14.20.

It is observed that 1D-CNN performs best with an overall accuracy of 94.37%followed by Stacked CNN (84.41%), CNN-LSTM (73.32%), Conv-LSTM (52.85%). Again, the RNN-LSTM model performs worst with an overall accuracy of 36%. Table 14.4 shows the activity-wise performance measures for the best case with the confusion matrix in Fig. 14.21.

TABLE 14.3 Performance measures in terms of Precision, Recall, and F1-score of each activity class using the 1D-CNN model on the MHEALTH dataset.

Activity class	Precision	Recall	F1-score	Support
Standing still	1.00	0.96	0.98	26
Sitting and relaxing	0.98	1.00	0.99	41
Lying down	0.97	1.00	0.99	35
Walking	1.00	0.94	0.97	31
Climbing stairs	0.96	0.98	0.97	47
Waist bends forward	1.00	1.00	1.00	41
Frontal elevation of arms	1.00	1.00	1.00	31
Knees bending (crouching)	1.00	1.00	1.00	33
Cycling	1.00	1.00	1.00	24
Jogging	1.00	0.97	0.98	33
Running	0.97	1.00	0.98	29
Jump front and back	1.00	1.00	1.00	12
Mean average	0.99	0.99	0.99	383

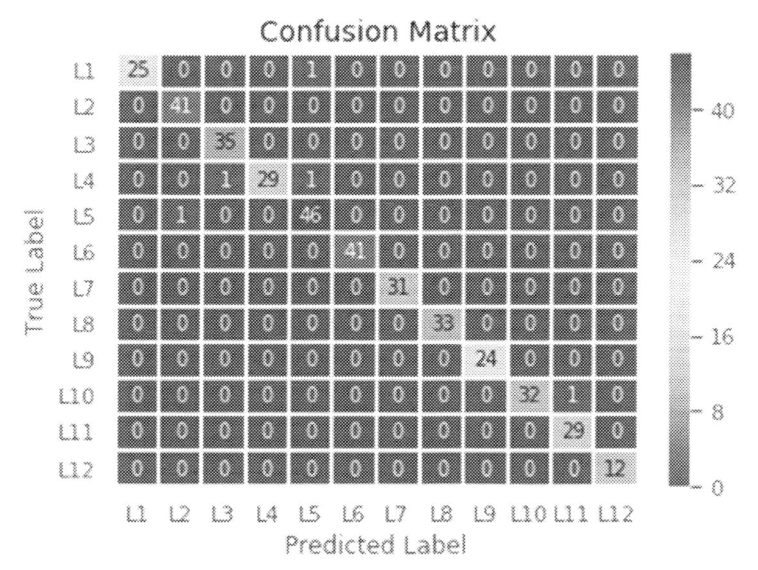

FIGURE 14.17 Confusion matrix of the last fold test set of the MHEALTH dataset for the 1D-CNN model.

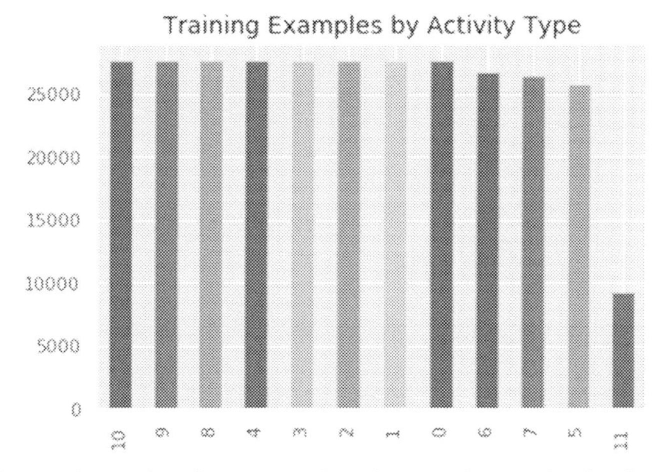

FIGURE 14.18 Showing the number of preprocessed window samples considered in the training phase for each type of activity in the MHEALTH dataset where the activities are labeled same as in Fig. 14.15.

5.3 UCI HAR dataset

The dataset has been split into a 7:3 ratio of training and validation (testing) sets based on the predefined split ratio provided in the dataset. The training set contains 7352 number of window samples, whereas the validation set consists of 2947 window samples. Each of the sets has a window size of length 128 and having nine distinct features. The distribution of movement of data per activity for a single subject (user) has been plotted for three sensor

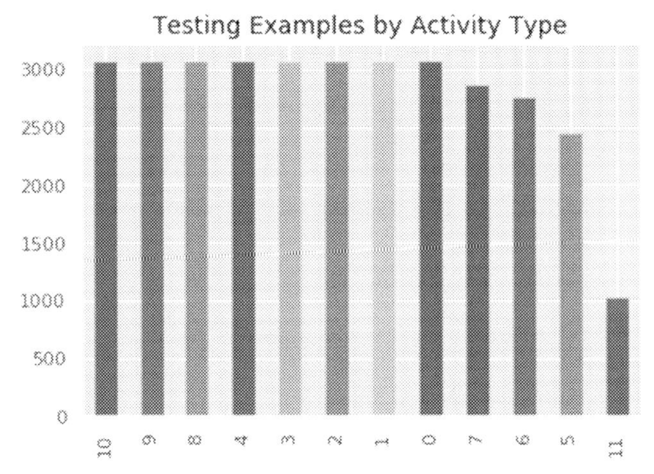

FIGURE 14.19 Illustration of the count of the number of preprocessed window samples considered in the testing (validation) phase for each type of activity in the MHEALTH dataset with label same as described in Fig. 14.15.

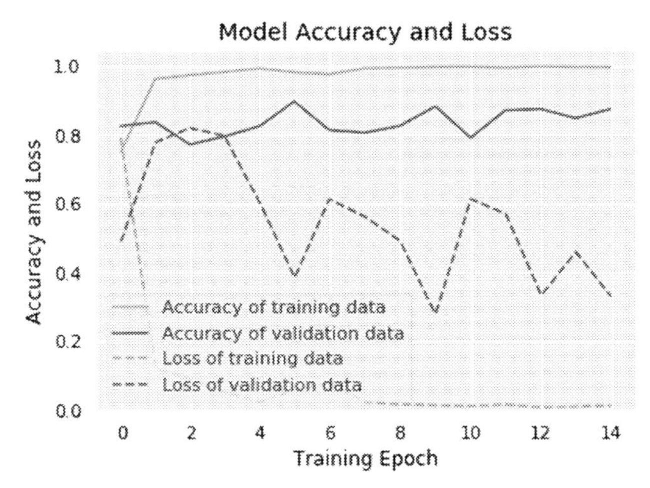

FIGURE 14.20 Training graph of 1D-CNN model for MHEALTH HAR dataset for the train with nine subjects and test with one subject.

types namely total body acceleration data calculated from linear and angular acceleration (illustrated in Fig. 14.22), body linear acceleration data (illustrated in Fig. 14.23) and body angular gyroscope data (shown in Fig. 14.24). In each of those plots, the distribution of respective sensor data for each axes (x, y, and z) has been demonstrated with three separate colors. The graphs of both training and validation on the UCI HAR dataset are depicted in Fig. 14.25.

TABLE 14.4 Performance measures like in terms of precision, recall, and F1-score of each activity class using the 1D-CNN model on MHEALTH dataset.

Activity class	Precision	Recall	F1-score	Support
Standing still	1.00	1.00	1.00	24
Sitting and relaxing	1.00	1.00	1.00	24
Lying down	1.00	1.00	1.00	24
Walking	1.00	1.00	1.00	24
Climbing stairs	0.71	0.62	0.67	24
Waist bends forward	0.95	1.00	0.97	19
Frontal elevation of arms	1.00	1.00	1.00	22
Knees bending (crouching)	0.73	1.00	0.85	22
Cycling	1.00	1.00	1.00	24
Jogging	1.00	0.96	0.98	24
Running	1.00	1.00	1.00	24
Jump front and back	0.67	0.25	0.36	8
Mean average	0.94	0.94	0.93	263

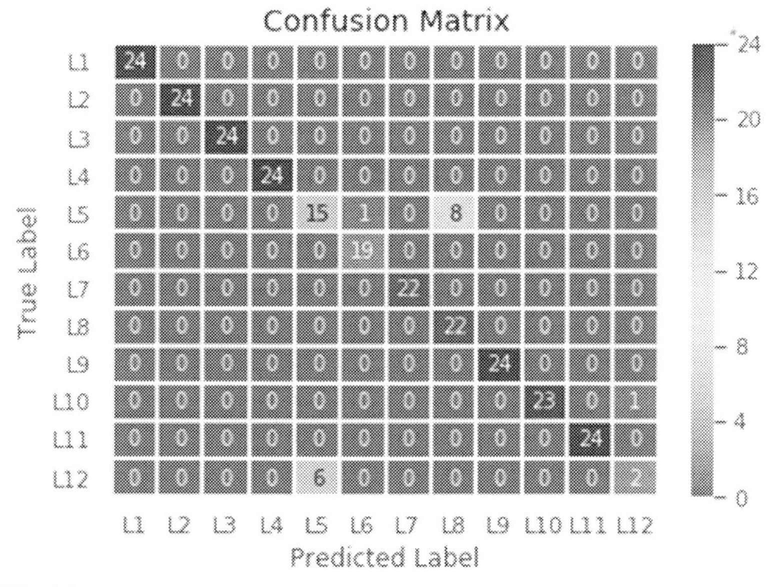

FIGURE 14.21 Confusion matrix of the test set of MHEALTH dataset for the 1D-CNN model.

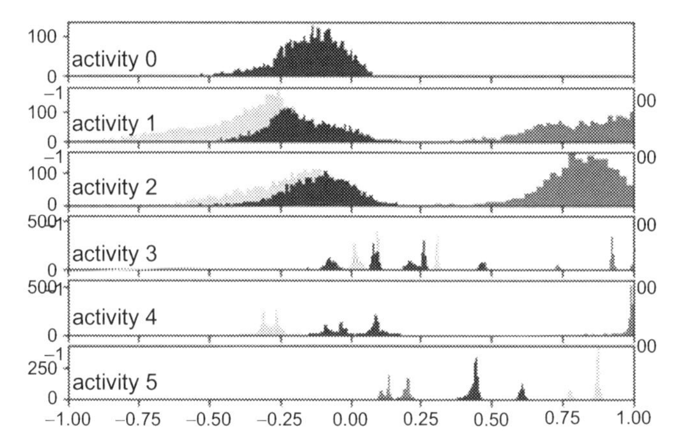

FIGURE 14.22 Histogram of total body acceleration data plotted per activity class for the UCI HAR dataset.

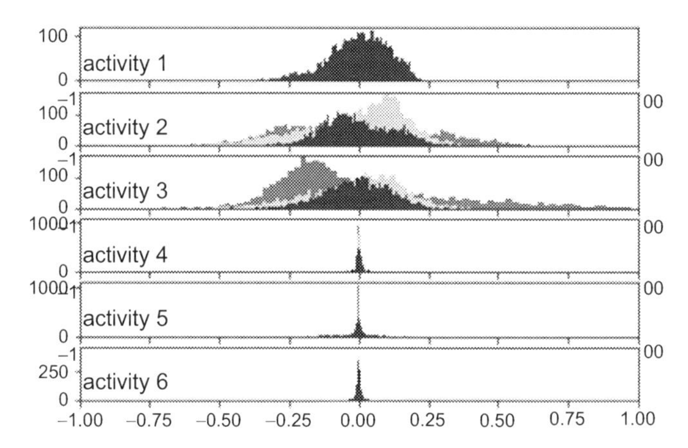

FIGURE 14.23 Histogram of body linear acceleration data plotted per activity class for the UCI HAR dataset.

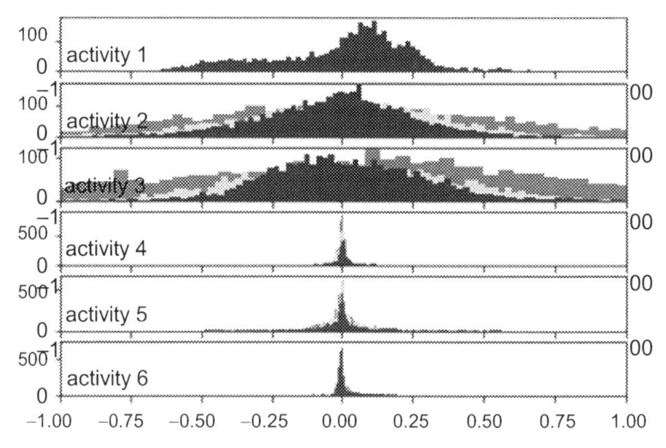

FIGURE 14.24 Histogram of body angular gyroscope data plotted per activity class for the UCI HAR dataset.

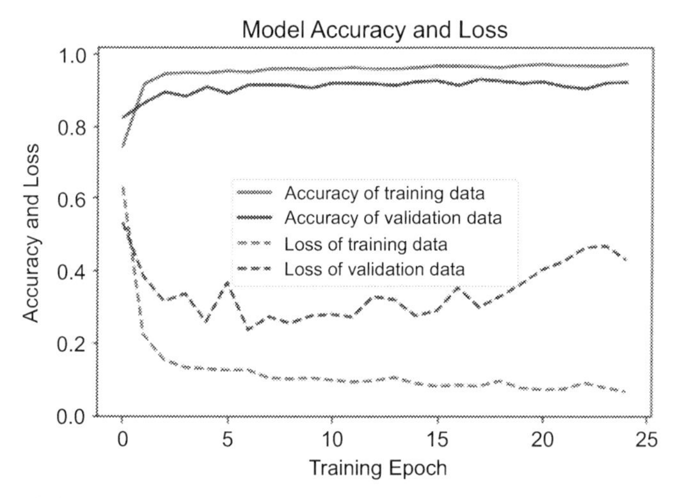

FIGURE 14.25 Training graph of CNN-LSTM model for UCI HAR dataset.

TABLE 14.5 Performance measures in terms of Precision, Recall, and F1-score of each activity class using the CNN-LSTM model on the UCI HAR dataset.

Activity class	Precision	Recall	F1-score	Support
Walking	0.98	0.99	0.98	496
Walking upstairs	0.92	0.94	0.93	471
Walking downstairs	0.93	0.99	0.96	420
Sitting	0.79	0.88	0.83	491
Standing	0.90	0.79	0.84	532
Laying	1.00	0.95	0.97	537
Mean average	0.92	0.92	0.92	2947

For the UCI HAR dataset, the CNN-LSTM model gives the best result with an accuracy of 91.89%, followed by 1D-CNN (90.90%), stacked-CNN (90.66%), Conv-LSTM (88.36%), and RNN-LSTM (88.32%). Table 14.5 shows the activity-wise performance measures for the best case and Fig. 14.26 shows the confusion matrix attained using the CNN-LSTM model.

From all the above experiments, it is clear that the simple RNN-LSTM performs the worst on all the three datasets as compared to CNN based models. The failure to remember the long state of information from the window sampled data with memory cell could be a reason for such poor performance.

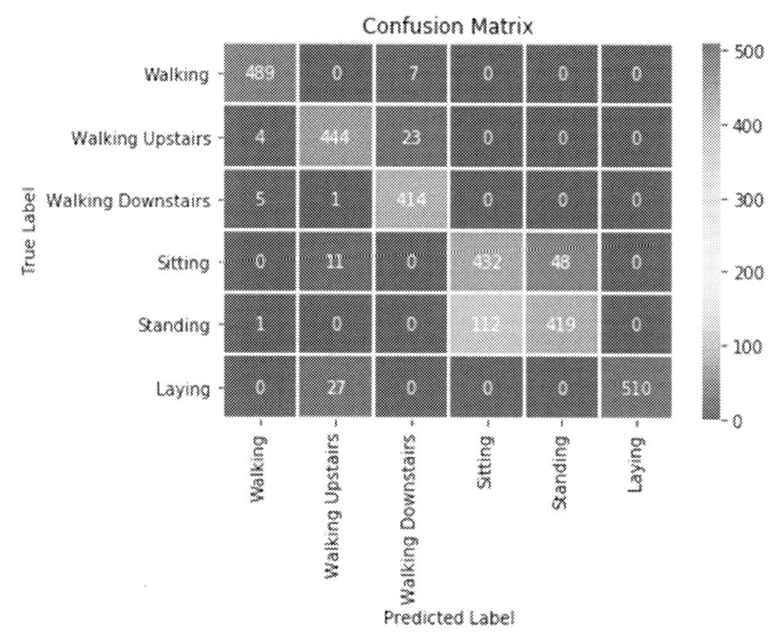

FIGURE 14.26 Confusion matrix of the test set of UCI HAR dataset for the CNN-LSTM model.

5.4 Comparison with other state-of-the-art HAR models

In the present work, we have also compared our results with some existing HAR models developed using both machine and deep learning-based models. Table 14.6 shows the comparison results obtained on the three datasets.

TABLE 14.6 Comparison of some of the existing models with our best performing deep learning models for HAR on RealWorld HAR, MHEALTH, and UCI HAR datasets.

Dataset	Work reference	Methodology	Accuracy (%)
RealWorld HAR	Sztyler et al. [29]	Random Forrest classifier	86
	Almaslukh et al. [37]	DNN Model	88
	Proposed method	**1D-CNN**	**92.7**
MHEALTH	Zdravevski et al. [38]	DBF feature selection	99.4
	Subasi et al. [39]	Data mining approach	99.08
	Proposed method	**1D-CNN**	**99.5**
UCI HAR	Anguita et al. [28]	Multi-class SVM	96.33
	Jain et al. [40]	Feature-level fusion with SVM, KNN classifiers	97.12
	Proposed method	**CNN-LSTM**	**91.89**

6. Conclusion and future work

Sensor-based HAR is a rapidly growing research topic in the domain of computer vision, where a sequence of data for a specified time span is collected from various sensors like accelerometer and gyroscope present in smart devices. In this paper, a comparative study of the performances of five different deep learning-based classification models, namely 1D-CNN, CNN-LSTM, Stacked-CNN, Conv-LSTM, and RNN-LSTM applied on HAR datasets has been made. These models have been applied to three benchmark datasets, namely RealWorldHAR, MHEALTH, and UCI HAR. Highest overall recognition accuracies of 92.07% on RealWorldHAR dataset using 1D-CNN, 99.50% on MHEALTH dataset using 1D-CNN, and 91.89% on UCI HAR dataset using CNN-LSTM have been attained. As the models considered here are supervised deep learning models, hence sometimes these misclassify the unseen test samples when its pattern is not present in the domain of training, or the samples are too noisy or complex.

However, unsupervised based deep learning models and reinforcement-based models could also be used, which we will consider in the future. In the 1D-CNN model, where we could vary the number of filters, size of the kernel, length of the stride to get a more accurate and optimized model. The introduction of deeper layers with more number of neurons in the dense layers might also improve the results. Similarly, different types of RNN layers like Gated Recurrent Unit (GRU) [35], Bidirectional LSTM (BLSTM) [36] can be used to create better models. This is a study of various deep learning models to recognize different human activities and we are also planning to use new concepts like transfer learning, ensembling of various models on different sensor datasets for our future work.

References

[1] B. Yao, L. Fei-Fei, Recognizing human-object interactions in still images by modeling the mutual context of objects and human poses, IEEE Trans. Pattern Anal. Mach. Intell. 34 (9) (2012) 1691–1703.

[2] J. Qin, L. Liu, Z. Zhang, Y. Wang, L. Shao, Compressive sequential learning for action similarity labeling, IEEE Trans. Image Proc. 25 (2) (2015) 756–769.

[3] M. Manic, K. Amarasinghe, J.J. Rodriguez-Andina, C. Rieger, Intelligent buildings of the future: cyberaware, deep learning powered, and human interacting, IEEE Ind. Electron. Mag. 10 (4) (2016) 32–49.

[4] D. Ravì, C. Wong, F. Deligianni, M. Berthelot, J. Andreu-Perez, B. Lo, G.Z. Yang, Deep learning for health informatics, IEEE J. Biomed. Health Inform. 21 (1) (2016) 4–21.

[5] T. Manoj, G.S. Thyagaraju, Active and assisted living: a comprehensive review of enabling technologies and scenarios, Int. J. Adv. Res. Comput. Sci. 9 (1) (2018).

[6] S. Guo, H. Xiong, X. Zheng, Y. Zhou, July. Indoor pedestrian trajectory tracking based on activity recognition, in: 2017 IEEE International Geoscience and Remote Sensing Symposium (IGARSS), IEEE, 2017, pp. 6079–6082.

[7] A. Avci, S. Bosch, M. Marin-Perianu, R. Marin-Perianu, P. Havinga, February. Activity recognition using inertial sensing for healthcare, wellbeing and sports applications: a survey, in: 23th International Conference on Architecture of Computing Systems 2010, VDE, 2010, pp. 1–10.

[8] S. Sadhukhan, S. Mallick, P.K. Singh, R. Sarkar, D. Bhattacharjee, A comparative study of different feature descriptors for video-based human action recognition, in: Intelligent Computing: Image Processing Based Applications, Springer, Singapore, 2020, pp. 35–52.

[9] F. Demrozi, G. Pravadelli, A. Bihorac, P. Rashidi, Human Activity Recognition Using Inertial, Physiological and Environmental Sensors: A Comprehensive Survey, 2020 arXiv preprint arXiv:2004.08821.

[10] W.H. Wu, A.A. Bui, M.A. Batalin, L.K. Au, J.D. Binney, W.J. Kaiser, MEDIC: medical embedded device for individualized care, Artif. Intell. Med. 42 (2) (2008) 137–152.

[11] A. Henpraserttae, S. Thiemjarus, S. Marukatat, May. Accurate activity recognition using a mobile phone regardless of device orientation and location, in: 2011 International Conference on Body Sensor Networks, IEEE, 2011, pp. 41–46.
[12] D. Micucci, M. Mobilio, P. Napoletano, Unimibshar: a dataset for human activity recognition using acceleration data from smartphones, Appl. Sci. 7 (10) (2017) 1101.
[13] D. Ravi, C. Wong, B. Lo, G.Z. Yang, A deep learning approach to on-node sensor data analytics for mobile or wearable devices, IEEE J. Biomed. Health Inform. 21 (1) (2016) 56–64.
[14] Z.Y. He, L.W. Jin, July. Activity recognition from acceleration data using AR model representation and SVM, in: 2008 International Conference on Machine Learning and Cybernetics, vol. 4, IEEE, 2008, pp. 2245–2250.
[15] M. Zeng, L.T. Nguyen, B. Yu, O.J. Mengshoel, J. Zhu, P. Wu, J. Zhang, November. Convolutional neural networks for human activity recognition using mobile sensors, in: 6th International Conference on Mobile Computing, Applications and Services, IEEE, 2014, pp. 197–205.
[16] P. Zappi, C. Lombriser, T. Stiefmeier, E. Farella, D. Roggen, L. Benini, G. Tröster, January. Activity recognition from on-body sensors: accuracy-power trade-off by dynamic sensor selection, in: European Conference on Wireless Sensor Networks, Springer, Berlin, Heidelberg, 2008, pp. 17–33.
[17] D. Roggen, A. Calatroni, M. Rossi, T. Holleczek, K. Förster, G. Tröster, P. Lukowicz, D. Bannach, G. Pirkl, A. Ferscha, J. Doppler, June. Collecting complex activity datasets in highly rich networked sensor environments, in: 2010 Seventh International Conference on Networked Sensing Systems (INSS), IEEE, 2010, pp. 233–240.
[18] J.W. Lockhart, G.M. Weiss, J.C. Xue, S.T. Gallagher, A.B. Grosner, T.T. Pulickal, August. Design considerations for the WISDM smart phone-based sensor mining architecture, in: Proceedings of the Fifth International Workshop on Knowledge Discovery from Sensor Data, ACM, 2011, pp. 25–33.
[19] M.M. Hassan, M.Z. Uddin, A. Mohamed, A. Almogren, A robust human activity recognition system using smartphone sensors and deep learning, Fut.Generat. Comput. Syst. 81 (2018) 307–313.
[20] C.A. Ronao, S.B. Cho, Human activity recognition with smartphone sensors using deep learning neural networks, Expert Syst. Appl. 59 (2016) 235–244.
[21] S. Matsui, N. Inoue, Y. Akagi, G. Nagino, K. Shinoda, User adaptation of convolutional neural network for human activity recognition, in: 2017 25th European Signal Processing Conference (EUSIPCO), IEEE, 2017, pp. 753–757.
[22] S.A. Rokni, M. Nourollahi, H. Ghasemzadeh, April. Personalized human activity recognition using convolutional neural networks, in: Thirty-Second AAAI Conference on Artificial Intelligence, 2018.
[23] A. Bulling, U. Blanke, B. Schiele, A tutorial on human activity recognition using body-worn inertial sensors, ACM Comput. Surv. 46 (3) (2014) 33.
[24] B. Almaslukh, J. AlMuhtadi, A. Artoli, An effective deep autoencoder approach for online smartphone-based human activity recognition, Int. J. Comput. Sci. Netw. Secur. 17 (4) (2017) 160–165.
[25] J.R. Kwapisz, G.M. Weiss, S.A. Moore, Activity recognition using cell phone accelerometers, ACM SigKDD Explor. Newsletter 12 (2) (2011) 74–82.
[26] Y.S. Lee, S.B. Cho, May. Activity recognition using hierarchical hidden markov models on a smartphone with 3D accelerometer, in: International Conference on Hybrid Artificial Intelligence Systems, Springer, Berlin, Heidelberg, 2011, pp. 460–467.
[27] C.A. Ronao, S.B. Cho, August. Human activity recognition using smartphone sensors with two-stage continuous hidden Markov models, in: 2014 10th International Conference on Natural Computation (ICNC), IEEE, 2014, pp. 681–686.
[28] D. Anguita, A. Ghio, L. Oneto, X. Parra, J.L. Reyes-Ortiz, April. A public domain dataset for human activity recognition using smartphones, in: Esann, 2013.
[29] T. Sztyler, H. Stuckenschmidt, March. On-body localization of wearable devices: an investigation of position-aware activity recognition, in: 2016 IEEE International Conference on Pervasive Computing and Communications (PerCom), IEEE, 2016, pp. 1–9.
[30] O. Banos, R. Garcia, J.A. Holgado-Terriza, M. Damas, H. Pomares, I. Rojas, A. Saez, C. Villalonga, December. mHealthDroid: a novel framework for agile development of mobile health applications, in: International Workshop on Ambient Assisted Living, Springer, Cham, 2014, pp. 91–98.
[31] A.F. Agarap, Deep Learning Using Rectified Linear Units (Relu), 2018 arXiv preprint arXiv:1803.08375.
[32] S. Hochreiter, J. Schmidhuber, Long short-term memory, Neural Comput. 9 (8) (1997) 1735–1780.

[33] S.H.I. Xingjian, Z. Chen, H. Wang, D.Y. Yeung, W.K. Wong, W.C. Woo, Convolutional LSTM network: a machine learning approach for precipitation nowcasting, in: Advances in Neural Information Processing Systems, 2015, pp. 802–810.

[34] D.P. Kingma, J. Ba, Adam: A Method for Stochastic Optimization. Proceedings of the 3rd International Conference on Learning Representations (ICLR), 2014, pp. 1–15. San Diego, CA, USA, arXiv preprint arXiv:1412.6980.

[35] R. Dey, F.M. Salemt, August. Gate-variants of gated recurrent unit (GRU) neural networks, in: 2017 IEEE 60th International Midwest Symposium on Circuits and Systems (MWSCAS), IEEE, 2017, pp. 1597–1600.

[36] M. Schuster, K.K. Paliwal, Bidirectional recurrent neural networks, IEEE Trans. Signal Proc. 45 (11) (1997) 2673–2681.

[37] B. Almaslukh, A.M. Artoli, J. Al-Muhtadi, A robust deep learning approach for position-independent smartphone-based human activity recognition, Sensors 18 (11) (2018) 3726.

[38] E. Zdravevski, P. Lameski, V. Trajkovik, A. Kulakov, I. Chorbev, R. Goleva, N. Pombo, N. Garcia, Improving activity recognition accuracy in ambient-assisted living systems by automated feature engineering, IEEE Access 5 (2017) 5262–5280.

[39] A. Subasi, M. Radhwan, R. Kurdi, K. Khateeb, February. IoT based mobile healthcare system for human activity recognition, in: 2018 15th Learning and Technology Conference (L&T), IEEE, 2018, pp. 29–34.

[40] A. Jain, V. Kanhangad, Human activity classification in smartphones using accelerometer and gyroscope sensors, IEEE Sensor. J. 18 (3) (2017) 1169–1177.

Index

Note: 'Page numbers followed by "*f*" indicate figures and "*t*" indicate tables.'

D

E

H

I

N

O

Printed in the United States
by Baker & Taylor Publisher Services